CELLS AND BIOMATERIALS IN REGENERATIVE MEDICINE

Edited by **Daniel Eberli**

Cells and Biomaterials in Regenerative Medicine
http://dx.doi.org/10.5772/58497
Edited by Daniel Eberli

Contributors

Joydeep Basu, Jan Oxholm Gordeladze, Dragica Maja Smrke, Danijela Semenič, Susanne Jung, Deana Georgieva Haralampieva, Simon Mensah Ametamey, Tullio Sulser, Daniel Eberli, Francine Goulet, Sara Bouhout, Jadson Moreira Pereira, Franck Simon, Stéphane Chabaud, Alex Therien, Éric Rousseau, Massimo Conti, Véronique Moulin, Jean-Pierre Lavoie, Michel Rouleau, Stéphane Bolduc, Stephane Bolduc, Cassandra R Goulet, Hazem Orabi, Julie Fradette, Manuel Monleon, Manuel Pérez Garnés, Ana Vallés Lluch, Juan Antonio Barcia, Gómez-Pinedo Ulises, Arnaldo Rodrigues Santos, Ralf Poertner, Regine Eibl, Dieter Eibl, Valentin Jossen, Herve Lesot, Tunay Kokten, Sabine Kuchler-Bopp, Alex Pontini

Published by InTech

Janeza Trdine 9, 51000 Rijeka, Croatia

Publishing Process Manager Iva Lipovic

Technical Editor InTech DTP team

Cover InTech Design team

First published December, 2014

Printed in Croatia

Additional hard copies can be obtained from orders@intechopen.com

Cells and Biomaterials in Regenerative Medicine, Edited by Daniel Eberli

p. cm.

ISBN 978-953-51-1731-5

For EU product safety concerns:
IN TECH d.o.o., Prolaz Marije Krucifikse Kozulić 3, 51000 Rijeka, Croatia,
info@intechopen.com or visit our website at intechopen.com.

Meet the editor

Daniel Eberli MD. Ph.D. is a scientific physician working in the translational field of urologic tissue engineering. He has a medical degree from the Medical School in Zurich, Switzerland, and a Ph.D. in Molecular Medicine from Wake Forest University, Winston Salem, NC. He currently has a faculty position at the Department of Urology at the University Hospital Zurich, where he devotes half of his time to patient care. He is a lecturer at the Medical School of Zurich and the Swiss Federal Institute of Technology. Together with his research team, he is working on novel biomaterials for bladder reconstruction, improving autonomic innervation, cellular treatment of incontinence and tracking of stem cells.

Contents

Preface

Tissue Engineering is one of the major approaches of regenerative medicine and a growing and exciting field of research. In combination with better understanding of structure, biology, physiology and cell culture techniques, Tissue Engineering may offer new treatment options for patients in need for replacement or repair of a deteriorated organ. The concept of Tissue Engineering has been applied clinically for a variety of disorders. The classical approaches of Tissue Engineering have not changed over the last three decades. The principle is to dissociate cells from a tissue biopsy, to expand these cells in culture, and to seed them onto the scaffold material in vitro in order to form a living tissue construct prior to implantation into the recipient's organism. In the appropriate biochemical and biomechanical environment these tissues will unfold their full functional potential and serve as native tissue equivalents.

The Tissue Engineering approach has major advantages over traditional organ transplantation and circumvents the problem of organ shortage. Tissues that closely match the patient's needs can be reconstructed from readily available biopsies and subsequently be implanted with minimal or no immunogenicity. This eventually conquers several limitations encountered in tissue transplantation approaches.

This book serves as a good starting point for anyone interested in the application of Tissue Engineering. It offers a colorful mix of topics, which explain the obstacles and possible solutions for TE applications.

Finally, I would like to thank all authors who have contributed to this book and hope that the readers will enjoy it.

Daniel Eberli MD. PhD.
University Zürich, Switzerland

Section 1

Cells

Chapter 1

Adipose-Derived Stem Cells — Are They the Optimal Cell Source for Urinary Tract Regeneration?

Hazem Orabi, Cassandra R. Goulet,
Julie Fradette and Stéphane Bolduc

Additional information is available at the end of the chapter

http://dx.doi.org/10.5772/59223

1. Introduction

The urinary system consists of two kidneys, two ureters, urinary bladder and urethra. The main function of the urinary system is to eliminate wastes and harmful materials from the body. Additionally, it adjusts blood volume and pressure, regulates electrolytes and blood pH. The urine passes from the kidneys via the muscular tubes of ureters to the urinary bladder where it is stored until it is socially convenient to be expelled outside through the urethra by the act of micturition.

The urinary tract (ureters, bladder and urethra) is composed of two main functional tissues: epithelial coverage (urothelium) and muscular coat (smooth muscle). A highly impermeable urothelial layer is necessary to prevent reabsorption of noxious materials present in urine despite its large osmotic and chemical gradients. Muscular coat creates coordinated waves of contractions necessary for urine transport and expulsion.

Since birth, the urinary system is subjected to a variety of diseases and pathologies including congenital and acquired disorders that threaten the patient`s life. Current therapeutic options to replace severely damaged urinary organs are associated with many complications and hazards on patient survival and quality of life. Regenerative medicine has emerged as potential replacement treatment. Although progenitor cells of the urinary tract have been used in experimental studies and clinical trials, however, they are not amenable options in cases of benign end-stage diseases and malignancy. Stem cells can be the source of cellular components in regenerating urinary organs. Among those stem cell types, Adipose-derived Stem Cells (ASCs) are the current most convenient source due to easiness of harvest in abundant quantity, potential differentiation into many cell types and lack of ethical problems.

ASCs had also been investigated to form matrix (scaffold) and differentiate into urothelial and smooth muscle cells. Both can be coupled together to form tissue-engineered constructs that can replace the wall of urinary tract including ureters, urinary bladder and urethra. When complete wall replacement is not needed in some situations, ASCs, as cellular therapy, had beneficial benefits reflected on pattern of voiding. Additionally, ASCs can enhance vascularity and improve survival of urinary flaps or grafts.

In this chapter, we will review the current and future applications of ASCs for regeneration and repair of urinary tract. Mechanism of actions, preclinical and clinical trials and forms for therapies will be addressed. We will highlight the challenges that face their use and potentials to overcome these obstacles.

2. ASCs characteristics and prospectives for regenerative medicine

In the past, little consideration has been given to adipose tissue since its main functions were mainly associated with the storage and release of lipids. For years, lipoaspirated material has been discarded as surgical waste. Over the past few years, scientists became interested in investigating this highly complex tissue. Adipose tissue is composed mainly of adipocyte cells organized into lobules. The most prominent fraction, volume-wise, are mature adipocytes, which store and hydrolyzed triglycerides in response to environmental signals. The stromal vascular fraction (SVF), which includes preadipocytes, endothelial cells, vascular smooth muscle cells, fibroblasts, resident monocytes/macrophages and lymphocytes [1] also, provides structural support to adipose tissue. In 2002, a novel adult stem cell population isolated from SVF was first identified by University of California,Los Angeles (UCLA) researchers and named processed lipoaspirate (PLA) cells [2,3].

A wide variety of terms have been used to describe the multipotent cells derived from white adipose tissues and there was no consensus on the nomenclature used, which can sometimes lead to confusion. At the Second Annual International Fat Applied Technology Society (IFATS) meeting in 2004, scientists concluded to refer to these cells as adipose-stem cells (ASCs) [53]. Accordingly, the term ASCs will be used throughout this chapter.

ASCs are mesenchymal stem cells (MSCs) and possess similar characteristics to those extracted from bone marrow or umbilical cord [4]. They are undifferentiated non-embryonic multipotent stem cells, which have the ability to divide and self-renew in undifferentiated state. When submitted to specific inductive signals, ASCs can mature into a broad spectrum of cell lineages. ASCs have been reported to express the MSCs surface protein markers CD10, CD13, CD29, CD44, CD54, CD73, CD90 and CD105 (Table 1) [5]. Major differences occur for the CD34 stem cell marker that is not express by MSCs, but is present in ASCs in early passages. Morphologically, ASCs are fibroblast-like and preserve their shape after expansion in vitro [6].

http://dx.doi.org/10.5772/59223

Cell surface markers	Level of expression
CD3	Low or negative
CD10	High
CD11b	Negative
CD14	Negative
CD29	High
CD34	Variable
CD44	High
CD45	Negative
CD73	High
CD90	High
CD105	High
HLA-DR	Negative

Table 1. Expression level of ASCs surface markers

2.1. Isolation of ASCs from adipose tissue

When compared to other stem cell populations and sources, ASCs can be easily harvested while providing higher yields upon processing of adipose tissue, with minimal discomfort under local anaesthesia [7]. The efficiency of the cell isolation process depends on the localisation of the white adipose tissue and the donor general condition. For example, ASCs from visceral deposits are more prone to apoptosis and, therefore, less proliferative than the same cells isolated from the subcutaneous deposits [8]. ASCs are ususally isolated from subcutaneous adipose tissue samples removed during liposuction, lipoplasty or lipectomy procedures, which are minimally invasive or painful. Adipose tissue is one of the richest sources of MSCs. From 1 gram of adipose tissue, 5×10^3 colony-forming stromal cells can be isolated, which is 500 times more cells than from an equivalent amount of bone marrow [9]. Current protocols used for release the cells rely on enzymatic digestion with collagenase, dispase, trypsin or related enzymes. Following the centrifugation and neutralization of the enzymes, the SVF fraction sediment is separated from the floating mature adipocytes. When seeded in culture flasks, the ASCs lack the intracellular lipid droplets seen in adipocytes and adhere to the plastic surface where they can be purified from other SVF cells using a combination of washing steps and culture expansion. Several exogenous supplements have been shown to have a stimulatory effect on the proliferation of ASCs. Platelet-derived growth factor, sphingosylphosphorylcholine, oncostatin M, and fibroblast growth factor (FGF) 2 have all been shown to increase ASC proliferation [9, 10, 11, 12].

2.2. ASCs potentials for regenerative medicine

The potential use of stem cell-based therapies for the repair and regeneration of various tissues and organs offers alternative therapeutic solutions for a number of diseases. Developments in stem cell research provided new cell source for regenerative medicine. An emerging body of literature suggests that adipose tissue may provide an abundant, readily accessible source of cells with similar potential to that described for other adult stem cells.

Adult stem cells are far more plastic than had previously been imagined. ASCs have been described as being able to give rise to several quite different mesenchymal cell phenotypes including osteocytes, adipocytes, neural cells, vascular endothelial cells, cardiomyocytes, pancreatic β cells, and hepatocytes [13, 14, 15, 16].

Cell lineage	Culture medium supplementation	References
Adipocyte	Insulin, dexamethasone, thiazolidinedione	17, 18
Chondrocyte	Ascorbic acid, BMP-6, FGF, IGF, dexamethasone, TGF-β	19, 20
Osteoblast	Ascorbic acid, BMP-2, valproic acid, dexamethasone	17
Myocyte	Dexamethasone, horse serum	21
Cardiomyocyte	Stem cell factor, transferrin, IL-3, IL-6, VEGF, TGF- β	22, 23
Neuronal-like	Valproic acid, insulin, butylated hydroxyanisole, hydrocortisone	24
Hepatocyte-like	HGF, FGF-1, FGF-4, oncostatin	24, 25

Table 2. Differentiation potential of ASCs

However, the ability to differentiate is not the only characteristic that makes these cells attractive for therapeutic purposes. Some reports have suggested that the therapeutic effects observed following ASCs administration, such as promotion of angiogenesis, reduction of inflammation, and functional recovery, are largely related to the trophic actions of their cytokines and growth factors secretion rather than by their differentiation into local tissue cell types [26]. The secretion of a broad range of bioactive molecules by ASCs, such as growth factors, cytokines and chemokines, constitutes their most biologically significant role under injury conditions [27]. Analyses of the soluble factors released from human cultured ASCs have revealed that, at relatively early passages, they secrete hepatocyte growth factor (HGF), vascular endothelial growth factor (VEGF), transforming growth factor-β (TGF-β), insulin-like growth factor (IGF)-1, fibroblast growth factor (bFGF), granulocyte-macrophage colony-stimulating factor (GM-CSF), tumour necrosis factor (TNF)-α, interleukin (IL)-6, 7, 8, and 11, adiponectin, angiotensin, cathepsin D, pentraxin, pregnancy zone protein, retinol-binding protein, and CXCL12 [28, 29, 30]. These factors promote an anti-inflammatory environment, angiogenesis and wound healing, which likely potentiate tissue repair.

ASCs possess immunomodulatory effects through their production of various soluble factors and rather low immunogenicity properties. In vitro, they have the ability to inhibit maturation of dendritic cells and suppress the proliferation of B cells, T cells, and natural killer cells following activation through both cell to cell binding and paracrine signalling [31, 32]. During inflammation, ASCs release TGF-β that promotes premature helper T-cell differentiation toward T regulatory cells promoting anti-inflammatory environment [33]. Moreover, ASCs promote tryptophan degradation, limiting the bioavailability of a crucial molecule for T-cell proliferation and function [34]. Low expression levels of class I Major Histocompatibilty Complex (MHC-I) and lack of MHC-II and co-stimulatory molecules on the cell surface (CD80 and CD86] makes ASCs immunoprivileged cells [35].

During the past decade, several studies have provided preclinical data on the safety and efficacy of ASCs administered in various animal models, supporting the use of these cells in future clinical applications. Furthermore, various clinical trials have shown the regenerative capability of ASCs in a variety of medical fields such as plastic surgery, autoimmune and inflammatory disorders, orthopedic surgery, oral and maxillofacial surgery, cardiac surgery and urology [36, 37, 38].

3. ASCs and kidney repair

Cell therapies using ASCs are highly promising in various clinical fields based on in vitro and in vivo research results. Important experimental findings in recent years suggest considerable therapeutic potential for cellular replacement in the context of kidney diseases (Figure 1).

3.1. Acute kidney injuries

Acute kidney injury (AKI), a multi-factorial syndrome that is common but often asymptomatic, is caused by such incidents as major infections, trauma, complications following major surgery and adverse reactions to drugs. It is characterised by the rapid loss of the kidney's functions, usually occurring within 48 hours or less, and represents a major treatment-resistant clinical problem with high mortality rates. In United States, AKI is an increasingly prevalent condition in hospitalized patients, a 10 percent increase per year, and the number of associated deaths has more than doubled over the last decade [39].

In the last few years, many studies have shown that stem cell transplantation is an efficient method for AKI repair. It was shown that ASCs drastically reduced mortality and improved renal structural and functional recovery using an animal ischemia-reperfusion model of AKI [56]. Indeed, intrarenal arterial injection of ASCs in ischemia-reperfusion-induced AKI rat model reduced blood urea nitrogen and creatinine levels [40]. Moreover, in a dose-dependent manner, ASCs reduced the tubular injury score 48 hours after ischemia-reperfusion [40]. The precise mechanism that attenuates kidney injury after ASCs injection remains to be clarified.

The renoprotective effects might be attributable to the replacement of damaged tubular cells by transdifferentiation of ASCs. The injured kidney expresses renotypic factors that may influence the differentiation of ASCs into a renal tubular epithelial lineage [41, 55]. A study using a murine model of AKI showed that differentiation and replacement of the dead cells at an early stage of injuriy by ASCs seems to be one of the major mechanisms in AKI kidney repair. However, most studies indicate that intravenously injected ASCs show minimal homing to renal tubules and have limited survival in the kidney environment [42]. To bypass this problem, scientists tried to inject ASCs directly at the injury site. However, more than 80% of grafted cells die within the first week after injection and the majority of cells leak out from the injection site. Recently, it was found that using an injectable scaffold could improve the survival of cells as well as enhance retention of ASCs at the injury site [43].

There are increasing evidences that other differentiation-independent mechanisms of ASCs play key role in promoting tissue repair. Beneficial effects could be partially explained by the anti-inflammatory and anti-oxydative properties of ASCs, as revealed by their capacity to reduce apoptosis and inflammation in the injured kidney tissue [56]. Indeed, the expression of inflammatory, oxidative stress and apoptotic biomarkers such as IL-10, glutathione peroxidase and TNF-α at both gene and protein levels, were significantly decreased in ASCs treated group in ischemia-reperfusion rat model [45].

3.2. Chronic kidney disease

Twenty-six million american adults have chronic kidney disease (CKD) with progression of the disease leading to kidney failure, which requires dialysis or a kidney transplant to maintain life.

Positive effects of bone marrow MSCs treatment in CKD animal models have been reported [46, 47]. However, bone marrow MSCs derived from CKD patients failed to provide the same beneficial effect [48, 49]. A recent study has shown that ASCs did not seem affected by renal disease [50]. Thus, the impact of ASC therapy has been studied in different CKD animal models. In the majority of induced models investigated, ASCs administration has resulted in beneficial therapeutic effects, as evidenced by decrease the speed of renal fibrosis progression and improvement in renal functional parameters such as plasma creatinin, proteinuria and renal filtration. Indeed, intraperitoneal injection of ASCs in ischemia-reperfusion CKD mouse model reduced renal dysfunction and tubular injuries after 24 h post-reperfusion [44]. At 6 weeks, ASCs-treated animals showed reduced renal fibrosis [44].

The mechanisms mediating the therapeutic effects described are still under investigation. ASCs would enhance the angiogenic process in CKD by increased expression of VEGF, a growth factor that stimulates new blood vessels formation, suggesting a paracrine mechanism. Moreover, the level of BMP-7 and Pax-2, two proteins that participate in cellular repair response to kidney damage, increased after ASCs administration in CKD models [52].

3.3. Graft tolerance

Kidney transplantation remains the best therapeutic option for patients with end stage renal disease. In 2012, the Scientific Registry for Transplant Recipients reported that approximately 17 000 kidney transplants were performed and over 92 800 patients were remaining on the waitlist at the end of the survey year [54]. Unfortunately, transplanted patients need life-long immunosuppressive drugs to prevent allograft rejection. This phenomenon may happen when transplantation of a kidney from a donor who differs genetically from the graft recipient induces an immune response in the recipient against alloantigens of the donor graft. Effector

T cells infiltrate the graft and orchestrate an inflammatory response leading to destruction of the tissue. The new immunosuppressive drugs have improved short-term patient survival but rejection remains a major problem. Effective treatments are necessary to effectively address the problem of transplanted organ rejection.

Because of their immunomodulatory properties, ASCs are believed to play a role in the induction of transplantation tolerance. Studies have shown that ASCs attenuated acute rejection in kidney transplantation by increasing graft survival and reducing rejection grade [60]. A kidney transplantation study in rats that received intravenous injection of autologous ASCs before intervention showed decreased infiltration of macrophages and lymphocytes into the allograft and suppression of alloreactive T cells [60]. Studies reported that graft tolerance can be induced by ASCs therapy. Indeed, donor-ASCs portal infusion in a 29 year-old recipient before renal transplantation induced production of T-regs cells [59]. At 2-years post-transplant, immunosuppression weaning was started and 6 months later, anti-rejection therapy was completely stop with normal graft function. An Indian study in 90 patients showed similar results [57]. Consequently, ASCs may serve as effective immunomodulators in clinical transplantation. Long-term studies in larger populations are required to confirm efficacy.

Indication	Disease model	Cells used	Functional Assessment	Ramakes	References
Acute renal injury (AKI)	Ischemia-reperfusion rat model	Human ASCs	Urea and creatinine level. Tubular injury score.	Improved renal recovery.	40
	Ischemia-reperfusion mouse model	Human ASCs	Dead cells replacement	Improved renal recovery.	51
	Ischemia-reperfusion mouse model	Mouse ASCs	-	Improved renal recovery.	58
	Ischemia-reperfusion rat model	Rat ASCs	Inflammation, oxidative stress and apoptose	Reduce inflammatory and oxidative stress reaction.	45
Chronic kidney disease (CKD)	5/6 NPX rat	Human ASCs	Creatinine level, structural analysis, angiogenesis	Improved renal recovery.	52
	Ischemia-reperfusion mouse model	Human ASCs	Renal fibrosis, inflammation	Improved renal recovery.	44
Graft Tolerance	Rat kidney transplantation	Rat ASCs	Inflammation, rejection	Improved graft tolerance	60
	Human kidney transplantation	Human ASCs	Inflammation, rejection	Improved graft tolerance	57

Table 3. Shows the potential of ACSs as Kidney cellular therapy.

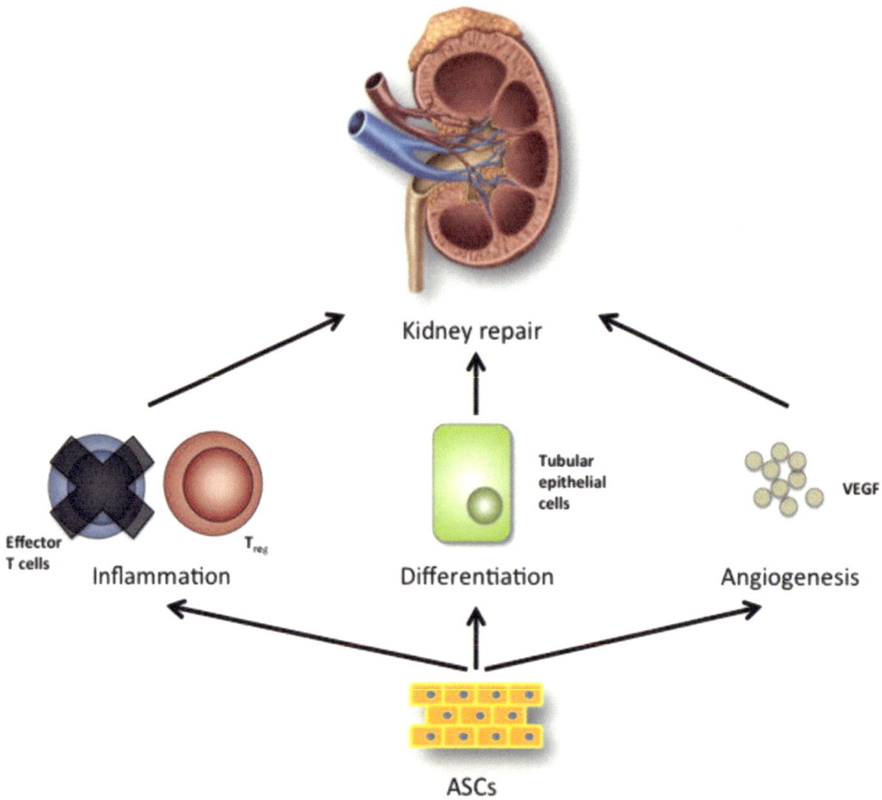

Figure 1. ASCs reduce inflammation by direct (cell–cell contact) and indirect (paracrine) actions, by suppression of T-cell and induction of T_{reg}. The differentiation of ASCs into tubular epithelial cells can contribute to kidney recovery. Microvascular damage is reduced by the secretion of the growth factor VEGF, which promotes angiogenesis.

4. ASCs and urinary tract

4.1. Urinary tract reconstruction

The urinary tract is anatomically composed of the renal pelvis, ureters, urinary bladder and urethra. It is concerned with urine transport from the kidneys, storage and expulsion outside of the body. Histologically, it is formed of 3 different layers including epithelium, connective tissue forming the submucosal layer and smooth muscle on the outside. The epithelial coverage is composed of transitional epithelium called urothelium, the function of which is to prevent the reabsorption of noxious materials present in urine despite its large osmotic and chemical gradients. Muscle layer creates coordinated waves of contractions necessary for urine transport and expulsion. The submucosa is formed of extracellular matrix containing blood vessels, lymphatics, nerves and variety of cells; predominantly fibroblasts.

The urinary tract is the target of many congenital and acquired insults that affect certain or all layers of the organ involved, resulting in partial or complete loss of function. This influences the patient survival and quality of life.

Although the urinary tract is naturally equipped with progenitor cells in the urothelium and muscle layers that help the regeneration of the urinary tract after any insult, severe disruption of the involved layer or the whole structure may exceed the capacity of the local progenitors or stem cells to compensate for the cellular loss. Hence, a necessity for repair with exogenous therapies emerges to restore the structure and hence the function, partially or completely.

Replacing the damaged urinary organs with non-native tissues is related to many problems including the lack of available tissues and postoperative complications including metabolic derangements. Regenerative medicine has evolved as a potential solution for replacing the damaged urinary tract. The patient's own cells that can be used in regenerating urinary tract are either local progenitors or stem cells. Local progenitor cells cannot be utilized in many instances as cancer or end stage benign diseases being the reasons for the defect. Consequently, stem cells should be harvested to build up new tissues for replacement.

Ideally, for clinical use, stem cells should be harvested in abundant quantities, by a minimally invasive procedure, and differentiated along multiple cell lineage pathways in a reproducible manner. Also, it can be manufactured in accordance with current Good Manufacturing Practice guidelines [61]. For these reasons, ASCs represent an ideal source for stem cell therapy especially for repairing the urinary system [62]. Additional advantages for the urogenital tract, ASCs populations likely contains vascular stem cells at various stages of differentiation toward becoming smooth muscle cells (SMCs) and endothelial cells (ECs), which are required components for regenerating the urinary tract [63]. Furthermore, ASCs have the ability to secrete many potentially synergistic proangiogenic growth factors delineating their angiogenic and antiapoptotic potential.

4.2. ASCs Contribution to urinary tract reconstruction

ASCs can be the source for epithelial and muscle cells through many established protocols of differentiation. Also, they can provide extracellular matrix (ECM) scaffold that sustains mechanical stress and supports the formation of epithelial coverage.

4.2.1. ASCs as cell source

1. Undifferentiated mesenchymal stem cells:

Unmodified ASCs can be seeded on different scaffolds to replace the urinary tract. In vivo differentiation of ASCs into SMCs has been observed after delivery of unmodified human ASCs within the bladder wall of nude mice [64], or autologous rat ASCs seeded on scaffold [65]. However, this phenotypic conversion occurs for a small percentage of the delivered ASCs.

2. Urothelial cell differentiation:

The main challenge in differentiating ASCs into urothelial cells (UCs) is due to the fact that the urothelium is derived from the endoderm [66], but ASCs are derived from the mesoderm, making cross-blastoderm induction difficult. In vitro differentiation of ASCs into UCs was achieved using either conditioned medium [67], growth factors as all-transretinoic acid [68] or direct (seeding both cell types together) or indirect (using Transwell system { Corning-Costar}) coculture with UCs [69,70]. Lineage specific markers such as keratin 18 and uroplakins were detected with high percentage among differentiated cells. ASCs were labeled with CM-DiI and mixed with immortalized human bladder urothelial cells in a collagen matrix to be implanted in subcutaneous tissue of athymic mice. After 4 weeks, the expression of uroplakin-Ia was 70% and the expression of uroplakin-II was 65%. However, there was a lack of organized stratified urothelium [70].

The difficulties of final phenotypic conversion of ASCs into UCs may be attributed to the lack of epithelial-specific microenvironment including 3D biomimetic culture conditions, mesenchymal component with its cross talk with UCs (organ-specific ECM and fibroblasts). Identifying the key factors responsible for induction of urothelial markers in previous experiments is another difficulty [71].

3. Smooth muscle cell differentiation:

ASCs are easier to differentiate into smooth muscle cells. A subpopulation of adipose SVF was isolated and proved to be different from other classes of adipose-derived cells after expanding SVF cells with DMEM (Dulbecco's Modified Eagle Medium) plus 10% FBS (fetal bovine serum). They were called adipose-derived SM-like cells as they consistently express SMCs markers (Smooth muscle alpha actin, Smooth muscle myosin Heavy chain, Myocardin, SM22 and calponin), independent of donor site and across multiple passages [72]. SMC differentiation could be achieved either with chemical induction using different protocols [73, 74] or with coculture with SMCs [64]. Mechanical extension stimulation could improve the feasibility of ASCs induction into SMCs [75]. When mature SMCs are seeded on collagen scaffolds and implanted in vivo, they not only regenerate SMCs needed for contraction but they also reduce inflammation and promote neovascularization [76].

4.2.2. ASCs as a possible source of scaffolds

The use of acellular matrices or synthetic scaffolds can be associated with immunologic reactions and possibility of infection transmission as exogenous ECM materials still retain a significant portion of residual DNA. Therefore, a new method was developed; the self-assembly method; to produce a tissue built by the cells themselves where a dense ECM is completely produced by fibroblasts. In opposition to all exogenous scaffold models, these models are autologous, which is a real advantage by eliminating the biocompatibility concerns. The absence of immunological response should reduce the inflammatory and fibrotic reactions and consequently improve the success rate of the procedure. Although this method has been developed for skin engineering in case of severe burn, however, its use has been extended to a wide variety of applications ranging from skin to blood vessels [77-79].

Our group has explored this unique technique for the reconstruction of urethra, urinary bladder and tunica albuginea [80-86]. As ASCs were able to lay down collagen-based matrix under the influence of ascorbic acid using the self-assembly technique [87], they would form another alternative for scaffolds for urinary tract regeneration. Additionally, they have been shown to allow the growth and stratification of epithelial cells when seeded on top [88]. ASCs have been utilized to form vesical equivalents and showed that there was no significant structural difference between a ASCs and a dermal fibroblasts equivalents. When compared to matrix synthesized from fibroblasts cultured under the same conditions, ASCs matrix was thicker and showed comparable failure strain [89].

4.3. Reconstruction of urinary tract with ASCs

4.3.1. Tissue engineering of the ureter

Ureteral damage may result from trauma or pathologies that lead to stricture formation obstructing the urine flow from the kidneys to the urinary bladder. The resulting ureteral defects may be short or long. While bridging short defects is usually surgically feasible, the reconstruction of long defects require extensive surgical repair that is not always possible and may carry complications including metabolic derangements and tissue harvest problems. Accordingly, new therapeutic options incorporating urothelium and avoiding ample tissue harvest are required.

Tissue engineering can offer these new therapeutic options. Unlike urethra and bladder, ureteral tissue engineering is only at its beginning. Unseeded synthetic or naturally derived biomaterials were used in few animal studies and resulted in ureterohydronephrosis due to the lack of normal tissue formation [90]. Cell seeded scaffolds present the ideal template for tubular ureteral regeneration. The seeded cells should include urothelial cells to prevent early contact of the scaffold with harmful urine and smooth muscle cells to regenerate the muscular layer necessary for urine transport through the ureter. Whenever autologous urinary tract cells are not available, ASCs can be the source of these cells. ASCs were differentiated into UCs through indirect coculture protocol and seeded on tubular polylactic acid (PLA)/collagen scaffolds [91]. Strong evidence of differentiation into urothelial lineage was detected with CK-18 and UP2. When implanted subcutaneously in athymic mice, the differentiated cells in the graft survived, stratified and exhibited urothelial markers. In another study, when ASCs were differentiated into SMCs and seeded onto decellularized rabbit aortas, at 16 weeks after implantation, radiography revealed patent ureter with no stricture associated with evidence of stratified epithelium and organized muscle bundles similar to the native tissue [92].

4.3.1.1. Tissue engineering of the urinary bladder

Cell-seeded scaffolds with autologous urinary tract cells represent the current strategy for tissue engineering of the urinary bladder [93]. Clinical trials using autologous urothelial and smooth muscle cells along with exogenous biomaterials have been performed [94]. Although organ specific cells would be an ideal cell source, however they are not available in many situations as in bladder cancer or benign end-stage bladder [95]. Stem cells derived from many tissues including bone marrow, muscle and adipose tissue are possible options to bridge the defective cell source in these situations. ASCs are more encouraging due to their previously mentioned advantages. ASCs can form ECM matrix and can be differentiated into urothelial cells and SMCs (Figure 2). Both the matrix and the differentiated cells can be used together or separately to reconstruct the urinary bladder for in vivo implantation.

ASCs have been used for in vivo studies for urinary bladder replacement or augmentation whether unmodified [96, 65] or after differentiating them into SMC [73]. Unmodified autologous ASCs were seeded on acellular matrices (bladder acellular matrix in rabbits or prepucial matrix in rats) and showed increase in smooth muscle content in the seeded group and improved bladder functional evaluation when compared to non-seeded scaffolds. However, ASCs were not labeled; therefore the origin of the regenerated muscle cells at the site of the seeded matrices was not known whether it is differentiated ASCs or growth of local bladder cells during normal healing process. In another study, ASCs were differentiated into SMCs and seeded on synthetic scaffold PLGA and implanted in rats to replace 50% of the bladder. Labeled SMC-differentiated ASCs persisted in vivo, became more organized with time and led to better bladder compliance when compared to non-seeded scaffolds. However, in either of the previous studies, the urothelial cell layer was lacking, exposing the scaffold to urine irritation in the early healing period in case of large defects, which may promote fibrosis and scarring of the graft highlighting the importance of the urothelial layer.

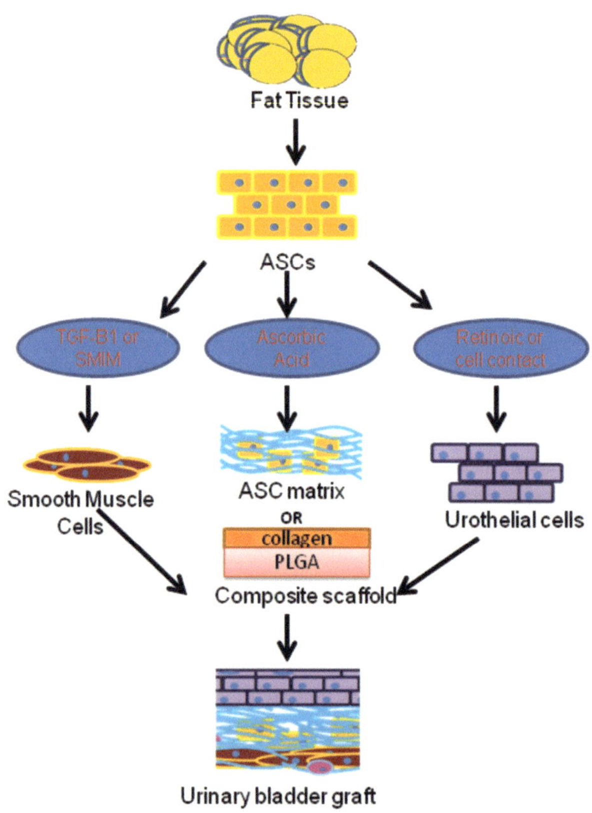

Figure 2. Shows the possible contributions of ASCs in tissue engineering of urinary bladder

http://dx.doi.org/10.5772/59223

4.3.1.2. Tissue engineering of the urethra

Tissue engineering of urethra, using seeded or unseeded scaffolds has been used with success in preclinical studies and clinical trials [93]. Among the available models, a biomaterial made by the self-assembly technique was fabricated from dermal fibroblasts and seeded with urothelial cells [85]. As preliminary formation of UC-seeded ASCs matrices for bladder replacement was achieved [89], it is expected that the same approach in tubularized form can be used for urethral regeneration.

ASCs have been employed to constitute urinary tract epithelium [97] and SMCs [75] in urethral tissue engineering. In the first study, ASCs were differentiated into urothelial cells with DMEM (supplemented with 2% FBS, all-trans retinoic acid, epidermal growth factor, hepatocyte growth factor, keratinocyte growth factor, hydrocortisone) at the air liquid interface to promote urothelial differentiation. Then, they were seeded on bladder acellular matrix to be implanted in rabbits. The urethral continuity was maintained with large caliber and the labelled differentiated urothelial cells survived and formed a stratified epithelial layer. In the second study, ASCs were induced chemically with 5-azacytidine (5-AZA) to differentiate into SMCs under mechanical extension stimulation. The autologous induced cells together with oral mucosal epithelium cells were seeded on PGA mesh to replace urethral defects in dogs. The mechanical extension-stimulated engineered urethras were developed into more normal architectures resembling nearby native urethra. Whether they are source of SMCs or UC, ASCs, with their advantages, can contribute significantly in urethral tissue engineering

4.3.1.3. Tissue engineering of urinary conduit

Urinary diversion after cystectomy with gastrointestinal segments is the current treatment for patients with bladder cancer or benign end-stage bladder diseases as bladder exstrophy or neurogenic bladder due to spina bifida. It can be either continent or incontinent diversion depending on the patient ability to control urine evacuation. Incontinent urinary diversion with ileal conduit is common; however it still carries the complications of the use of gastrointestinal segments. That is why a new approach for construction of tissue engineered urothelium-lined urinary conduit was developed.

A limited number of studies incorporated the use of adult cells from the urinary tract to construct neourinary conduit (NUC) [98]. Nevertheless, autologous cells in these disease states are not appropriate source for formation of NUC. Stem cells differentiated to mature urinary tract cells may be a suitable alternative for the development of NUC. ASCs as a source of easily obtained autologous stem cells can be an ideal option. A study led by Tengion™ [99] used tubular scaffolds made of PLGA seeded with adipose-derived SMCs in a porcine cystectomy model. Implantation of these constructs led to the formation of patent tubular neo-organ that was histologically comparable to native urinary bladder. They had well developed urothelial and smooth muscle layers with minimal collagen deposition. As a result, Tengion has started Phase I clinical trials of NUC constructs in human patients requiring urinary diversion. This Phase I study "Incontinent Urinary Diversion Using an Autologous NeoUrinary Conduit" (http://www.clinicaltrials.gov/ct2/show/NCT01087697) is currently recruiting patients using autologous adipose-derived SMC with biodegradable biomaterial.

4.3.1.4. *Tissue engineering of sling materials for urinary incontinence and pelvic organ prolapse*

Synthetic slings used in treatment of urinary incontinence (SUI) and pelvic organ prolapse (POP) have serious complications such as tissue erosion [100]. Biodegradable biological materials; whether xenografts or allografts; are more likely to undergo tissue remodeling and less likely to cause erosion [101]. They are designed to provide a scaffold of acellular material to facilitate autologous cell infiltration and subsequent replacement of the graft tissue with regenerated functional host tissue. However, they have the drawbacks of poor integration into the host and rapid degradation with early recurrence [102]. A new approach was developed to use a cell-seeded bioabsorbable material to attain a long-term repair. Addition of autologous cells will help the formation of new ECM, leading to long-term mechanical integrity while the degradable nature of the scaffold should avoid a harmful chronic inflammatory response, allowing integration, neovascularization, and remodeling.

In two in vitro studies, ASCs were seeded onto biodegradable materials and the mechanical properties and cell attachment were observed. In the first one, human ASCs were seeded on a collagen mesh from porcine dermis. An increase in the mechanical properties after cell seeding was observed [103]. In the second study, the authors compared the use of human oral fibroblasts (OF) and ASCs seeded on poly-L-lactic acid (PLA) scaffolds as candidate cell types for the development of a pelvic floor tissue engineered repair material (TERM). They found that both cells were well attached and proliferated on scaffolds. The addition of any of the two cell types led to improvement in the mechanical properties compared to non-seeded scaffolds in vitro. In addition, ASCs produced more total collagen and a denser homogenous ECM than OF with unrestrained scaffolds. The authors concluded that OF and ASC both appeared to be suitable cell types to combine with biodegradable scaffolds, in the development of a TERM for the treatment of SUI and POP [104]. Morever, autologous ASCs have the other advantages of modulating the inflammatory response and promoting a weaker foreign body reaction, which could have a beneficial impact in patients treated with cellularized meshes. Also, ASCs decrease the degradation rate of the collagen meshes and enhance angiogenesis. However, it remains to examine the in vivo response to these cell seeded meshes in animal studies as well as their degradation rate and changes in mechanical properties over time.

Preclinical studies for urinary reconstruction are included in table 4.

Involved organ/ disease	Nature of the study/ disease model	Cells used	Functional Assessment	Notes	References
Ureteral replacement	Athymic mice	Human ASCs differentiated into urothelial cells	Not available	The grafts was implanted subcutaneously for 14 days	91
	Normal rabbits	Autologous rabbit ASCs differentiated into smooth muscle cells	Intravenous Urography (IVU)	IVU demonstrated no ureteral stricture or hydroureteronephrosis.	92

http://dx.doi.org/10.5772/59223

Involved organ/ disease	Nature of the study/ disease model	Cells used	Functional Assessment	Notes	References
Bladder replacement	In vitro study	Human cultured unmodified ASCs	Not available	ASCs formed matrix graft with good mechanical properities	89
	Normal rabbits	Autologous cultured ASCs were seeded on bladder acellular matrix.	Cystography. Normal bladder capacity was acquired.	Exogenous scaffold was used.	96
	Normal Rats	Autologous rat ASCs were seeded on decelluarized prepuce dynamically	Cystometry. It showed better parameters with seeded group.	ASCs were not labeled. Small number of animals were used.	65
	Athymic rats	Human ASCs differentiated into SMCs were seeded on PLGA	Cystometry. It showed better compliance with ASCs-SMC seeded group.	Labeled ASCs survived and formed organized muscle tissue.	73
Urethral replacement	Normal rabbits	Autologous cultured ASCs and urothelial - differentiated ASCs were seeded on bladder acellular matrix.	Urethrography. It revealed restoration of urethral continuity with only urothelial-differentiated cell seeded constructs	BrdU-labeled cells survived in vivo transplantation.	97
	Normal canine model	Autologous SMC-differentiated ASCs and oral epithelial cells were seeded on PGA scaffolds	Urethrography. It showed slight strictures at the site of implantation	The use of bioreactor improved the characters and outcome of engineered graft	75
Urinary conduit	porcine cystectomy model	Autologous Adipose-derived SMCs were seeded on PLGA scaffolds	Not available	Urothelial and smooth muscle regeneration with no fibrosis	99
Pelvic sling material	In vitro study	Human unmodified ASCs seeded on porcine dermis	Not available	Increase in mechanical properties after cell seeding	103
	In vitro study	Human oral fibroblasts and	Not available	Both cell types had good results regarding	104

Involved organ/ disease	Nature of the study/ disease model	Cells used	Functional Assessment	Notes	References
		unmodified ASCs were seeded on PLA scaffolds		cell attachment and mechanical properties	

Table 4. Shows the different studies involving the use of ASCs in urinary tract reconstruction.

5. Cellular therapy

5.1. Voiding dysfunction

There are approximately 400 million patients with bladder diseases, and a lot of them might eventually need bladder reconstruction [66]. The International Continence Society and International Urogynaecological Association define voiding dysfunction as abnormally slow and/or incomplete micturition (voiding) based on symptoms and urodynamic investigations [105]. Voiding dysfunction usually presents in 2 forms; lower urinary tract symptoms (LUTS) and urinary tract decomposition [106]. Both affect the patient's quality of life and survival. It can be classified functionally into failure to store or failure to empty urine or both; and anatomically, bladder dysfunction and bladder outlet dysfunction or both [107]. Bladder outlet obstruction (BOO) causes voiding dysfunction through increased collagen deposition, detrimental changes in ultrastructure of bladder SMCs and decrease blood flow [108]. All lead to impaired smooth muscle function and decreased bladder compliance. The inadequate efficacy of current pharmacological treatment and invasiveness of other modalities has supported the search for new stable therapeutic modalities for voiding dysfunction including bladder overactivity or underactivity. Additionally, none of the current treatments are able to modify the pathologic effects in the diseased bladders.

Possible mechanisms for a role of ASCs for the treatment of voiding dysfunction

1. Engraftment:

When delivered either locally or systemically, ASCs are recruited to the affected tissue with the effect being more pronounced with the local delivery. Homing cytokines, such as stromal derived factor-1, has been shown to attract ASCs to the site of injury [109]. After an acute injury, however, homing of transplanted stem cells may not be achieved because of the minimal amount of tissue damage, which subsequently leads to less cytokine expression [110].

2. Differentiation

Few studies have shown in vivo differentiation of ASCs after transplantation [64, 65]. SMC differentiation may occur due to the plasticity of ASCs and the effect of local bladder micro-environment rather than cell fusion [64]. The low evidence of in vivo SMC differentiation can be explained by the fact that labelled DNA in dividing cells is quickly diluted by cell divisions whereas dilution takes much more time for slow dividing stem cells [111] and the rapid wash of the cells from the desired sites into circulation with lack of time needed for phenotype change. That is why in vitro differentiation of ASCs before delivery may be needed if SMC regeneration in vivo is strongly required.

3. Paracrine effect

Growing evidence has been shown that the beneficial effects of ASCs are largely due to paracrine actions with the release of cytokines and growth factors by the transplanted cells or neighbouring cells [112]. These cytokines and growth factors result in anti-inflammatory, musculotropic, angiogenic, antifibrotic and antiapoptotic actions. These actions cause modulation of local and systemic inflammatory responses and mobilization, stimulation and differentiation of local stem cells, promotion of vascularisation of regenerating tissues and reduction of fibrosis [113]. The secretome of ASCs includes: hepatocyte growth factor (HGF), insulin-like growth factor-1 (IGF-1), vascular endothelial growth factor (VEGF) and basic fibroblast growth factor (bFGF) [114]. In a rat model of BOO, human ASCs increased sequence-specific transcription of Oct4, Sox2, and Stella in the submucosal and muscle layer of the rat bladders. These are markers for primitive pluripotent stem cells. In addition, ASCs enhanced the expression of several genes responsible for stem cell trafficking, including SDF-1/CXCR4, HGF/cMet, PDGF/PDGFR, and VEGF/VEGFR signaling axis. Through these paracrine effects, ASCs caused the stimulation and mobilization of endogenous stem cells [115].

5.2. Preclinical studies

ASCs could potentially reverse many of the bladder pathologic changes in different animal models [113]. ASCs alleviated the symptoms of bladder dysfunction in various animal models of detrusor overactivity [115, 116] or underactivity [117] or variable spectrum of voiding dysfunction [112]. Also, ASCs seemed to preserve the bladder vascularity and decrease apoptosis. Human ASCs decreased the frequency and irregularity of detrusor contractions and slightly increased their amplitude when injected into the rat bladders subjected to outlet obstruction [112]. This suggests the possibility of allogenic stem cells transfer for people with perturbed stem cell depot as in diabetic or geriatric populations. There is no known human trial incorporating the use of ASCs for treatment of voiding dysfunction.

ASCs differentiated into SMC before local injection; have been shown to survive and increase SMC content at the injury site. However, no record on the improvement of bladder function after injection was reported [118]. Although systemic injection of ASCs has improved voiding dysfunction in animals, as seen with local injection into urinary bladder, like other MSCs, it may have serious side effects such as hemodynamic compromise, respiratory distress and impeding of pulmonary gas exchange that hinder its adoption as a regular route of delivery [119].

It is important to note that ASCs can be useful in early stages of voiding dysfunction before severe affection of the bladder wall happens. This beneficial effect may be preventive (arrest of further pathologic effects) or ameliorative (correct existing pathologic effects) or both. The exact underlying mechanisms, the magnitude and type of positive outcomes and durability need to be furtherly investigated.

5.3. Urinary incontinence

There are 17 million people in the USA and more than 200 million people worldwide who live with urinary incontinence [120, 121]. Stress urinary incontinence (SUI) is the most common type of urinary incontinence. SUI is involuntary leakage of urine with sudden increase in the intra-abdominal pressure. It happens when intra-abdominal pressure causes the bladder pressure to exceed the urethral closure pressure. SUI is classified into three conditions, including intrinsic sphincter deficiency (ISD), urethral hypermobility, or a combination of both [122]. Most of the patients have both disorders in varying degrees [123]. SUI affects both males and females and decreases quality of life [124]. Many injectable bulking agents are minimally invasive but have a poor long-term efficacy and complications such as chronic inflammatory reactions, particle migration, periurethral abscess and erosion [125]. More invasive approaches, like sling procedures, bladder neck suspensions or artificial urinary sphincter implantation are more effective but have higher morbidity [126, 127]. More importantly, none of these therapies replace the deficient urethral sphincter. The ideal strategy for treating SUI using stem cell therapy besides being a bulking agent would be to allow for the regeneration of functional periurethral tissues, provide adequate mucosal coaptation and restore or improve resting urethral closure pressures [128].

5.4. Cell source for injection therapy

The ideal cells for cell therapy should be easily procured from minimally invasive procedures, proliferate quickly in a well-controlled manner, provide sufficient quantities of cells, exhibit capabilities of differentiation to regenerate multiple tissues, and be able to be transplanted into an autologous host [129]. Currently, bone marrow-derived stem cells (BMSCs), adipose-derived stem cells (ASCs), and muscle-derived stem cells (MDSCs) are the stem cell sources applied in SUI therapy. ASCs carry future special importance in this regard due to its reported myoblast and neuronal-like differentiation capacity and neovascularization potential in addition to their ease of harvest and high stem cell content [130].

5.5. Preclinical studies and clinical trials

ASCs were used in many animal studies as SVF or cultured ASCs or differentiated into myoblasts or coupled with biomaterials or with growth factors. Unmodified ASCs were labelled and injected periurethrally and systemically into female rats after induction of SUI postpartum vaginal balloon dilation and bilateral ovariectomy. There was improvement in cystometric parameters in 2/3 of the rats treated locally or systemically in relation to control animals with increase in muscle and elastin content. The ASCs beneficial effects were attributed mainly to growth factors that supported host tissue regeneration as most of the delivered ACSs remained undifferentiated after injection [131]. ASCs were also used in another model of SUI after injury to pudendal nerve injury. ASCs were labelled and injected at 3, 9, and 12 o'clock around the mid and distal urethra. Urodynamic evaluations revealed considerable improvements in maximum bladder capacity, abdominal leak point pressure, maximum urethral closure pressure and functional urethral length. Morphologic changes and significant improvement in urination control were consistent over time. Labeled cells gradually migrated in vivo toward the urethra and its lumen from urethra's edges [132].

ASCs were differentiated into myoblasts using 5-AZA and injected in the posterior urethra after induction of SUI in rats. Maximal bladder capacity and abdominal leak point pressure (ALLP) significantly increased 1 and 3 months after implantation with unmodified and differentiated rat ASCs with better results in case of differentiated ASCs [133].

ASCs coupled with biodegradable microbeads as carriers improved in abdominal leak point pressure (ALPP) and retrograde urethral perfusion pressures (RUPP) in a rat model of SUI [134]. ASCs in combination with nerve growth factor (NGF) and PLGA resulted in significant improvements in ALPP and RUPP as well as the amount of muscle and ganglia when compared to ASCs alone [135].

Few clinical trials are incorporating the use of ASCs for treatment of SUI (www.clinicaltrials.gov). In a clinical trial, 11 male patients with persistent post-prostatectomy SUI received ASCs in 2 fractions; ASCs alone and mixed with fat. SUI improved progressively in eight patients during the 1-year follow up, as determined by a 59.8% decrease in the leakage volume in the 24h pad test, decreased frequency and amount of incontinence, and improved quality of life. One patient achieved total continence up to 12 months after stem cell injection [136].

In a pilot study, 5 female patients with SUI were included to be treated with ASCs combined with bovine collagen gel and saline. The ASCs mixture with collagen was injected endoscopically through the urethra. The effect of the treatment was assessed objectively with cough test as a primary end point and subjectively with validated questionnaire. At 1 year, the cough test was negative for three patients; two of them were satisfied with the treatment and did not wish further treatment for SUI. Validated questionnaires showed some subjective improvement in all five patients [137].

Preclinical studies and clinical trials for cellular therapy are included in table 5.

Involved disease	Nature of the study/ disease model	Cells used	Functional Assessment	Notes	References
Voiding dysfunction	BOO	Cultured Human ASCs injected into rat bladder wall	UDS. Decrease bladder overactivity (frequency and irregularity of contractions) with increase in bladder voiding pressure.		115
		Autologous cultured ASCs and muscle precursor cells	UDS. Micturiting pressure (maximum and threshold) and		117

Involved disease	Nature of the study/ disease model	Cells used	Functional Assessment	Notes	References
		(MPCs) injected into rat bladder.	voided volumes increased.		
	Diabetes Mellitus	Autologous cultured ASCs injected in bladder wall or tail vein of diabetic type II rats.	UDS. It showed Diabetic Voiding dysfunction improvement in 40-60 %.	Improvement with local (bladder) injection is more effective than systemic (tail vein) injection	112
	Hyperlipidemia	Autologous cultured ASCs injected into bladder or tail vein of hyperlipdemic rat	Improved micturition frequency and voided volumes	Improvement with direct (bladder) injection is more efficient than systemic (tail vein) injection	117
	Cryo-injury	Human ASCs differentiated into SMCs and injected into cryo-injured bladder wall of mice.	Not available	There was an Increase in the ASMA positive area of injured Bladder. The injected labeled cells were detected in vivo.	118
Urinary incontinence	Stress urinary incontinence (SUI) -rat model	Autologous unmodified ASCs	UDS with different measures including ALPP, RUPR and bladder capacity	SUI was induced by vaginal balloon dilation and bilateral Ovariectomy.	131
		Unmodified ASCs and ASCs differentiated into myoblasts			133

Involved disease	Nature of the study/ disease model	Cells used	Functional Assessment	Notes	References
		Autologous unmodified ASCs		SUI was induced by pudendal nerve injury.	135
		Cultured ASCs with PLGA microbeads		SUI was induced by urethrolysis.	134
		Autologous cultured ASCs with PLGA or NGF or both		SUI was induced by bilateral pudendal nerve transection	135
	Postprostatectomy urinary incontinence – clinical trial	11 patients received autologous ASCs with and without fat.	Frequency and amount of incontinence, daily leakage volume, UDS and ICIQ-SF	The cells were injected endoscopicall y into the region of external urethral sphincter and submuocal space.	136
	Female SUI-clinical trial	5 female patients received ASCs combined with bovine collagen gel and saline	Cough test and questionnaire	The mixture was injected transurethrall y via cystoscope	137

BOO=bladder outlet obstruction SMC=smooth muscle cells / ASMA=smooth muscle α-actin /PGA=Poly-Glycolic acid

PLGA=poly(lactic-co-glycolic acid) /NGF=Nerve growth factor

ICIQ-SF=The International Consultation on Incontinence Questionnaire-Short Form (ICIQ-SF)

UDS=Urodynamic study

ALPP=abdominal leak point pressure

RUPP=retrograde urethral perfusion pressure

Table 5. Shows the different animal studies and human trials with ASCs as a cellular therapy for lower urinary tract.

6. Hurdles and future directions

There are many hurdles that face the broad utilization of ASCs in clinical therapies for urinary tract diseases. First of all, is the limited ASCs efficiency for the treatment of chronic diseases. This can be explained by the fact that in chronic pathologies, there are less release of cytokines and so minimal attraction of ASCs to the site of the disease [138]. Research is needed to improve the accuracy of predictions regarding disease progression. Similarly, the time delay between injury and the initiation of treatment represents an obstacle as after an acute injury, homing signals are often diminished with time. Therefore, it is important to determine the duration of time after injury that equates to the optimal release of homing bioactive factors and to develop innovative methods for up-regulating the expression of these cytokines. Methods such as electrical stimulation or local injection of homing cytokines could help the direct recruitment of systemically delivered stem cells to the target organ [139].

The second is the form of delivery of ASCs; whether differentiated or undifferentiated cells would allow better regeneration of the urinary tract. It would be interesting to investigate whether pre-differentiation of ASCs into the targeted tissue cell types would increase their benefits and help engraftment without affecting their secretomes. A mix of differentiated and undifferentiated cells may be a good option.

Moreover, there is no final agreement on the preferred type of cells to use (SVF cells or cultured ASCs), total number of cells for the treatment or the number of cells for single injection. Although recent studies showed that after cell labelling, ASCs migrate into the bone marrow after systemic or local delivery [140,109,141], however, more precise imaging studies are required to to observe the fate of the implanted ASCs regarding cell survival, proliferation, migration and formation of functional tissues [142]. Live image tracing might be a good choice for that purpose [143]. Neovascularization and reinnervation of engineered tissues are critical obstacles to overcome for future application to bladder reconstruction. Ingrowth of native vessels and nerves—which is stimulated by MSCs bioactive factors—is achievable for smaller grafts; however, complete regeneration with functional integration is far more challenging.

ASCs are able to secret some cytokines, which can modulate the host inflammatory response, and guards against severe inflammatory and fibrotic responses damage in the urinary tract [144]. The contributions of each of the cytokines and growth factors to the repair of urinary tract need to be evaluated. Finally, the safety of ASCs-based treatment needs to be carefully checked as ASCs have the potential risk of tumorigenic transformation [145]. Hence, more chronic animal models, consistent protocols and many clinical trials are required to make sure of ASCs therapeutic efficacy and safety.

7. Conclusions

Tissue Engineering and Regenerative Medicine is an emerging field of research for organ and tissue replacement to compensate for the deficiency of organ donation and complications of immunosuppression. In the search of new therapeutic options for urinary tract disorders, both SVF and cultured ASCs have been the focus of numerous in vitro studies, various animal models and few clinical trials. Depending on the type of dysfunction to be treated, ASCs can be used either as cellular therapies or combined with scaffolds for tissue-engineering applications. Although few clinical trials showed promising results, however, more future clinical studies are required to prove their efficacy for those particular applications while exploring the mechanisms ensuring their functional activity. Clinical trials are listed for the use of ASCs in different diseases including urinary incontinence, renal failure, ischemic nephropathy and urethral strictures in males (http://www.clinicaltrials.gov/).

Acknowledgements

We acknowledge the Canadian Institutes of Health Research (grant #79437), Canadian Male Sexual Health Council (CMSHC), Fonds de la Recherche du Québec en santé (FRQ-S).

Author details

Hazem Orabi, Cassandra R. Goulet, Julie Fradette and Stéphane Bolduc*

*Address all correspondence to: stephane.bolduc@fmed.ulaval.ca

Centre de recherche en organogénèse expérimentale de l'Université Laval / LOEX, Division of Regenerative Medicine, CHU de Québec Research Centre. Québec City, Canada

References

[1] Traktuev DO, Merfeld-Clauss S, Li J, Kolonin M, Arap W, Pasqualini R, Johnstone BH, March KL. A population of multipotent CD34-positive adipose stromal cells share pericyte and mesenchymal surface markers, reside in a periendothelial location, and stabilize endothelial networks. Circ Res 102(1):77-85, 2008.

[2] Zuk PA, Zhu M, Ashjian P, De Ugarte DA, Huang JI, Mizuno H, Alfonso ZC, Fraser JK, Benhaim P, Hedrick MH. Human adipose tissue is a source of multipotent stem cells. Mol Biol Cell 13(12):4279-4295, 2002.

[3] Zuk PA, Zhu M, Mizuno H, Huang J, Futrell JW, Katz AJ, Benhaim P, Lorenz HP, Hedrick MH. Multilineage cells from human adipose tissue: implications for cell-based therapies. Tissue Eng 7(2):211-228, 2001.

[4] Witkowska-Zimny M, Walenko K. Stem cells from adipose tissue. Cell Mol Biol Lett 16(2): 236-257, 2011.

[5] Romanov YA, Darevskaya AN, Merzlikina NV, Buravkova LB. Mesenchymal stem cells from human bone marrow and adipose tissue: isolation, characterization, and differentiation potentialities. Bull Exp Biol Med 140(1):138-143, 2005.

[6] Zannettino AC, Paton S, Arthur A, Khor F, Itescu S, Gimble JM, Gronthos S. Multipotential human adipose-derived stromal stem cells exhibit a perivascular phenotype in vitro and in vivo. J Cell Physiol 214(2):413-421, 2008.

[7] Casteilla, L. & Dani, C. Adipose tissue-derived cells: from physiology to regenerative medicine. Diabetes Metab. 32, 393–401, 2006.

[8] Cawthorn WP, Scheller EL, MacDougald OA, Adipose tissue stem cells meet preadipocyte commitment: going back to the future, J Lipid Res, 53(2):227–246, 2012.

[9] Fraser J K, Wulur I, Alfonso Z, Hedrick M H. Fat tissue: an underappreciated source of stem cells for biotechnology. Trends Biotechnol. 24 150-154, 2006

[10] Choi Y S, Cha S M, Lee Y Y, Kwon S W, Park C J, Kim M. Adipogenic differentiation of adipose tissue derived adult stem cells in nude mouse. Biochem Biophys Res Commun. 345 631-637, 2006.

[11] Jeon E S, Moon H J, Lee M J et al.. Sphingosylphosphorylcholine induces differentiation of human mesenchymal stem cells into smooth-muscle-like cells through a TGF-beta-dependent mechanism. J Cell Sci 119 (Pt 23) 4994-5005, 2006.

[12] Eto H, Suga H, Matsumoto D et al.. Characterization of structure and cellular components of aspirated and excised adipose tissue. Plast Reconstr Surg. 124 1087-1097, 2009.

[13] Planat-Benard, V., Menard, C., Andre, M., Puceat, M., Perez, A., Garcia-Verdugo, J. M., Penicaud, L. & Casteilla, L. Spontaneous cardiomyocyte differentiation from adipose tis-sue stroma cells. Circ. Res. 94, 223–229, 2004.

[14] Awad, H. A., Wickhan, M. Q., Leddy, H. A. Gimble, J. M. & Guilak, F. Chondrogenic differentiation of adipose-derived adult stem cells in agarose, alginate, and gelatin scaffolds. Biomaterials 25, 3211–3222, 2004.

[15] Okura, H., Komoda, H., Fumimoto, Y., Lee, C. M., Nishida, T., Sawa, Y. & Matsuyama, A. Transdifferentiation of human adipose tissue-derived stromal cells into insulin-producing clusters. J. Artif. Organs 12, 123–130, 2009.

[16] Konno, M., Hamazaki, T. S., Fukuda, S., Tokuhara, M., Uchiyama, H., Okazawa, H., Okochi, H. & Asashima, M. Efficiently differentiating vascular endothelial cells from adipose tissue-derived mesenchymal stem cells in serum-free culture. Biochem. Biophys. Res. Commun. 400, 461–465, 2010.

[17] Neupane M, Chang CC, Kiupel M, et al. Isolation and characterization of canine adipose-derived mesenchymal stem cells. Tissue Eng Part A 14(6): 1007-15, 2008.

[18] Senesi S, Marcolongo P, Manini I, et al. Constant expression of hexose-6-phosphate dehydrogenase during differentiation of human adipose-derived mesenchymal stem cells. J Mol Endocrinol 41(3): 125-33, 2008.

[19] Pilgaard L, Lund P, Duroux M, et al. Transcriptional signature of human adipose tissue-derived stem cells (hASCs) preconditioned for chondrogenesis in hypoxic conditions. Exp Cell Res 315(11): 1937-52, 2009.

[20] Ogawa R, Mizuno S, Murphy GF, et al. The Effect of hydrostatic pressure on 3-D chondroinduction of human adipose-derived stem cells. Tissue Eng Part A 15(10): 2937-45, 2009.

[21] Mizuno H, Zuk PA, Zhu M, et al. Myogenic differentiation by human processed lipoaspirate cells. Plast Reconstr Surg 109(1): 199-209; discussion 210-1, 2002.

[22] Gwak SJ, Bhang SH, Yang HS, et al. In vitro cardiomyogenic differentiation of dipose-derived stromal cells using transforming growth factor-beta1. Cell Biochem Funct 27(3): 148-54, 2009.

[23] Van Dijk A, Niessen HW, Zandieh Doulabi B, et al. Differentiation of human adipose-derived stem cells towards cardiomyocytes is facilitated by laminin. Cell Tissue Res 334(3): 457-67, 2008.

[24] Banas A, Teratani T, Yamamoto Y, et al. Adipose tissue-derived mesenchymal stem cells as a source of human hepatocytes[J]. Hepatology 46(1): 219-28, 2007.

[25] Sgodda M, Aurich H, Kleist S, et al. Hepatocyte differentiation of mesenchymal stem cells from rat peritoneal adipose tissue in vitro and in vivo. Exp Cell Res 313(13): 2875-86, 2007.

[26] Van Poll D, Parekkadan B, Borel Rinkes I et al. Mesenchymal stem cell therapy for protection and repair of injured vital organs. Cell Mol Bioeng 1:42–50, 2008.

[27] Luis Fernando Arenas da Silva, Lionel Micol, Dorien Tiemessen, Toin H. van Kuppevelt, Peter Frey, Egbert Oosterwijk, Paul Geutjes, and Wout F. Feitz. Is There a Need for Smooth Muscle Cell Transplantation in Urethral Reconstruction? TISSUE ENGINEERING: Part A Volume 20, Numbers 9 and 10, 2014.

[28] Rehman J, Traktuev D, Li J et al. Secretion of angiogenic and antiapoptotic factors by human adipose stromal cells. Circulation 109:1292–1298, 2004

[29] Salgado AJ, Reis RL, Sousa NJ et al. Adipose tissue derived stem cells secretome: Soluble factors and their roles in regenerative medicine. Curr Stem Cell Res Ther 5:103–110, 2010.

[30] Kilroy GE, Foster SJ, Wu X et al. Cytokine profile of human adipose-derived stem cells: Expression of angiogenic, hematopoietic, and pro-inflammatory factors. J Cell Physiol 212:702–709, 2007.

[31] McTaggart SJ, Atkinson K. Mesenchymal stem cells: Immunobiology and therapeutic potential in kidney disease. Nephrology (Carlton) 12:44–52, 2007.

[32] McIntosh. K.R. Evaluation of cellular and humoral immune responses to allogeneic adipose-derived stem/stromal cells. Methods Mol Biol, 702;133–150, 2011.

[33] Li, M.O., Flavell. R.A. Contextual regulation of inflammation: a duet by transforming growth factor-x and interleukin-10. Immunity, 28;468–476, 2008.

[34] Meisel, R. et al. Human bone marrow stromal cells inhibit allogeneic T-cell responses by indoleamine 2,3-dioxygenase–mediated tryptophan degradation. Blood 103, 4619–4621, 2004.

[35] Uccelli, A., et al. Immunoregulatory function of mesenchymal stem cells. Eur. J. Immunol., 36;2566–2573, 2006.

[36] Fang, B. Song, Y. Liao, L. Zhang, Y. Zhao. R.C. Favorable response to human adipose tissue-derived mesenchymal stem cells in steroid-refractory acute graft-versus-host disease. Transplant Proc, 39; 3358–3362, 2007.

[37] Gonzalez MA, Gonzalez-Rey E, Rico L et al. Treatment of experimental arthritis by inducing immune tolerance with human adipose-derived mesenchymal stem cells. Arthritis Rheum 60:1006–1019, 2009.

[38] Mesimäki, K. Lindroos, B. Törnwall, J. et al. Novel maxillary reconstruction with ectopic bone formation by GMP adipose stem cells. Int J Oral Maxillofac Surg, 38; 201–209, 2009.

[39] Hsu, R. K. McCulloch, C. E. Dudley, R. A. Lo, L. J. Hsu, C.-Y. Temporal Changes in Incidence of Dialysis-Requiring AKI. Journal of the American Society of Nephrology, 24, 37-42, 2012.

[40] Shih YC, Lee PY, Cheng H, Tsai CH, Ma H, et al. Adipose-derived stem cells exhibit antioxidative and antiapoptotic properties to rescue ischemic acute kidney injury in rats. Plast Reconstr Surg 132: 940e–951e, 2012.

[41] Morigi M, Imberti B, Zoja C et al. Mesenchymal stem cells are renotropic, helping to repair the kidney and improve function in acute renal failure. J Am Soc Nephrol 15: 1794–1804, 2004.

[42] Burst VR, Gillis M, Putsch F et al. Poor cell survival limits the beneficial impact of mesenchymal stem cell transplantation on acute kidney injury. Nephron Exp Nephrol 114: e107–e116, 2010.

[43] Gao et al. The use of chitosan based hydrogel for enhancing the therapeutic benefits of adipose-derived MSCs for acute kidney injury. Biomaterials 33, 2012.

[44] Donizetti-Oliveira et al. Adipose Tissue-Derived Stem Cell Treatment Prevents Renal Disease Progression. Cell Transplantation, Volume 21, Number 8, 1727-1741(15), 2012.

[45] Chen, Y.-T. Sun, C.-K. Lin Y.-C. et al., "Adipose-derived mesenchymal stem cell protects kidneys against ischemia-reperfusion injury through suppressing oxidative stress and inflammatory reaction," Journal of Translational Medicine, vol. 9, article 51, 2011.

[46] Asanuma, H., Meldrum, D. R. and Meldrum, K. K. Therapeutic applications of mesenchymal stem cells to repair kidney injury. J. Urol. 184, 26–33, 2010.

[47] Pino, C. J. and Humes, H. D. Stem cell technology for the treatment of acute and chronic renal failure. Transl. Res. 156, 161–168, 2010.

[48] De Groot, K., Bahlmann, F. H., Sowa, J., Koenig, J., Menne, J., Haller, H. and Fliser, D. Uremia causes endothelial progenitor cell deficiency. Kidney Int. 66, 641–646, 2004.

[49] Noh, H., Yu, M. R., Kim, H. J., Jeon, J. S., Kwon, S. H., Jin, S. Y., Lee, J., Jang, J., Park, J. O., Ziyadeh, F. et al. Uremia induces functional incompetence of bone marrow-derived stromal cells. Nephrol. Dial. Transplant. 27, 218–225, 2012.

[50] Roemeling-van Rhijn, M., Reinders, M. E., de Klein, A., Douben, H., Korevaar, S. S., Mensah, F. K., Dor, F. J., IJzermans, J. N., Betjes, M. G., Baan, C. C. et al. Mesenchymal stem cells derived from adipose tissue are not affected by renal disease. Kidney Int. 82, 7 48–758, 2012.

[51] Li K, Han Q, Yan X, Liao L, Zhao RC. Not a process of simple vicariousness, the differentiation of human adipose-derived mesenchymal stem cells to renal tubular epithelial cells plays an important role in acute kidney injury repairing. Stem Cells Dev. Aug;19(8):1267-75, 2010.

[52] Villanueva S, Carreño JE, Salazar L, Vergara C, Strodthoff R, Fajre F, Céspedes C, Sáez PJ, Irarrázabal C, Bartolucci J, Figueroa F, Vio CP: Human mesenchymal stem cells derived from adipose tissue reduce functional and tissue damage in a rat model of chronic renal failure. Clin Sci (Lond) 125:199–210, 2013.

[53] Daher, S. R., Johnstone, B. H., Phinney, D. G. and March, K. L. Adipose Stromal/Stem Cells: Basic and Translational Advances: The IFATS Collection. STEM CELLS, 26: 2664–2665, 2008.

[54] Matas, A. J., Smith, J. M., Skeans, M. A., Lamb, K. E., Gustafson, S. K., Samana, C. J., Stewart, D. E., Snyder, J. J., Israni, A. K. and Kasiske, B. L. OPTN/SRTR 2011 Annual Data Report: Kidney. American Journal of Transplantation, 13: 11–46, 2013.

[55] Singaravelu K, Padanilam BJ. In vitro differentiation of MSC into cells with a renal tubular epithelial-like phenotype. Renal Failure. 31:492–502, 2009.

[56] Feng B, LaPerle JL, Chang G, and Varma MV. Renal clearance in drug discovery and development: molecular descriptors, drug transporters and disease state. *Expert Opin Drug Metab Toxicol* 6:939–952, 2010.

[57] Trivedi H. L., Vanikar A. V., Kute V. B., Patel H. V., Gumber M. R., Shah P. R., Dave S. D. Pretransplant Infusion of Donor Stem Cells Open Gateway to Tolerance

through Induction of Regulatory T Cells and Activation of Other Allosuppressive Immune Mechanisms – Single Centre Experience in Living Donor Renal Transplantation. CellR4 1 (3): e528, 2013.

[58] Furuichi, K. Shintani, H. Sakai Y. et al., "Effects of adipose-derived mesenchymal cells on ischemia-reperfusion injury in kidney," Clinical and Experimental Nephrology, vol. 16, no. 5, pp. 679–689, 2012.

[59] Vanikar et al. Pre-transplant co-infusion of donor-adipose tissue derived mesenchymal stem cells and hematopoietic stem cells may help in achieving tolerance in living donor renal transplantation. Ren Fail, 36(3): 457–460, 2014.

[60] Kato T, Okumi M, Tanemura M, Yazawa K, Kakuta Y, Yamanaka K, Tsutahara K, Doki Y, Mori M, Takahara S, Nonomura N. Adipose Tissue-Derived Stem Cells Suppress Acute Cellular Rejection by TSG-6 and CD44 Interaction in Rat Kidney Transplantation. Transplantation. 2014 Jun 30.

[61] Gimble JM. Adipose tissue-derived therapeutics. Expert Opin Biol Ther 3:705–13, 2003.

[62] Lin G, Banie L, Ning H, Bella AJ, Lin C-S, Lue TF. Potential of adipose-derived stem cells for treatment of erectile dysfunction. J Sex Med 6(3 suppl):320–7, 2009.

[63] Lin G, Garcia M, Ning H, Banie L, Gio YL, Lue TF, Lin CS. Defining stem and progenitor. Stem Cells Dev. 6:1053-63, 2008.

[64] Rong Zhang, Gregory S. Jack, Nagesh Rao,Patricia Zuk,Louis J. Ignarro, Benjamin Wu, Larissa V. Rodriguez. Nuclear Fusion-Independent Smooth Muscle Differentiation of Human Adipose-Derived Stem Cells Induced by a Smooth Muscle Environment. Stem Cells 30:481–490, 2012.

[65] . Abdol-Mohammad Kajbafzadeh, Ali Tourchi, Amir-Abbas Mousavian, Leila Rouhi, Seyyed Mohammad Tavangar, Nastaran Sabetkish. Bladder muscular wall regeneration with autologous adipose mesenchymal stem cells on three-dimensional collagen-based tissueengineered prepuce and biocompatible nanofibrillar scaffold. J Pediatr Urol. May 2. pii: S1477-5131(14)00127-2, 2014.

[66] Staack A, Hayward SW, Baskin LS, et al. Molecular, cellular and developmental biology of urothelium as a basis of bladder regeneration. Differentiation. 73:121-133, 2005.

[67] Jian-Guo Shi, Wei-Jun Fu, Xiao-Xiong Wang, Yong-De Xu, Gang Li, Bao-Fa Hong & Kun Hu, Fu-Zhai Cui, Yan Wang, Xu Zhang. Transdifferentiation of human adipose-derived stem cells into urothelial cells: potential for urinary tract tissue engineering. Cell Tissue Res 347:737–746, 2012.

[68] M. Brzoska, H. Geiger, S. Gauer, and P. Baer, "Epithelial differentiation of human adipose tissue-derived adult stem cells," Biochemical and Biophysical Research Communications, vol. 330, no. 1, 142–150, 2005.

[69] J. Liu, J. Huang, T. Lin, C. Zhang, and X. Yin, "Cell-to-cell contact induces human adipose tissue-derived stromal cells to differentiate into urothelium-like cells in vitro," Biochemical and Biophysical Research Communications, vol. 390, no. 3, 931–936, 2009.

[70] Ming Zhang, Yubing Peng, Zhe Zhou, Juan Zhou, Zhong Wang, and Mujun Lu. Differentiation of Human Adipose-derived Stem Cells Co-cultured With Urothelium Cell Line Toward a Urothelium-like Phenotype in a Nude Murine Model. UROLOGY 81: 465.e15e465.e22, 2013.

[71] Hongbin Li,Yuemin Xu, Qiang Fu, and Chao Li. Effects of Multiple Agents on Epithelial Differentiation of Rabbit Adipose-Derived Stem Cells in 3D Culture. TISSUE ENGINEERING: Part A Volume 18, Numbers 17 and 18, 2012.

[72] Basu J, Genheimer C, Guthrie KI, Sangha N, Quinlan SF, Bruce AT, Reavis B, Halberstadt CR, Ilagan R, Ludlow JW. Expansion of the human adipose-derived stromal vascular cell fraction yields a population of smooth muscle-like cells with markedly distinct phenotypic and functional properties relative to mesenchymal stem cells. Tissue Eng Part C Methods 17:843–60, 2011.

[73] Jack GS, Zhang R, Lee M, Xu Y, Wu BM, et al. Urinary bladder smooth muscle engineered from adipose stem cells and a three dimensional synthetic composite. Biomaterials. Jul;30(19):3259-70, 2009.

[74] Harris LJ, Abdollahi H, Zhang P, McIlhenny S, Tulenko TN, DiMuzio PJ. Differentiation of adult stem cells into smooth muscle for vascular tissue engineering. J Surg Res 168:306–14, 2011.

[75] Qiang Fu a,Chen-Liang Deng,Ren-Yan Zhao, Ying Wang, Yilin Cao. The effect of mechanical extension stimulation combined with epithelial cell sorting on outcomes of implanted tissue-engineered muscular urethras. Biomaterials 35 105-112, 2014.

[76] Luis Fernando Arenas da Silva, Lionel Micol, Dorien Tiemessen, Toin H. van Kuppevelt, Peter Frey, Egbert Oosterwijk, Paul Geutjes, and Wout F. Feitz. Is There a Need for Smooth Muscle Cell Transplantation in Urethral Reconstruction? TISSUE ENGINEERING: Part A Volume 20, Numbers 9 and 10, 2014.

[77] F. A. Auger, M. R´emy-Zolghadri, G. Grenier, and L. Germain, "A truly new approach for tissue engineering: the LOEX self assembly technique," Ernst Schering Research Foundation Workshop, no. 35, 73–88, 2002.

[78] F. A. Auger, F. Berthod, V. Moulin, R. Pouliot, and L. Germain, "Tissue-engineered skin substitutes: from in vitro constructs to in vivo applications," Biotechnology and Applied Biochemistry,vol. 39, part 3, 263–275, 2004.

[79] N. L'Heureux, J. C. Stoclet, F. A.Auger, G. J. Lagaud, L.Germain, and R.Andriantsitohaina, "Ahuman tissue-engineered vascular media: a new model for pharmacological studies of contractile responses," The FASEB Journal, vol. 15, no. 2, 515–524, 2001.

[80] S. Bouhout, E. Perron, R. Gauvin et al., "In vitro reconstruction of an autologous, watertight, and resistant vesical equivalent" Tissue Engineering A, vol. 16, no. 5, 1539–1548, 2010.

[81] S. Bouhout, R. Gauvin, L. Gibot, D. Aub´e, and S. Bolduc, "Bladder substitute reconstructed in a physiological pressure environment" Journal of Pediatric Urology, vol. 7, no. 3, 276–282, 2011.

[82] V. Cattan, G. Bernard, A. Rousseau et al., "Mechanical stimuli induced urothelial differentiation in a human tissue-engineered tubular genitourinary graft," European Urology, vol. 60, no. 6, 1291–1298, 2011.

[83] A. Imbeault, G. Bernard, G. Ouellet, S. Bouhout, S. Carrier, and S. Bolduc, "Surgical option for the correction of Peyronie's disease: an autologous tissue-engineered endothelialized graft" Journal of Sexual Medicine, vol. 8, no. 11, pp. 3227–3235, 2011.

[84] M.Magnan, F. Berthod,M. Champigny, F. Soucy, andS.Bolduc, "In vitro reconstruction of a tissue-engineered endothelialized bladder from a single porcine biopsy," Journal of Pediatric Urology, vol. 2, no. 4, pp. 261–270, 2006.

[85] M. Magnan, P. L´evesque, R. Gauvin et al., "Tissue engineering of a genitourinary tubular tissue graft resistant to suturing and high internal pressures," Tissue Engineering A, vol. 15, no. 1, pp.197–202, 2009.

[86] G. Ouellet, J. Dub´e, R. Gauvin, V. Laterreur, S. Bouhout, and S.Bolduc, "Production of an optimized tissue-engineered pig connective tissue for the reconstruction of the urinary tract" Tissue Engineering A, vol. 17, no. 11-12, pp. 1625–1633, 2011.

[87] Fortier GM, Gauvin R, Proulx M, Vallée M, Fradette J. Dynamic culture induces a cell type-dependent response impacting on the thickness of engineered connective tissues. J Tissue Eng Regen Med. Apr;7(4):292-301, 2013.

[88] Trottier V, Marceau-Fortier G, Germain L, Vincent C, Fradette J. IFATS collection: Using human adipose-derived stem/stromal cells for the production of new skin substitutes. Stem Cells. Oct;26(10):2713-23, 2008.

[89] Rousseau A, Fradette J, Bernard G, Gauvin R, Laterreur V, et al. Adipose-derived stromal cells for the reconstruction of a human vesical equivalent. J Tissue Eng Regen Med. Apr 11, 2013

[90] Vasileios Simaioforidis, Paul de Jonge,Marije Sloff, Egbert Oosterwijk, Paul Geutjes, and Wout F.J. Feitz. Ureteral Tissue Engineering: Where Are We and How to Proceed? TISSUE ENGINEERING: Part B, Volume 19, Number 5, 2013.

[91] Shi, J.G., Fu, W.J., Wang, X.X., Xu, Y.D., Li, G., Hong, B.F., Wang, Y., Du, Z.Y., and Zhang, X. Tissue engineering of ureteral grafts by seeding urothelial differentiated hADSCs onto biodegradable ureteral scaffolds. J Biomed Mater Res A 100, 2612, 2012.

[92] Zhao, Z., Yu, H., Xiao, F., Wang, X., Yang, S., and Li, S. Differentiation of adipose derived stem cells promotes regeneration of smooth muscle for ureteral tissue engineering. J Surg Res 178, 55, 2012.

[93] Orabi H, Bouhout S, Morissette A, Rousseau A, Chabaud S, et al. Tissue Engineering of Urinary Bladder and Urethra: Advances from Bench to Patients. Scientific World Journal. 154564, 2013.

[94] Atala, S. B. Bauer, S. Soker, J. J. Yoo, and A. B. Retik. Tissue engineered autologous bladders for patients needing cystoplasty. The Lancet, vol. 367, no. 9518, 1241–1246, 2006.

[95] R. Subramaniam, J. Hinley, J. Stahlschmidt, and J. Southgate, "Tissue engineering potential of urothelial cells from diseased bladders," Journal of Urology, vol. 186, no. 5, 2014–2020, 2011.

[96] Zhu WD, Xu YM, Feng C, Fu Q, Song LJ, Cui L. Bladder reconstruction with adipose-derived stem cell-seeded bladder acellular matrix grafts improve morphology composition. World J Urol. Aug;28(4):493-8, 2010.

[97] Li H, Xu Y, Xie H, Li C, Song L, Feng Cet al. Epithelial-differentiated adipose-derived stem cells seeded bladder acellular matrix grafts for urethral reconstruction: an animal model. Tissue Eng Part A. Feb;20(3-4):774-84, 2014.

[98] Marije Sloff, Rob de Vries, Paul Geutjes, Joanna in't Hout, Merel Ritskes-Hoitinga, Egbert Oosterwijk, Wout Feitz. Tissue Engineering in Animal Models for Urinary Diversion: A Systematic Review. PLoS One. Jun 25;9(6):e98734, 2014.

[99] Basu J, Jayo MJ, Ilagan RM, Guthrie KI, Sangha N, Genheimer CW, Quinlan SF, Payne R, Knight T, Rivera E, Jain D, Bertram TA, Ludlow JW. Regeneration of native-like neo-urinary tissue from nonbladder cell sources. Tissue Eng Part A. May; 18(9-10):1025-34, 2012.

[100] Chen HY, Ho M, Hung YC, et al. Analysis of risk factors associated with vaginal erosion after synthetic sling procedures for stress urinary incontinence. Int Urogynecol J Pelvic Floor Dysfunct 19:117–21, 2008.

[101] Song PH, Kim YD, Kim HT, et al. The 7-year outcome of the tension-free vaginal tape procedure for treating female stress urinary incontinence. BJU Int 104:1113–7. 2009.

[102] Nazemi TM, Kobashi KC. Complications of grafts used in female pelvic floor reconstruction: Mesh erosion and extrusion. Indian J Urol 2007;23:153–160.

[103] Ochoa I, Peña E, Andreu EJ, Pérez-Ilzarbe M, Robles JE, Alcaine C, López T, Prósper F, Doblaré M. Mechanical properties of cross-linked collagen meshes after human adipose derived stromal cells seeding. J Biomed Mater Res A. 2011 Feb;96(2):341-8.

[104] Roman S, Mangera A, Osman NI, Bullock AJ, Chapple CR, MacNeil S. Developing a tissue engineered repair material for treatment of stress urinary incontinence and pelvic organprolapse-which cell source? Neurourol Urodyn. Jun;33(5):531-7, 2014.

[105] Haylen BT, de Ridder D, Freeman RM, et al: An International Urogynecological Association (IUGA)/International Continence Society (ICS) Joiunt Report on the Terminology of Remale Pelvic Floor Dysfunction. International Urogynecology Journal and Pelvic Floor Dysfunction 21:5-26, 2010

[106] Victor W. Nitti and Michael Ficazzola. Chapter 2 Voiding Dysfunction: Diagnostic Evaluation. Current Clinical Urology: Voiding Dysfunction: Diagnosis and Treatment. Edited by: R. A. Appell. Humana Press Inc., Totowa, NJ 0 2000 Huniana Press Inc.

[107] Wein AJ. Classification of neurogenic voiding dysfunction. J Urol 125:605–607, 1981.

[108] Ahmad Elbadawi, Seife Hailemariam, Subbarao V. Yalla and Neil M. Resnick. Structural Basis Of Geriatric Voiding Dysfunction. Vii. Prospective Ultrastructurcul/Urodynamic Evaluation Of Its Natural Evolution. Journal of Urology, Vol. 157, 1814-1822, May 1997.

[109] Fandel TM1, Albersen M, Lin G, Qiu X, Ning H, Banie L, Lue TF, Lin CS. Recruitment of intracavernously injected adipose-derived stem cells to the major pelvic ganglion improves erectile function in a rat model of cavernous nerve injury. Eur Urol. Jan;61(1):201-10, 2012.

[110] Vaegler M, Lenis AT, Daum L, Amend B, Stenzl A, Damaser MS, et al. Stem cell therapy for voiding and erectile dysfunction. Nat Rev Urol 2012.

[111] Rando TA. The immortal strand hypothesis: segregation and reconstruction. Cell 129:1239-43, 2007.

[112] Zhang H, Qiu X, Shindel AW, Ning H, Ferretti L, et al. Adipose tissue-derived stem cells ameliorate diabetic bladder dysfunction in a type II diabetic rat model. Stem Cells Dev. Jun 10;21(9):1391-400, 2012.

[113] Kim JH, Lee SR, Song YS, Lee HJ. Stem cell therapy in bladder dysfunction: where are we? And where do we have to go? Biomed Res Int. 930713, 2013.

[114] Baer PC, Geiger H. Adipose-derived mesenchymal stromal/stem cells: tissue localization, characterization, and heterogeneity. Stem Cells Int. 812693, 2012.

[115] Song M, Heo J, Chun JY, Bae HS, Kang JW, et al. The Paracrine Effects of Mesenchymal Stem Cells Stimulate the Regeneration Capacity of Endogenous Stem Cells in the Repair of a Bladder-Outlet-Obstruction-Induced Overactive Bladder. Stem Cells Dev. Mar 15;23(6):654-63, 2014.

[116] Huang YC, Shindel AW, Ning H, Lin G, Harraz AM, Wang G, Garcia M, et al. Adipose derived stem cells ameliorate hyperlipidemia associated detrusor overactivity in a rat model. J Urol. Mar;183(3):1232-40, 2010.

[117] Tremp M, Salemi S, Largo R, Andersson KE, A Plock J, et al. Adipose-derived stem cells (ADSCs) and muscle precursor cells (MPCs) for the treatment of bladder voiding dysfunction. World J Urol. 2013.

[118] Sakuma T, Matsumoto T, Kano K, Fukuda N, Obinata D, et al. Mature, adipocyte derived, dedifferentiated fat cells can differentiate into smooth muscle-like cells and contribute to bladder tissue regeneration. J Urol. Jul;182(1):355-65, 2010.

[119] J. Gao, J. E. Dennis, R. F.Muzic, M. Lundberg, and A. I. Caplan. "The dynamic in vivo distribution of bone marrow-derived mesenchymal stem cells after infusion," Cells Tissues Organs, vol. 169, no. 1, pp. 12–20, 2001.

[120] Levy R, Muller N. Urinary incontinence: economic burden and new choices in pharmaceutical treatment. Adv Ther 23:556–573, 2006.

[121] Norton P, Brubaker L. Urinary incontinence in women. Lancet 367:57–67. 2009.

[122] McGuire EJ, Lytton B, Pepe V, Kohorn EI. Stress urinary incontinence. Obstet Gynecol 47:255–264, 1976.

[123] Kayigil O, Iftekhar Ahmed S, Metin A The coexistence of intrinsic sphincter deficiency with type II stress incontinence. J Urol 162:1365–1366, 1999.

[124] Corcos J, Beaulieu S, Donovan J, Naughton M, Gotoh M. Quality of life assessment in men and women with urinary incontinence. J Urol 168:896–905, 2002.

[125] McGuire EJ: Urethral bulking agents. Nat Clin Pract Urol 3: 234–235, 2006.

[126] Sharifi-Aghdas F. Surgical management of stress urinary incontinence. Urol J 2:175–182., 2005.

[127] Lai HH, Hsu EI, Teh BS, Butler EB, Boone TB: Thirteen years of experience with artificial urinary sphincter implantation at Baylor College of Medicine. J Urol 177: 1021–1025, 2007.

[128] Staack A, Rodriguez LV: Stem cells for the treatment of urinary incontinence. Curr Urol Rep 12: 41–46, 2011.

[129] Atala A. Regenerative medicine and tissue engineering in urology. Urol Clin North Am 36:199–209, 2009.

[130] Régis Roche, Franck Festy, Xavier Fritel. Stem cells for stress urinary incontinence: the adipose promise. J. Cell. Mol. Med. Vol 14, No 1-2, 135-142, 2009.

[131] Lin G, Wang G, Banie L, Ning H, Shindel AW, et al: Treatment of stress urinary incontinence with adipose tissue-derived stem cells. Cytotherapy. 12(1): 88–95, 2010.

[132] GuiZhu Wu, YanFeng Song, Xiu Zheng, ZhongQing Jiang. Adipose-derived stromal cell transplantation for treatment of stress urinary incontinence. Tissue and Cell 43 246– 253, 2011.

[133] Fu Q, Song XF, Liao GL, Deng CL, Cui L. Myoblasts differentiated from adipose-derived stem cells to treat stress urinary incontinence. Urology. Mar;75(3):718-23, 2010.

[134] Zeng X, Jack GS, Zhang R. Treatment of SUI using adipose derived stem cells: restoration of urethral function. J Urol 175:291, 2006.

[135] Zhao W, Zhang C, Jin C, Zhang Z, Kong D, et al. Periurethral injection of autologous adipose-derived stem cells with controlled-release nerve growth factor for the treatment of stress urinary incontinence in a rat model. Eur Urol. Jan;59(1):155-63, 2011.

[136] Gotoh M, Yamamoto T, Kato M, Majima T, Toriyama K, et al. Regenerative treatment of male stress urinary incontinence by periurethral injection of autologous adipose-derived regenerative cells: 1-year outcomes in 11 patients. Int J Urol. 2013.

[137] Kuismanen K, Sartoneva R, Haimi S, Mannerström B, Tomás E, Miettinen S, Nieminen K. Autologous Adipose Stem Cells in Treatment of Female Stress Urinary Incontinence: Results of a Pilot Study. Stem Cells Transl Med. 2014

[138] Hakim L, Van der Aa F, Bivalacqua TJ, Hedlund P, Albersen M. Emerging tools for erectile dysfunction: a role for regenerative medicine. Nat. Rev. Urol. 9:520–36, 2012.

[139] YANG Bin, ZHENG Jun-hua and ZHANG Yuan-yuan. Myogenic differentiation of mesenchymal stem cells for muscle regeneration in urinary tract. Chin Med J 126 15, 2013.

[140] Lin G, Qiu X, Fandel TM, Banie L,Wang G, Lue TF, Lin CS. Tracking intracavernously injected adipose-derived stem cells to bone marrow. Int J Impot Res 23:268–75, 2011.

[141] H. Orabi, G. Lin, L. Ferretti, C. Lin, and T. F. Lue, "Scaffoldless tissue engineering of stem cell derived cavernous tissue for treatment of erectile function," Journal of Sexual Medicine, vol. 9, no. 6,1522–1534, 2012.

[142] Yang B, Peng B, Zheng J. Cell-based tissue-engineered urethras. The Lancet 378: 568-569, 2011.

[143] Zhu J, Zhou L, XingWu F. Tracking neural stem cells in patients with brain trauma. N Engl J Med 355: 2376-2378, 2006.

[144] Prockop DJ, Oh JY. Mesenchymal stem/stromal cells (MSCs): role as guardians of inflammation. Mol Ther 20: 14-20, 2012.

[145] Ning H, Liu G, Lin G, Garcia M, Li L-C, Lue TF, Ching-Shwun L. Identification of an aberrant cell line among humanadipose tissue-derived stem cell isolates. Differentiation 77:172–80, 2009.

Chapter 2

Mechanisms of Cell Regeneration — From Differentiation to Maintenance of Cell Phenotype

Arnaldo Rodrigues Santos Jr,
Vitor Andrade Nascimento,
Selma Candelária Genari and
Christiane Bertachini Lombello

Additional information is available at the end of the chapter

http://dx.doi.org/10.5772/59150

1. Introduction

In adult organisms the regenerative capacity of certain organs or tissues can be limited, resulting in an important clinical challenge for physicians and scientists [1-3].

Regeneration involves the capacity for renewal or recomposition of tissues, organs or even organisms, after considerable physical injury or damage, resulting from pathologies, tumors, congenital diseases or traumas, for example. As a consequence of tissue regeneration, both the composition and the tissue properties are restored, and the newly formed tissue is highly similar to the original tissue. The regenerative capacity is directly related to the presence of stem cells or progenitor cells, which are capable of proliferation and differentiation [4,5]. Tissues that maintain a high proliferative capacity, such as the hematopoietic system, have regenerative capacity even in adult organisms [6].

Cell proliferation occurs in repair processes in general, accompanied by intense production of extracellular matrix, with large amounts of collagen, resulting in the formation of fibrous tissue to occupy the injured area. Although there is lesion filling, both the composition and the tissue properties are different from the original tissue, and the tissue organization pattern is not restored, leading to an altered performance of its functions [2]. Skin healing processes with the presence of scars are examples of tissue repair [3].

Besides the natural processes of regeneration and repair, it is possible, through medical intervention, to fill lesions with natural or synthetic materials, aiming at the recovery of the compromised area, and conferring certain properties to the tissue, avoiding, for example, exacerbation of the initial lesion or the evolution of degenerative processes [1,7].

The three approaches can be used in tissue engineering, targeting regenerative medicine, as they allow the recovery of compromised areas in different degrees. However, the primary objective is regeneration, recomposition of the original tissue and resumption of the biomechanical and molecular properties, with the normal performance of their functions [1,7-8].

Tissue regeneration involves cell recruitment, growth, proliferation and differentiation, with the latter representing a crucial stage for the success of regeneration, avoiding the formation of fibrous tissue characteristic of the repair [9-12]. Tissues with greater regenerative capacity, such as the skin and liver, intrinsically present cells able to migrate to occupy the affected region, and the same cells maintain the proliferative capacity, enabling occupation of the lesion [3]. In other tissues the regenerative capacity is even more impaired. In the cartilage, for example, the cells remain embedded in the extracellular matrix, and the absence of blood vessels inhibits the presence of other types of component cell in the tissue; even the cell migration and proliferation processes are compromised. In general, regeneration and repair processes do not occur naturally in these cases, requiring surgical intervention to stimulate the subchondral bone marrow, thus enabling the presence of cells capable of tissue repair in the compromised area [13]. Other scientific techniques and methodologies seek alternatives to enable the processes both of repair and of tissue regeneration [9-12,14-16].

Anyhow the final stage of the abovementioned processes, cell differentiation, is critical. An understanding of the mechanisms that lead to the differentiation process in adult organisms allows the proposition of improvements in existing technologies and of alternatives geared towards the optimization of guided tissue regeneration processes, in regenerative medicine.

2. Mechanisms of differentiation

In higher organisms, such as mammals, organs and tissues have limited regeneration capacity. Cell fate determination occurs during the embryonic development of these animals, from the zygote. The zygote, classified as totipotent, has the potential to originate any type of cell, among more than two hundred cell types found in the adult man, for example [12,17]. However, during embryonic development, especially in the gastrulation phase, a more intense and differentiated form of gene expression begins. This phase of embryonic development is still marked by the decrease in cell proliferation capacity and by morphogenetic movement, or cell migration. From this stage on, the morphogens, soluble inductive factors, play a vital role in cell differentiation [17]. The cells gradually exhibit changes in the gene expression pattern, resulting in the development of a specific type of cell, i.e., cell differentiation. The differentiated cells keep the gene expression pattern stable, allowing the expression of its characteristic phenotype [12,14].

The relationship between potentiality and cell differentiation is therefore inversely proportional. With cell differentiation, its potentiality gradually becomes more restricted, and the more differentiated cells have less capacity to originate other cell types [17-18]. In cell culture systems it is possible to clearly observe situations that favor the proliferative capacity of the cells, such as growth in monolayers with fetal bovine serum supplementation, while generally three-dimensional frameworks, and specific supplementation with growth factors, are inductors for cell differentiation [19-20].

In the adult organism the cells are mostly specialized, or differentiated, form part of complex tissues and organs, and have a low proliferative capacity, limiting the regenerative possibility. It is important to emphasize that some vertebrates maintain more significant regenerative capacities that, in a number of cases, extend to the substitution of complete limbs. Some strategies allow the organism to maintain, albeit in a restricted manner, the tissue regeneration potential, through the: 1) presence of progenitor cells or stem cells; 2) reversal of the cell differentiation process, though limited, by means of dedifferentiation, transdifferentiation or cell reprogramming, the latter with certain restrictions as it is essentially obtained using the artificial manipulation of cells and rarely occurs naturally in the adult organism [8,12].

2.1. Aspects relating to stem cells

The considerable potentiality of embryonic stem cells is lost over the course of the cell differentiation process during the development of the organism. In the adult organism the stem cells, called somatic, maintain the self-renewal capacity but have less potentiality, and are found in some specific tissues, such as the hematopoietic, mesenchymal, nervous, adipose, liver, pancreas and skin tissues as well as others [21-22]. These cells are maintained in the adult organism by two main mechanisms: deterministic and stochastic. In the deterministic model, the division of a stem cell produces another stem cell and a cell with the ability to differentiate, or a progenitor cell. In the stochastic, or random model, some stem cells generate only stem cells, while others generate differentiated cells [6,23-24].

Even in adult organisms the somatic stem cells diminish over time, and older individuals have a smaller amount of somatic stem cells [21,25]. With aging, mechanisms such as cell senescence and apoptosis are apparently related to a decrease in the proliferative capacity of stem cells, resulting from the accumulation of intrinsic (DNA mutations) and extrinsic factors (changes in the specific microenvironment, or niche, in which the stem cells are located) [24,26]. Circumjacent factors (both internal and external), or mediators, are essential for the maintenance of these cells, and can lead to quiescence or activation of the stem cells [24-25,27]. These mediators can be: 1) autocrine or paracrine factors produced by the somatic stem cells, present in the niches; 2) paracrine factors originating from adjacent mesenchymal or stromal cells; 3) molecules present in the ECM, or cell-cell adhesion molecules; 4) factors originating from distant sources, such as blood circulation, immune cells or neurons. The balance between the presence, secretion and expression of these factors modulates cell activity [28].

Some tissues with high cell renewal capacity, such as the hematopoietic tissue, maintain the activity of a population of fast-cycling stem cells so as to maintain the characteristic complexity and their functions. It is proposed that slow-cycling stem cells coexist in these tissues for recruitment in the case of injury and consequent need for tissue regeneration, characterizing inductive and quiescent niches in these tissues. In the case of tissues with a low cell renewal rate, such as in the brain, muscle and liver, only quiescent slow-cycling stem cells are maintained for activation in the event of injury [28]. In spite of recent advances in the characterization of these niches, not all the modulation factors of somatic stem cells have yet been fully understood [24,28].

2.2. Morphogens and growth factors

Anyhow, it has been established that inductive soluble factors, such as morphogens, are of crucial importance in the cell differentiation process, during embryonic development and in somatic, stem or differentiated cells, since these factors can lead to the abovementioned dedifferentiation and transdifferentiation processes [8]. Morphogens are signaling molecules that induce concentration-dependent cell-specific response. A concentration gradient is formed from the synthesis site of these molecules, and interferes directly in cell response. The initial identification of morphogens in *Xenopus* resulted in the identification of molecules involved in the control of cell growth, belonging to the Fibroblast Growth Factor (FGF) and Transforming Growth Factor Beta (TGFβ) families [29-30]. Some examples of morphogens studied are Sonic hedgehog (Shh), in mice and Decapentaplegic (Dpp), in wing development in *Drosophila* [30].

During embryonic development, FGF is responsible for the differentiation of the ventral and posterior mesoderm, while TGFβ acts in the epithelial-mesenchymal interactions and BMP (bone morphogenetic protein), a class of TGFβ, acts in mesodermal and bone differentiation [12]. In adult organisms the same molecules act as growth factors; for example, FGF acts in the proliferation of chondrocytes [31] while TGFβ and BMP act in the differentiation of the cartilaginous and bone tissue, besides other specific functions [32-33]. The cultivation of mesenchymal stem cells in the presence of TGFB allows the reduction of self-renewal markers (Oct4, Stella, Nanos3, and Abcg2), besides inducing the expression of osteoblast dedifferentiation markers (Runx2, Opn, and Col1) [34].

2.3. Inhibition of cyclin-CDK

Since differentiation is inversely proportional to proliferation, both processes occur in a controlled fashion. The high proliferative capacity is maintained in slightly differentiated cells, such as stem cells and progenitor cells. With the differentiation, or specialization process, the proliferative capacity is diminished or even absent, as in tissues with low replacement activity, such as the nervous tissue [35]. Control of the different phases of the cell cycle is a key point both in embryonic development, and in the adult organism. The transition between the G1, S and G2 phases of the interphase, and M phase (mitosis) occurs through a balance of specific signals. Some cells exit the cell cycle, and remain in G0, and in mammals both the G1 sequence, and the escape to G0, occur at the so-called restriction point R, where the cellular metabolic conditions and the gene expression pattern, influenced by extrinsic factors, determine quiescence (G0), growth, differentiation or cell proliferation [36].

The protein p53 is one of those responsible for blocking the cell cycle, especially in the presence of DNA damage. The increase in protein p53 induces the expression of p21 protein, which in turn binds, through the amino-terminal region, to the cyclin-CDK (cyclin-dependent kinase) complex, usually responsible for cell cycle progression [37-38]. The inhibition of cyclin-CDK enables DNA repair prior to the progression of the cell cycle, since besides the inhibitory action, p21 also binds, through the carboxy-terminal region, to the proliferating cell nuclear antigen (PCNA*)*, blocking DNA replication [39]. The same regulation mechanism can determine the apoptosis process, when there is no repair of the DNA damage. Other proteins that act together with p21, inhibiting CDK, are the proteins p27 and p57 which are both from the KIP (kinase inhibitor protein) family. Besides interfering in cell cycle progression, there is evidence of the action of these proteins in cell differentiation processes [12]. Protein p21 can be expressed independently of p53, and is responsible for the differentiation process in adult organisms and in cell culture systems, being stimulated in the latter by the induction of systems with fetal bovine serum [40].

Another family of inhibitor proteins, INK4 (inhibitors of CDK-4), composed of the proteins p15, p16, p18 and p19, also acts in the control of cell behavior. The cyclin-CDK complex promotes the phosphorylation of the retinoblastoma (pRb) protein, inactivating it and allowing the action of the E2F factor, besides directing the cell to the division process. It may also be related to the progression of tumors in the case of deregulated activity [36]. In the presence of proteins from the INK4 family, this process is inhibited, and the cell cycle does not proceed. However, their performance in cell differentiation is not clearly determined [12,17]. For somatic stem cells, it is suggested that p16 has a repressive action in older organisms, i.e., the reduction of the self-renewal capacity of these cells over time would be due mainly to the regulation of repressive pathways, and not to the loss of permissive capacity for self-renewal, which in spite of reducing the tissue regeneration capacity of these cells, would avoid potentially tumorigenic cell behavior [25].

2.4. Control of gene expression

Besides control of the cell cycle, via inhibition of cyclin-CDk, control of gene expression in non-proliferating cells is an essential cell differentiation control mechanism, maintaining preserved universal patterns in different organisms, both in *Drosophila* and in higher organisms, such as mammals [12, 41].

Certain genes are expressed throughout cell differentiation, and the main point of control of gene expression occurs precisely in DNA transcription. Despite universal gene expression patterns, the greater the complexity of the organism, the more complex the molecular mechanism that acts in this control [42]. Transcription factors are proteins that bind specifically to DNA sequences, inducing or repressing gene expression, and together with chromatin remodeling pattern, influence cell differentiation [42]. The transcription factors act specifically at DNA sites, more specifically the following sequences: promoters, or binding sites for the RNA polymerase; enhancers, binding site for regulatory proteins; and silencers, binding sites for regulatory proteins that block the gene expression [12, 43].

Changes that lead to chromatin remodeling basically occur through the processes of acetylation, methylation and phosphatation of histones, resulting in the removal of positive charges from these proteins, in the case of acetylation and methylation, or adding negative charges to the phosphate groupings, as a result of phosphorylation. Changes caused by methylation in general hinder gene expression, while acetylation favors access to DNA [44]. In all cases, structural changes modify the access of transcription factors, regulating gene expression, even for cells with the same cell differentiation pattern, i.e., the same genotype can correspond to different phenotypes, due to reversible changes in gene expression, without any changes in the gene content itself [42,45-47].

Control of cell phenotype can be extended beyond gene expression, in post-transcriptional controls, by modulation of the mRNA transcription process, and in post-traductional controls, with changes in the proteins for their biological activation. The protein phosphorylation process is one of the most common post-traductional changes, and results in the formation, for example, of signal transduction mediators [12,35].

2.5. Influence of the extracellular matrix and cell-cell interaction

The influence of the extracellular matrix (ECM) in the differentiation process involves the presence of factors immobilized in the ECM, such as morphogens, secreted growth factors and chemokine gradient. The ECM can be defined as an interconnected network of macromolecules composed of adhesive proteins (such as fibronectin, vitronectin, tenascin and laminin), structural fibrous proteins (collagen and elastin), glycoproteins and specialized proteins (such as growth factors) [35,48]. The modulation of behaviors such as adhesion, growth, migration, proliferation and differentiation, occurs as a result of the direct influence of the contact of cells with the ECM, or with adjacent cells [6-7,28,48]. The bond with the ECM elements occurs via membrane receptors and results in cytoskeletal reorganization, which in turn triggers a cascade of intracellular events through signal transduction [35,48].

An example of differentiation control is the interaction of cells with fibronectin. Fibronectin is a multiadhesive protein of the ECM. At least 20 isoforms of this protein obtained through the alternative processing of the transcript of a single gene are known. They are proteins that act both in migration and in cell differentiation, during embryogenesis and in adult organisms [35]. The intracellular events of binding with fibronectin area result of their action with transmembrane receptors, the integrins. Integrins are dimeric glycoproteins, formed by pairs of different combinations of the subunits α and β [48]. In the organization of epithelia, such as of the skin, the basal layers are kept in contact with the ECM through the bond with fibronectin. As the cells migrate to the upper stratums, this bond is lost and cell differentiation occurs [12]. In the keratinocytes, for example, the differential expression of receptors for fibronectin was observed throughout the cell differentiation process. These receptors are the integrins: $\alpha5\beta1$, which recognize RGD sequences of fibronectin and act in cell growth and initial migration; $\alpha v\beta6$, not characteristic of cells from the basal layer. It can activate the growth factor TGFβ, directing differentiation, and is present mainly in the tissue repair process; and $\alpha v\beta1$, which is a low-affinity receptor that facilitates the final stages of cell migration. The alternation in the expression of the integrins allows the migration and differentiation of these cells [48]. Similarly, in the pulmonary epithelium the tissue response to pathological condition of pulmonary fibrosis to integrin $\alpha v\beta6$ is responsible for the activation of the constituent expression of TGFβ, acting as epithelial transdifferentiation regulators during fibrogenesis [49-50].

Not only the fibronectins, but also the proteoglycans, act in tissue organization and direct differentiation. These glycoproteins have repetitive chains of disaccharides that are generally acid, bonded to a protein nucleus. Besides the structural function and the high hydrophilicity of this molecule, these molecules have the ability to bind to other diffuse proteins in the ECM, acting direct or indirectly in cell differentiation. Betaglycan and decorin bind to TGFβ, and heparan sulfate to several other morphogens, such as FGF and Wnt [12]. There have been reports of the non-differentiation of mice embryo stem cells due to the reduction of heparan sulfate sulphatation, impeding the response to FGF [51].

Collagen is the most abundant component in the ECM, and is directly linked to cell differentiation processes. This structural fibrous protein begins to be secreted during embryonic development, in the gastrulation stage, concomitant to differentiation of the three germ layers [52]. In the adult organism the action of this protein in maintaining differentiation and during the tissue repair and regeneration processes has been described for different cells such as fibroblasts, hepatocytes, pancreatic acinar, thyroid epithelial, mammary epithelial and others [12]. Mammary epithelial cells of mice exhibit cuboid dispersed morphology, when cultivated on surfaces with type I collagen, preventing their differentiation and functional activation in the production of β-casein [48]. Similarly, hepatocyte cultures also tend to firmly adhere to surfaces coated with a fine film of type I collagen, are polygonal, dispersed and with suppression of specific function [48]. However, three-dimensional collagen gels induce the cells to cytoskeletal reorganization and the resumption of differential gene expression, leading to the specialized phenotype of the cells. Mammary epithelial cells in collagen gels can form duct-like structures and secrete milk proteins, such as β-casein. Hepatocytes cultivated in three-dimensional collagen gels present albumin secretion [12,48].

Studies reporting the *in vitro* differentiation of chondrocytes are also demonstrative of the action of the ECM elements in this process. Chondrocytes cultivated in two-dimensional monolayers appear dedifferentiated, with flat morphology and expressing mainly type I collagen, characteristic of fibrocartilage, or fibrous tissues. With the maintenance of three-dimensional culture systems, the chondrocytes resume the expression pattern characteristic of type II collagen, and synthesis of proteoglycans [53-55]. The binding of chondrocytes to the ECM allows the modulation and maintenance of the differentiated phenotype of the cells, being one of the principles of the chondrocyte implantation technique, a tissue engineering method applied to chondral regeneration. This binding occurs via integrin, collagen receptors and laminins, which in turn signal the modulation of cell behavior via Wnt, nitric oxide, retinoic acid and protein kinase C [55]. Most integrins expressed in chondrocytes present β1 chain, and its absence brings about important changes in the cartilage phenotype, being one of the factors of apoptosis, together with integrin αnβ5 [55].

ECM degradation also directly influences cell differentiation, from embryonic morphogenesis to adult tissues. The proteolytic cleavage of the ECM elements, through the action of metalloproteinases, serine proteases and cysteine proteases, and the consequent solubilization of their components, signal cell behaviors. Morphogens and growth factors such as TGFβ which are embedded in the ECM, often bound to proteoglycans, are made available, signaling the tissue repair or regeneration process. During salivary gland formation the development of ramifications occurs in the presence of collagenase inhibitors, favoring the interaction of the cells with the ECM [12]. Mammary gland involution results in the cleavage of laminin, whose fragments bind to EGF receptors, resulting in increased cell migration, which together with collagen cleavage and release of soluble factors such as Wnt, TGFβ and FGF, favor tissue remodeling [12].

Tissue remodeling in bone formation, starting from endochondral ossification, is another example of differentiation regulation by the specific degradation of the ECM. The substitution of chondrocytes in the tissue maturation process involves the presence of a lecithin, galectin-3, located in the ECM and with antimitotic action. During the remodeling of the ECM, this lecithin is cleaved and inactivated, allowing a sequence of events that leads to ossification. In cases where this lecithin is not degraded, chondrocyte apoptosis does not occur, resulting in bone formation defects [56-57].

Besides the interaction directly with ECM elements, cell-cell interaction also directs the differentiation process, but expressively in embryonic development and epithelial tissues. Among the Cell Adhesion Molecules (CAMs), the cadherins, a superfamily of calcium–dependent transmembrane proteins, are important cell differentiation mediators. N-cadherin is responsible for cell migration, both in embryonic development and in adult tissues, while E-cadherin is expressed mainly in the embryonic and epithelial tissues [35]. Returning to the example of mammary epithelial cells, anti-integrin β1 antibodies resulted in the blocking of β-casein synthesis, while anti-E-cadherin antibodies do not interfere in protein synthesis, demonstrating the influence of ECM in the transduction of signals independent of cell-cell interaction [48].

Therefore cell differentiation mechanisms are obtained by complex relations between intrinsic factors and extrinsic influences, which trigger signaling reactions and modulate cell behavior as well as gene expression. In general the factors that act in cell differentiation from receptors on the cell surface lead to transduction of signals via protein kinase, activating intracellular phosphorylation cascades and culminating in gene expression regulation, besides cytoskeletal reorganization [4,12,35].

It is important to stress that the factors responsible for cell behavior modulation during differentiation, do not only apply to somatic stem cells. Other types of cell can be involved in dedifferentiation and transdifferentiation processes. The use of animal models is essential in establishing the concepts presented and the understanding of cell differentiation mechanisms, as described below.

3. Regeneration in animal models

Given the importance of differentiation for the acquisition of cell functions concerning the formation and maintenance of the organism, it is advisable to gather information on the regeneration process that, as mentioned previously, involves the capacity for recomposition and renewal of parts through the remodeling of somatic tissue [10,14]. Regeneration in an adult

http://dx.doi.org/10.5772/59150

animal appears to be a noteworthy example of postembryonic morphogenesis. It involves recognition of tissue loss or injury, followed by mechanisms of reconstruction or recovery of the structure in question [58].

One of the mechanisms associated with natural regeneration is **dedifferentiation**, which involves a terminally differentiated cell returning to a less differentiated phase of its own lineage. This process allows the cell to proliferate again before redifferentiation, which leads to the substitution of these cells that were lost [59]. Cell dedifferentiation changes a program that directs the specific function of a somatic cell to another program or to proliferation, regardless of whether the destination is the same as the origin [60].

Transdifferentiation is another natural mechanism that was observed for the first time in the regeneration of the salamander lens over 100 years ago. As mentioned above, this process involves the conversion of a differentiated cell type into another [12]. This occurs because the cells recede to a point where their change of lineage becomes possible [59].

We should designate another process, which aims to induce differentiated cells to revert their pluripotency: **reprogramming.** From this point on, they can differentiate into almost any type of cell. Although reprogramming occurs naturally during fertilization to produce totipotent cells that can differentiate into any type of cell, it has not yet been formally shown as a true regenerative response. Moreover, reprogramming avoids the need to use embryos for regenerative therapies using differentiated cells created from a patient. From the clinical point of view, this has the additional bonus of circumventing the immunological problems associated with grafting (such as transplant rejection and graft versus host diseases) [59].

3.1. Regeneration in different animals

Some animals, such as starfish, planaria (flatworms) and *Hydra* have significant regenerative capacity, and can originate complete specimens from small fragments [10,14]. Is interesting to see the comparison between regeneration and fragmentation, fission and budding, which are forms of propagation in invertebrates. Fragmentation is the simplest form of agamic reproduction and is essentially identical to regeneration after cross sectioning. Fragmentation can occur in animals that can be broken by external forces and that regenerate completely from each lost part. Fission, in contrast to fragmentation, is generally an endogenous process [58,61]. Returning to the topic of regeneration, we present below some model systems:

- **Zebrafish:** Fins and tail of zebrafish can regenerate after amputation. The phenomena called appendage regeneration have been studied at length. This fish can also regenerate cardiac tissue after resection or the destruction of cardiomyocites. In the regeneration of the zebrafish fin after amputation, there are four stages called "epithelialization, regeneration or healing of wounds", "blastema formation", "regenerative consequence" and "termination". In epithelialization, the proximal epidermis migrates to cover the stump and to form a 3-4-cell thick layer. Inflammation then proceeds to clean the clotted plasma and cell debris. For blastema formation, there is histolysis and remodeling of the extracellular matrix; cells to be released by histolysis start to dedifferentiate in this period. Then blastema aggregation is achieved through an accentuated increase in the mitosis, which is dependent on factors from the wound and regeneration of the epidermis and nerves. A notable feature is the rare presence of apoptotic cells in the blastema despite the presence of avascular tissue, which could present hypoxia and potentially be susceptible to apoptosis.

- **Xenopus:** Fins and tail of *Xenopus* larvae can regenerate after amputation.
- **Hydra:** When cross sectioned, *Hydra* will regenerate a head (from the lower piece resulting from the sectioning) or a foot (from the upper piece), which will always depend on the relative position of the regenerated structure in the piece being regenerated. In other words, the sectioning surface closest to the head forms a head. The same applies to the foot, which shows well defined general polarity. Regeneration in *Hydra,* starting from a small fragment of the spine, does not involve an initial increase in size. This results in a small animal, which will have its normal size reestablished with feeding [10].
- **Urodeles:** Salamanders and tritons (urodeles) have major regenerative capacity. Actually, so far as we known, the Urodeles are the only adult vertebrates able to regenerate their limbs. The regenerative capacity of the adult triton covers different regions, such as maxilla, crystalline lens, retina, large sections of the heart, and its limbs and tails, in response to tissue damage or even amputation. The salamander can regenerate its limbs and tail, ocular tissues, the intestine, and small sections of the heart. Limb regeneration also takes place in salamanders, where amputated limbs are covered by the epidermis, and the immature cells accumulate and proliferate below them, forming the "blastema". The blastema, which is coated by the newly formed epidermis, involves undifferentiated cells that are restricted to the musculoskeletal cells, chondrocytes, Schwann cells, and mesenchymal fibroblasts. Heart tissue can also regenerate after resectioning or the destruction of cardiomyocites.

There is a large gap between the examples presented previously and the regenerative capacity in other vertebrates. With the exception of Urodeles, regeneration in vertebrates is classified as very limited. In mammals, it is much more restricted, being limited to regeneration of the liver, when part of this is removed, or even by a bone fracture healing process. The regeneration of lost limbs is not possible [10,14]. Cardiomyogenesis was observed in murine hearts only at less than one week of age. The regeneration of vertebrates implies dedifferentiation. However, the factors distinguishing vertebrates with or without regenerative capacity still need to be clarified [60].

An interesting view is that regeneration is an essential attribute of metazoa, and was secondarily lost in closely related species or more distant groups [58]. In analyzing the mechanisms that form the basis of regenerative responses in Urodeles, a comparative study with mammals was allowed. Thus it is possible to identify the primary differences between Urodeles and mammals, especially with regards to the evolutionary bases for regeneration [58]. Nevertheless, the reasons that lead to the understanding of the animals having such a different regenerative capacity are not clear. Several hypotheses have been raised to this effect [58], which illustrates the complexity of the problem: (1) some species have a fixed number of segments in adults, and a very large number of non-regenerative species share this characteristic; (2) if amputation removes a structure that is not critical for survival, then the regenerative capacity could be lost, as it would not be a selective advantage; (3) if some species have low rates of amputation of their segments, this could lead to a loss of regenerative capacity; (4) if fission is present in a given species, this may diminish the selective advantages for regeneration.

On the other hand, some observations indicate that these regeneration responses have not been totally lost in these species. This idea is supported by the fact that many species have good regeneration capacity as embryos (including humans), but this capacity is gradually lost (or silenced) during development and aging [62]. Many aspects of this matter still need to be discussed.

3.2. Response to injury

The regenerative response is initiated through recognition of the loss of tissue or local wound. It is not yet understood which factors initiate this response, as there are potentially lots of signals involved. Nowadays, events such as *bioelectric signaling, thrombin activation* and its hemostasis, the possible *influence of the immune response* in regeneration and the formation of a *wound epithelium* are considered relevant.

- **Bioelectric signaling** involves electric currents in lesion that flow after amputation of limbs and reflects the geometry of the altered tissue and consequently, the electrical resistance. The amputation of a salamander limb produces derivation of low resistance at the end of the stump, through which the ionic current flows during the first days [63]. Wounds are generally electrically positive in relation to the more proximal uninjured areas. Recent experiences in the caudal regeneration of *Xenopus* larvae indicate the existence of important activity of the V-ATPase proton pump [64-65]. The inhibition of this pump blocked regeneration, while maintenance of the V-ATPase expression maintained the regenerative capacity. The need for the presence of nerves for effective regeneration in limbs has already been described. It was proposed that the need to provide a nerve for regeneration is the aspect influenced by circuits of internal currents produced by this nerve [66]. Another important point in regeneration in planaria, in which the regeneration polarity can be manipulated by applied electrical fields, so that the animal heads are always formed when facing the cathode [67].

- **Thrombin activation** is an essential regulator of the response to injury in vertebrates and has become a strong candidate to initiate the regenerative response in salamanders as observed in the repair of the intestine [68], limbs [69] and heart [70]. Clots are formed as a result of the action of thrombin protease on the plasma fibrinogen, and thrombin activation from prothrombin can be regulated by hemostasis. The possibility that local thrombin activation could be a regeneration signal originated from studies in which salamander myotubes cultivated *in vitro* returned to the cell cycle through thrombin-linked activity [71]. After the removal of the crystalline lens in tritons, thrombin activity appears transitorily in the dorsal pupillary margin of the iris and can be blocked by the introduction of inhibitors in the ocular chamber [72]. This intervention reduces cell cycle reentry by the pigment epithelial cells (PECs) on the dorsal margin and inhibits crystalline lens regeneration [73]. In the same model, cell proliferation is dependent on the activity of the fibroblast growth factor 2 (FGF2), while activation of the Wnt signaling system determines the crystalline lens regeneration site [74]. During regeneration of the liver in mammals, some evidence suggests that, after hepatectomy, the release of serotonin by the platelets is a fundamental signal for the onset of hepatocyte proliferation [75]. Platelet activation is another aspect of the thrombin-dependent response to injury.

- **Formation of the scar epithelium on wounds** is an early response to the injury and consists of the migration of epithelial cells to the amputation plane or tissue lesion [67]. The wound epithelium assumes a specialized identity and plays an important role in subsequent regeneration events. In some cases, the formation of epithelium on the wound does not occur. In crystalline lens regeneration, epithelial transdifferentiation occurs at the site without involving the formation of a wound epithelium. In cardiac regeneration in zebrafish there is early and generalized activation of the epicardium, which can perform a role similar to that of the wound epithelium [76]. The role of the wound epithelium is not yet fully understood, but some points can be raised. The formation of wound epithelium in salamanders can be avoided by suturing skin over the extremity of a limb or amputated tail. This procedure allows the wound to heal, but prevents limb regeneration. The wound epithelium can provide a distal limit to standardize mechanisms during regeneration, even though positional identity is usually considered a function of the mesenchymal cells [77]. It has been suggested that the epidermal cells of different circumferential identities can migrate and form a functional wound epithelium [78]. The formation of this wound epithelium is a target for a variety of regulatory events. The formation of the wound epithelium has been shown to be related to FGF20a activity in zebrafish. This can be regulated by Wnt 10a and Wnt5b [79], where it was shown that inhibition of signaling via Wnt/β-catenin leads to a decrease in FGF20a expression.

Following removal of the apical region of the triton ventricle, the heart seals, by contraction, around the clot. Adult cardiomyocites re-enter the cell cycle and the division of a peripheral zone of the clot. If the animal is injected with tritiated thymidine to identify the cells that are in S phase, about 10% of the cardiomyocites in this region are marked in the period of a day. In experimental comparisons with the heart of an adult mammal, very few cells are marked after injury [22,80]. In tritons, after crystalline lens removal, the population of regenerative cells involves pigment epithelial cells, which are invariably located on the dorsal pupillary margin of the iris. These cells re-enter the cell cycle, lose their pigment granules and are converted into crystalline lens cells, a process known as transdifferentiation [22,80].

After the amputation or tissue injury, there is the regeneration and supply of the nerve to the damaged region. Normal limbs require the presence of nerves to regenerate. Limbs that have had nerves removed prior to amputation do not regenerate. However, aneurogenic limbs can regenerate normally [10]. In most cases, this regeneration involves only the axonal extension, so as to reestablish functional contacts with the newly formed tissue. In the case of tail regeneration in salamanders, or head regeneration in *Hydra*, the generation of new nervous cells can also be involved. It has been observed that regeneration is dependent on and concomitant to the nerve supply [67]. This dependence is widely conserved phylogenetically, being observed not only in several contexts in vertebrates, but also in examples of echinoderms [81] and annelid regeneration [82]. This dependence shows the clearest example of the difference between development and regeneration. The dependence on nerve regeneration may be a means of ensuring that the regenerated tissue is functional.

The **plasticity** of cell differentiation-capacity to differentiate into a cell range featuring distinct functionalities-provides a convenient cell assay for the comparison of a differentiated cell in Urodeles with its congener in mammals. The plasticity of differentiated cells is a remarkable characteristic in heart regeneration in salamanders, as it depends on the capacity of the cardiomyocytes close to the lesion to re-enter the cell cycle. Cardiomyocytes of adult salamander ventricles when dissociated in culture re-enter the S phase. About one third of these cells progress through mitosis and undergo successive cell divisions, in contrast to their counterparts in mammals. This is accomplished without major loss of differentiated properties and cells resume their heartbeats after cytokinesis [83].

It is important to recognize that there are examples of regeneration in mammals that involve plasticity. An example in a mammal that depends on the plasticity of the differentiated cells is in the liver that appears to be comparable to cardiac regeneration in salamanders, as the hepatocytes split without loss of differentiated function [84]. The regeneration of the myelinated peripheral nerves requires the reentry of Schwann cells in the cell cycle, with the loss of their differentiated properties, such as myelin expression, and the acquisition of a phenotype that facilitates axon regeneration, before redifferentiating together with the regeneration axon [83].

In appendages (limb and tail) of urodeles under regeneration, multinucleated myotubes or striated myofibers undergo cellularization to produce mononucleated cells capable of division. The nuclei of multinucleated muscle cells can also reenter in the S phase, although this is not apparently necessary for cellularization to occur. Thus multinucleated skeletal myotubes are formed by the fusion of mononucleated precursor cells. The myotube enters a post-mitotic imprisonment state in which it is totally refractory to the growth factors that stimulate the division of its precursors. The change in the cell, from mononucleated to multinucleated, together with the stationary phase after mitosis, provides two indices for the reversal of the myogenic phenotype (dedifferentiation) [22,80]. In tritons, after limb amputation, epidermal cells migrate to the surface of the wound, which is extremely important for subsequent growth. A bastema forms from cells beneath the wound epidermis, which lose their differentiated nature and begin to divide, thus producing a regenerated limb. As the limb regenerates, these cells differentiate into cartilage, muscles and connective tissue. The question regarding this phenomenon is whether the cells that differentiate into cartilage and muscles in the blastema remain faithful to their original types, i.e., whether previously muscle cells are, necessarily, those that will produce the muscle cells after regeneration.

In vitro, triton myotubes clearly differ from their vertebrate counterparts, in which these enter and cross the S phase after stimulation with fetal serum in the culture. The nuclei in the myotubes duplicate their DNA content and are detained stably in G2. The response to the serum is not observed for other myotubes of vertebrates, with the exception of rat cells in which both copies of the retinoblastoma gene are missing. pRb plays a familiar and essential role in regulating the G1-S phase transition. Several lines of evidence indicate that this is crucially important to maintain the differentiation state in myotubes of vertebrates, not only for the stable imprisonment of the cell cycle, but also for transcription starting from certain muscle promoters that depend on the activation of limbs of the *myocyte enhancer factor 2* (MEF2), a family of transcription factors. Triton myotubes express pRb, but the serum stimulates a pathway that leads to its inactivation by phosphorylation, and consequently, causes progression from G1 to S phase.

Serum is the soluble fraction of clotted blood, and results from prothrombin activation to generate the serine protease thrombin. Thrombin activates the coagulation cascade and other events to mediate the response to the injury. When crude prothrombin is activated *in vitro*, the resulting thrombin preparations contain a distinct activity, which acts directly on the triton myotubes in a serum-free medium [80]. Vascular prothrombin activation following injury occurs in relation to a protease complex known as *tissue factor*, which is mounted on the cell surface. Thrombin formation is subject to strict spatial and temporal regulation, as it is essential for clot formation to be restricted to the area of the wound, and not to spread. It is speculated that regeneration in Urodeles-in the heart, limb, or even eye-is linked to acute events of injuries or the removal of tissue from the thrombin activity site. Thrombin activity is locally increased in the early mesenchymal blastema of the limb, and prothrombin has been seen to be selectively activated over the dorsal margin of the iris after injury in recent discoveries [80].

A critical contribution to tissue standardization can also come from the fibroblasts of the connective tissue of the dermis, and the degree of alteration of its differentiated state is not yet clear [85].

3.3. Morphallaxis, epimorphosis and tissue regeneration

Typically, the regeneration process can be distinguished in two types: *Morphallaxis*-Regeneration occurs through the re-standardization and delimitation of existing tissue, with little growth of new tissues-and *Epimorphosis*-where there is the growth of correctly standardized new structures[10,14]. There has also been talk of a third type, called *tissue regeneration*.

Morphallaxis [*morph* (form)+*allaxis* (change/substitution)] involves remodeling of the without proliferation. The absence of cell division is not currently regarded as an essential criterion to define morphallaxis. The important criterion is the extreme remodeling of the remaining structures, as can be seen in the hydrozoans, planarian <??>, some annelids and other invertebrate animals. In *Hydra* regeneration occurs dynamically, with stable and continuous cell proliferation. As the tissues grow, the cells shift along the body column. The adult *Hydra* needs to lose cells continually to maintain its size. The loss of cells occurs at the ends of the tentacles and in the basal disc of the budding foot, and most of the excessive production of cells is used in the asexual budding of new *Hydra* from the body column [10]. Such dynamic re-standardization of the cells occurs through mechanisms that confer regenerative capacity to this animal.

Hydra has two organizing regions (one at each end, namely, hypostome and basal disc) which confer its general polarity. There is also interaction between the head inhibition gradient – which prevents the formation of other heads in other regions-and a positional value gradient – which determines the threshold at which head inhibition occurs-in the regeneration process. Such organization and interaction confer a dimensional arrangement that precludes (or hinders) malformations, or even incoherence between their positioning. It is assumed that the head inhibitor is a secreted factor, produced by the head itself, which spreads downwards along the body column and is degraded at the basal extremity. It is also assumed that the positional value gradient is an intrinsic property of the cells. Both are linear with values at a constant rate of decrease as they move further away from the head. Accordingly, the specification of a region of the head on the surface of the section is the first basic stage in this morphallactic regeneration when the head region is removed [10].

As regards **Epimorphosis** [*epi* (on/over)+*morph* (form)] it is defined today as the method of regeneration, where a blastema, a mass of proliferative cells located at the tip of the amputation stump, is formed [67,86]. As pointed out previously, there is standardized regeneration of new structures which leads us to reflect on the fact that cells completely differentiated from the mature limb return to the cell cycle, dedifferentiating to redifferentiate in the regeneration process, forming different types of cell, with restructuring of the injured limb [10].

Epimorphic regeneration occurs in Urodele amphibians. As a problem in cellular and molecular biology, regeneration in Urodeles provides information of considerable importance about the reversal and the plasticity of the differentiated state of cells. Limb regeneration thus adds up to a system in which the key for studying the positional identity in cells is established. Notwithstanding the usefulness of studies about development, evolution and phylogeny, it is a widespread concern of regeneration research to understand the material differences between species that regenerate and those that do not. In salamanders, although the tissue restoration process occurs differently in the heart and in the limbs, the result appears to depend on the plasticity of the differentiated cells that remain after tissue removal [22,86]. Salamanders can regenerate an entire limb from a blastema. In mammals, digit-tip regeneration does not originate from a blastema, but instead from progenitor cells in the ungueal bed [87]. The blastema of a limb consists of a group of mesenchymal stem cells at the end of the stump.

Regeneration always occurs distally in relation to the sectioned surface, which allows the replacement of the lost part of the limb. The development of the blastema, as well as the nature of the structures that it originates, depend on the amputation site and not on the nature of the nearest tissues. The limb undergoing regeneration identifies the positional value at the amputation site, regenerating all the positional values distal to this site. Note that epimorphic regeneration involves the retention of embryonic processes, such as the ability to specify new positional values, which are encoded as a property in gradient, partly on the cell surface, and whose relevant behavior for axial specification (growth, movement and adhesion) is a function of the expression of this property in relation to the neighboring cells [10].

The strategy used by Urodeles in regenerating most structures is, therefore, the re-specification of differentiated cells into local progenitor cells, rather than the activation of a pluripotent cell. This is called dedifferentiation. Thus if the epithelial cells of the iris are transplanted to the blastemal of the limb, they produce a crystalline lens, and blastemas of limbs always produce a limb after transplantation, even after transfer to the anterior chamber of the eye. This contrasts with the recent results of considerable plasticity of stem cells following transplantation in mammals. At present there is no evidence that adult stem cells can contribute to regeneration in limbs of Urodeles, although it is not possible to completely rule out this possibility. An advantage of the regeneration mechanism in Urodeles may be that it allows the progenitor cells to derive local signals from their differentiated parental cells [22,86].

Tissue regeneration is another type of reparative regeneration that involves proliferation, close to the cut surface, but unlike epimorphosis, there is no formation of a blastema [67,86]. It involves the restoration of tissues damaged by the removal of dead cells, the proliferation of progenitor cells and of functional restoration of the tissues [67]. Classical examples of this type of regeneration in mammals are the skin, bones and regeneration of the skeletal muscle, although the degree of regeneration in mammals is very limited. Liver regeneration, in turn, involves enlargement of the remaining lobules by proliferation of cells throughout the organ to make up for the lost mass, without this resulting in the restoration of the initial morphology. Once removed the lobules do not grow back [88].

It is possible for regeneration to occur in the same animal through more than one model. Thus the limits between these definitions are often not particularly clear.

3.4. Positional identity and polarity in regeneration

We now know that this positional identity is encoded in the cell membrane by a PD gradient of glycoproteins from the cell surface [89].

An important issue in regeneration is how the blastema cells identify their spatial position. Blastema cells are derived by dedifferentiation of adult mesenchymal cells at the amputation plane. If a blastema is removed from its limb by transection to the amputation plane and is transplanted to an adequate site, such as the anterior chamber of the eye or of a lesion in the connective tissue of the dorsal fin, they form a normal regenerated appendage/tissue. The cells derived from the blastema after amputation, at any level on the proximodistal axis, will produce the distal structures. This property is called positional memory. Blastema cells are derived by dedifferentiation from adult mesenchymal cells at the amputation plane, and they derive critical suggestions about their identity and potentiality of their precursors. When the regenerative cells are transplanted in tissues in the salamander, they keep their original identity. Position memory is a critical aspect for limb regeneration autonomy, because it specifies the initial population of blastema cells in relation to the extension of the axis to be regenerated. An understanding of its molecular basis is generally important for our appreciation of how stem cells are specified to produce different structures, instead of different cell types [83].

It is accepted that gradients of morphogens associated with cell-cell interactions provide the blastema cells with information on their three-dimensional position, which is similar to what happens during morphogenesis [71]. However, unlike development, the blastema cells need to know their proximodistal level to allow regeneration to occur from the correct portion. Interaction studies between proximodistal portions of the limbs in salamander blastemas have suggested that cell adhesion, movement and division can be important in the positional identity expressed in cells [89]. The hypothesis was that the proximodistal identity is encoded by a molecule or molecules to the cell surface, possibly as a level of gradual expression along the proximodistal axis. These considerations led to the identification of *Prod 1*. *Prod 1* encodes a small protein that is bound to the cell surface by a glycosylphosphatidylinositol (GPI) anchor, is apparently analogous to the mammalian CD59 protein and is implicated in positional identity determination. The mammalian CD59 protein is associated with inhibition of the terminal phase of activation of the complement system, and it is also able to mediate the activation of intracellular tyrosine kinases. The distal cells of the blastema of axalotl larvae are converted experimentally into proximal cells when expressing *Prod 1* [5,90]. The *Nag* protein was also identified as a binder to *Prod 1. Nag* acts as a mitogenic factor for blastema cells [90].

The head formation process, which occurs during the regeneration of *Hydra,* involves apoptosis, proliferation, and re-specification of cells at the amputation plane [91]. This leads to the formation of a transitory organizer in the transient head, which apparently involves the signaling pathways wnt and β-catenin. There are data showing that regeneration in planaria also occurs with the participation of β-catenin, as mentioned above.

3.5. Mobilization of regenerative cells

The determination of the origin of blastema cells has been one of the main concerns of regenerative biology for a long time. Theoretically, the blastema cells could originate from: (1) mature cells dedifferentiated from the stump of the limb, (2) the stem cells or reserve cells of the stump, (3) the plasticity of mature cells of the stump and (4) progenitor cells from the rest of the body. Transplantation experiments using cells marked with green fluorescent protein (GFP) of several tissues of the axalotl showed that grafted cells dedifferentiate, proliferate, and redifferentiate into cells restricted to their lineage of origin (5). The tracking of osteoblasts in the regeneration of zebrafish fins demonstrated the dedifferentiation of osteoblasts and their subsequent redifferentiation [92]. However, it is possible that the resident stem cells are involved in appendage regeneration [5], especially in the case of the skeletal muscle, which are accompanied by a population of stem cells called satellite cells [93].

All the tissues from the distal stump of the limb appear to contribute somehow to the formation of the blastema, except the epidermis. The dedifferentiation of mature cells is believed to be an important mechanism for blastema formation [86]. Reports have confirmed that dedifferentiated cells acquire proliferative and migrant capacities, necessary for the formation of the blastema [59,90]. There is a dichotomy between a reserve of progenitor cells in relation to the local plasticity of the differentiated cells. Previous studies suggested that the regeneration of the zebrafish myocardium occurs through reactivation of the cell cycle of adult cardiomyocytes [94]; more recently, other data has indicated that the new myocardium originates from progenitor cells [76].

Besides dedifferentiation, there is evidence of the contribution of adult stem cells to the blastema [95-96]. It is believed that Urodeles, and perhaps some fish, may contain a different muscle progenitor cell, which originates in the muscle fibers that dissociate through the dedifferentiation process. The multinucleated muscle fibers fragment into mononuclear cells that recover their proliferative capacity and migrate to the blastema. These progenitor cells presumably produce a new muscle together with the activated satellite cells[5].

In vitro, skeletal salamander myotubes are able to re-enter the S phase in response to a thrombin-derived binder present in the serum [71]. Myotubes of mammals and other vertebrates do not respond to stimulation with serum; hence there is a distinct property of differentiated cells in the salamander. A second aspect of the plasticity of skeletal muscle in salamander is the conversion of multinucleated myotubes and myofibers into viable mononuclear cells. In tritons, muscle satellite cells, during regeneration of the limbs, express *Pax-7* [97] when activated. In this model, there appears to be a balance between the satellite cells and cellularization originating from myogenic cells. A recent study of mandibular regeneration in tritons suggested that both sources are important [98]. On the other hand, in the regeneration of the tail of *Xenopus* tadpoles, regeneration appears to occur exclusively from satellite cells [95]. Triton myotubes can be prevented from reentering the cell cycle both by X-radiation and by p16INK4. Such myotubes, when marked and implanted into a blastema of a limb under regeneration, were effectively converted to the mononucleated cells. Hence although reentry and cellularization occur at the same time, after implantation, they are not mechanically linked [80].

An important impetus for studying celullarization came from the recent discovery of two methods that induce this process in rat myotubes. In an approach, the mononuclear cells were transfected with the homeobox gene *Msx-1.* Several studies have previously indicated that *Msx* genes promote cell proliferation, and that their expression is inversely correlated with differentiation. After the fusion of the transfected cells, the expression of Msx-1 was induced in myotubes and this led to a decrease in the expression of myogenic regulatory genes. About 5% of myotubes were induced to cleave into viable fragments and another 5% fragmented into mononuclear cells, which proliferated. In some cases, the clonal progeny of a single myotube was isolated, propagated and showed itself capable of chondrogenic, adipogenic or myogenic differentiation, depending on the culture conditions. It has been proposed that Msx-1 is a master regulator of the program for cellularization that is expressed in regeneration in Urodeles, and that can also induce this program in mammal myotubes [80].

A second impetus for the analysis of these matters comes from the application of "chemical genetics". A large combinatorial library of tritiated and substituted purines was exhibited to identify a compound that induces mammalian myotubes to regenerate. A compound, with substituents methoxybenzyl and isopropylic, effectively fragments the myotubes over 24 hours to produce viable mononuclear cells that can divide and also fuse together once more for myotubes to reform. This compound, called *myoseverin,* appears to have two activities in mammalian myotubes. Firstly, it depolymerizes microtubules. Secondly, it induces changes in the expression of a specific set of genes that are involved in repair, wound healing and regeneration. Although there are other agents that can fragment microtubules, they do not tend to produce viable mononuclear cells that can fuse into myotubes again. Therefore, it is possible that both activities, i.e., depolymerization of microtubules and changes in the gene expression, are important for the generation of viable mononuclear cells [80].

A possibility is that the compound responsible for the fragmentation of myotubes can also regulate genes that are related in tissue remodeling and repair. The alternative and most appealing hypothesis is that *myoseverin* can activate the expression of a program that acts as mediator of cellularization and other functions that are important for regeneration. DNA microarray analysis has identified approximately 90 genes so far. Many of these genes belong to categories that are regulated in fibroblasts in response to the serum [80].

The results with myoseverin raised the issue of the identity of the endogenous signal that activates this response in the muscle fibers of Urodeles during the initial phase of regeneration. A recent study presented evidence proving that binders present in extracts from the start of the regeneration in triton limbs can induce the myotubes of tritons and of rats to submit to cell cycle reentry and cellularization. Cellularization indicates the presence of myoseverin as an activity both in triton cells and in mouse cells. However, reentry in S phase of rat myotubes is surprising in view of the previously discussed evidence pointing to a clear difference between the capacity for response to serum and thrombin between triton and rat myotubes. A more in-depth analysis of the factors that are present in the extracts should provide further information [80].

Crystalline lens regeneration is an example in which the regenerative cells appear through transdifferentiation of pigmented epithelial cells of the iris [99]. It is possible to aggregate pigmented epithelial cells of salamander iris from the dorsal iris, to implant these cells in an eye whose crystalline lens has been previously removed and to obtain the formation of a new crystalline lens in the aggregate. Pigmented epithelial cells of the ventral iris do not normally produce a crystalline lens, but the activity of the *Six*-3 gene induces lens formation through the transdifferentiation of ventral cells. Inhibition of the signaling pathway of the bone morphogenetic protein (BMP) also leads to induction of crystalline lens formation [100]. There is the theory that the BMP signaling pathway maintains ventral identity and that the cells assume a dorsal identity after its inhibition.

Planarian regeneration appears to occur almost exclusively through the contribution of neoblasts, although the current evidence does not necessarily exclude a contribution of differentiated cells [58]. In hydra regeneration, the activity of the epithelial stem cells implies budding and regeneration [101-102]. Studies with the jellyfish *Podocoryne* showed an example of transdifferentiation and its regulation by signaling [58]. When portions of striated muscle were explanted, they could be activated by enzymes that degrade the extracellular matrix, and that are later differentiated into smooth muscle and other types of cell. The impression left by these studies is that control of the differentiated state is prepared to allow considerable plasticity, which may be related to the important role of agamic reproduction and regeneration in these animals [58].

In view of evolutionary and biomedical problems, it is interesting to compare the relevant examples of plasticity in myotubes and in the pigmented epithelial cells of Urodeles and mammals. Although salamanders are the only adult vertebrates that can regenerate their crystalline lens, the pigmented epithelial cells of several vertebrates can be converted into crystalline lens cells in culture. It will be necessary to identify the blastemal signals that trigger these responses from differentiated cells, as well as the molecular basis of any differences between the differentiated cells of the two species. It is already evident that this approach to regeneration – i.e., the investigation of the plasticity of differentiated cells-is a productive and informative study, particularly in the absence of a complete genetic analysis [80].

Nowadays, the main method for regenerative medicine in mammals is the isolation of stem cells, followed by manipulations that aim to direct differentiation towards the morphogenesis of complex structures. Although this is attracting considerable interest at present, many of the applications are sufficiently problematic to justify the consideration of alternative and supplementary approaches. The strategy in Urodeles – the limited re-specification of residual differentiated cells-is so successful that it would come as a surprise if it were not potentially regarded as a therapeutic approach in some regeneration contexts in mammals. The example of myoseverin shows that the responses that are discussed in this review are a potential target for therapies geared towards regulating the stability of the differentiated state [80].

Finally, in theory, differentiated cells could also contribute to the formation of the blastema, without losing their differentiation status. This would apply to cells such as fibroblasts. Fibroblasts are mature cells that synthesize the extracellular matrix, but still maintain some of the characteristics of progenitor cells, such as elevated motility and proliferation capacity. Fibroblasts can be regarded as one of the main contributors to the blastema and they can achieve this without the occurrence of dedifferentiation [62]. Therefore, the processes that lie behind the formation of the blastema and cell origin are still being studied.

4. Cellular differentiation and environmental insults or stress

The cells are susceptible to changes arising from normal physiological processes as well as from changes in the external environment units. To avoid subsequent damage to these cells, there are homeostatic mechanisms that enable the adaptation of cells and tissues to changing conditions. Such mechanisms are activated not only under physiological conditions, but also to limit the damage imposed in response to injury and disease processes. Hence certain factors that may be present in the tissue microenvironment can lead to cellular stress, which in turn triggers mechanisms involving cell survival and adaptation strategies as well as injuries and consequent pathological process, or even cell death [103].

Stress-inducing conditions caused by changes in the cell microenvironment may affect the eukaryotic cells exposed either acutely, when they occur for a transitory period, or chronically, when they persist for a long time, prompting the cells to respond in a number of different ways, e.g., changing their metabolism, secretion and gene expression etc. [104].

Stress-inducing agents include thermal shock caused by hyperthermia or hypothermia [105, 106], osmotic shock [107, 108], nutrient [109-111] and ATP depletion, low oxygen concentration and oxidative stress [112,113], changes in cell metabolism induced by cancerous transformation, mechanical stress, DNA damage due to exposure to UV radiation, baric stress and others. Therefore the cells may be exposed to a wide variety of agents, which will trigger different responses, affecting their function, activity and differentiation [103].

Once exposed to stress agents, the initial response of cells is generally a decrease in the synthesis processes of macromolecules such as normal proteins, RNA, DNA and fatty acids, and an increase in the synthesis of proteins belonging to a specific group called stress proteins, probably regulated by changes occurring in the cytoskeleton during stress. These proteins are associated with the mechanism of cell cycle control, biosynthesis reactions and processing, including protein folding and oligomerization, translation, secretion, and repair of damaged proteins, and oxidative function [104]. This group of proteins is evolutively conserved, is present in the Archaea, Eubacteria and Eukarya domains, including the human proteome, and includes the so-called heat shock proteins, or Hsp, thus named because they were initially observed by Ritossa in 1962, due to the rise in temperature, but are also associated with other types of stress. Their high level of conservation suggests the performance of an essential role in cell metabolism [105].

The Hsps are part of the large group of proteins known as molecular chaperones, and have the ability to interact reversibly with other proteins, assisting in formation, folding and transmembrane transport [116]. In adverse conditions, such as increased temperature, osmotic or oxidative stress, Hsp levels are increased, remodeling the proteins that suffered injury, activating the synthesis and maturation of new proteins that will replace those affected by metabolic stress [117], or allocating them to an adequate proteolytic system, facilitating their elimination when the damage is not reparable [118]. The Hsps can be classified according to their molecular weight into five families or groups: HSP100, HSP90, HSP70, HSP60, and low-molecular-weight HSPs between 12 and 43 kDa [119].

Stress may also result in epigenetic modifications involving DNA and histone methylation, phosphorylation and acetylation, which change the expression of the genes without altering the DNA. Genomic changes may occur due to structural rearrangements and these alterations can be passed on to daughters and may lead to some changes which can be maintained in many cell generations [103].

The HSPs may interact with different apoptotic proteins, and can block essentially all apoptotic pathways, most of them involving the activation of cysteine proteases called caspases [120]. Some mediator proteins may alternate adaptive responses and direct the cells to apoptosis due to more severe stress, when this causes, for instance, DNA breakages. This mechanism involves checkpoint kinase 2 (Chk2) phosphorylation via p21 with cell cycle arrest and the action of proapoptotic p53 genes, activating apoptosis and cell death. Thus the cells may respond to stress, depending on its level and intensity, in various ways ranging from the activation of survival and adaptation pathways in milder situations, to channeling to cell death, in the case of more severe damage [103]. The degree of susceptibility to environmental changes varies in different cell types. Brain neurons, for example, are extremely sensitive to changes in their environment, dying quickly when environmental conditions are beyond the normal physiological patterns. On the other hand, fibroblasts exhibit extreme resistance to damage, surviving drastic metabolic changes, such as the complete deprivation of oxygen.

The characteristics and functions of stem cells make them particularly susceptible to certain stress factors such as oxidative stress, mechanical stress, growth factors, and cytokines signal affecting the cell differentiation and division regulation mechanisms. The long lifespan and high proliferative potential of stem cells makes them prone to cellular transformation (cancer), and the stress responses that lead to senescence and apoptosis can in fact act as anticancer protection mechanisms, by preventing the continuity of the cell division process, thus limiting their lifespan [121].

4.1. Stress proteins, differentiation and cancer

During cellular transformation in tumor growth, the cells are experiencing drastic shifts in their intracellular and extracellular environment, because they are frequently exposed to stress conditions that include hypoxia, acidosis, nutrient deprivation and immune system attacks. Accordingly, the tumor cells must be able to adapt to a variety of stress conditions to survive and proliferate, so these stress condition may act as a driving force behind evolution (oncogenesis) and adaptation (acquired treatment resistance) [122]. In various types of tumor, the expression of molecular chaperones including the Hsps is increased, reflecting an effort of the tumor cells to maintain tissue homeostasis in an unsuitable environment. This mechanism can facilitate the survival of tumor cells from these injury conditions, thus representing a possible selecting agent for further mutations, and the management of the oncogenic process [123].

However, the Hsp levels are useful biomarkers for carcinogenesis in some tissues and signal the degree of differentiation and the aggressiveness of some cancers. Therefore, these levels can be useful as an adjunct in tumor diagnosis, but are not informative at the diagnostic when considered alone. Several Hsps are implicated with the prognosis of specific cancers; for example Hsp70 is correlated with poor prognosis in breast, endometrial, uterine, cervical, and bladder carcinomas, while Hsp27 expression is associated with poor prognosis in gastric, liver, and prostate carcinoma, and osteosarcomas. On the other hand, increased Hsp expression may also predict the response to some anticancer treatments. For example, Hsp27 and Hsp70 are implicated in resistance to chemotherapy in breast cancer, Hsp27 predicts a poor response to chemotherapy in leukemia patients, while Hsp70 expression predicts a better response to chemotherapy in osteosarcomas [123]. As Hsp70 and Hsp27 are the most active chaperones in the subversion of the programmed cell death pathway (apoptosis) and play major roles in cancer, because these chaperones are also able to inhibit senescence pathways (e.g., p21 and p53 dependent), allowing cell division limits to be surpassed [119,124], these implications of Hsp in tumor progression and response to therapy have led to the possibility of Hsp use in anticancer vaccines, exploiting their ability to act as immunological adjuvants [125,126].

Based on the fact that most cancer cells abundantly express HSP70 at the basal level to resist various insults at different stages of tumorigenesis and during anti-cancer treatment, some studies have investigated HSP70 and HSP90 inhibitors for their use in anti-cancer therapy [127,128].

5. Conclusion

The goal of regenerative medicine is to restore the cells, tissues and structures that are lost or damaged after disease, injury or aging. Regenerative medicine currently uses three main approaches (4): the use of stem cells to build new structures, the implantation of cells preconditioned to differentiate in a particular direction, and the stimulation of endogenous cells to replace structures that have disappeared. In these strategies, we can choose to use or disregard biomaterials with carrier and/or stimulator agents. The regeneration of organs and limbs that occurs in several groups of animals provides another important viewpoint, as it demonstrates that under some conditions, complex adult tissues can be rebuilt. Biological regeneration lessons have not been widely assimilated, partly because this attribute seems minor when we focus on mammals.

It is not understood why some animals are able to regenerate and others are apparently not, but even from our current limited perspective, there appears to be a series of differences between mammals and Urodeles that prevent or limit regeneration. One of the aspects of wound healing in adult mammals is the reduction of the regenerative capacity and the occurrence of fibrosis and inflammatory responses [83].

We understand that it is necessary to glean a better idea of the cell differentiation, transdifferentiation, dedifferentiation and redifferentiation mechanisms that occur in different contexts, such as during regeneration in animals. The participation of stem cells and progenitor cells in this process is also of considerable relevance. The understanding of these mechanisms can facilitate highly relevant therapeutic strategies in regenerative medicine and tissue engineering.

Gene/Factor	Effect/Importance
pRb	Essential in regulating G1-phase S transition. Evidence indicates major importance for the maintenance of the state of differentiation in vertebrate myotubes.
Prothrombin	Activated selectively on the dorsal margin of the iris after injury. Generator of thrombin protease.
Thrombin	Activates the coagulation cascade and other events to mediate the response to the injury.
Prod 1	Encodes a small protein that is bound to the cell surface by a GPI anchor. Apparently analogous to the mammalian CD59 protein, implicated in positional identity determination.
Nag	Identified as a binder to *Prod 1*. Acts as a mitogenic factor for blastema cells.
β-catenin	Signaling pathway for the regenerative process in the formation of the *Hydra* head. Participation in regeneration in planaria.
Pax-7	Expressed from the activation of muscle satellite cells, during the regeneration process in tritons.
$p16^{INK4}$	Causes an impediment to cell cycle reentry.
Homeobox Gene Msx-1	Promotes cell proliferation. Its expression is inversely correlated to differentiation. It has been proposed that Msx-1 is a master regulator of the program for cellularization in regeneration in Urodeles.
Myoseverin	Depolymerizes microtubules, an activity observed in myotubes of mammals. Induces changes in the expression of genes involved in repair, wound healing and regeneration. There is a theory that *myoseverin* can activate the expression of a program that acts on cellularization and other functions that are relevant to regeneration;
Six-3 Gene	Induces the formation of the crystalline lens through the transdifferentiation of ventral cells of the iris in mice.

Gene/Factor	Effect/Importance
Bone Morphogenetic Protein (BMP)	The inhibition of its signaling pathway leads to the induction of crystalline lens formation. It can maintain ventral identity and the cells assume a dorsal identity after its inhibition.
RA gradient	Operates in the development of rat limbs, as a consequence of its synthesis, close to the midline of the animal and its degradation by the product of the CYP 26 gene, expressed at the distal extremity of the limb bud.
Canonical Wnt signaling pathways	Involved in the appendage regeneration process
Fibroblast growth factor (FGF)	Involved in the appendage regeneration process.
Signaling of activin; retinoic acid; non-canonical Wnt.	Involved in the regulation of regeneration in the sea urchin.
SALL4	Involved in the maintenance of pluripotency. Expressed during blastema formation.
ATF3, JUN3, EGR1, NR4A2 and FOS	An increase is observed in these oncogenes during the epithelialization and blastema phases.
MSP1	Mitotic kinase essential for cardiac regeneration in the zebrafish.
GATA4	Heart development transcription factor essential for cardiac regeneration in the zebrafish.

Table 1. Some biochemical parameters to the regenerative responses and their consequences.

Acknowledgements

We would like to thank Instituto Nacional de Ciência e Tecnologia em Biofabricação (INCT-Biobabris) from Brazilian government.

Author details

Arnaldo Rodrigues Santos Jr[1*], Vitor Andrade Nascimento[1], Selma Candelária Genari[2] and Christiane Bertachini Lombello[3]

*Address all correspondence to: arnaldo.santos@ufabc.edu.br

1 Centro de Ciências Naturais e Humanas (CCNH), Universidade Federal do ABC, São Bernardo do Campo, SP, Brazil

2 Centro Estadual de Educação Tecnológica Paula Souza, Faculdade de Tecnologia de Jacareí (FATEC), Jacareí, SP, Brazil

3 Centro de Engenharia e Ciências Sociais Aplicadas (CECS),Universidade Federal do ABC, São Bernardo do Campo, SP, Brazil

References

[1] Vacanti J.Tissue engineering and regenerative medicine: from first principles to state of the art. Journal of Pediatric Surgery 2010; 45(2) 291–294.

[2] Baddour JA, Sousounis K, Tsonis PA. Organ repair and regeneration: An overview. Birth Defects Research C: Embryo Today 2012; 96(1) 1-29.

[3] Hoffmann A, Tsonis PA. Pattern formation in regenerating tissues. In: Capasso V, Gromov M, Harel-Bellan A, Morozova N, Pritchard L (ed.) Pattern formation in morphogenesis. Springer Proceedings in Mathematics (Vol. 15). Springer-Verlag Berlin Heidelberg; 2013. p7-16.

[4] Yannas JV. Tissue and organ regeneration in adults. New York: Springer; 2001.

[5] Tanaka EM, Reddien PW. The cellular basis for animal regeneration. Dev Cell 2011; 21(1) 172-185.

[6] Sephel GC, Woodward SC. Chapter 3: Repair, Regeneration and Fibrosis In: Rubin's Pathology: Clinicopathologic Foundations of Medicine. 6th Ed. Rubbin R, Strayer DS (ed.); 2011. p71-98.

[7] Lanza R, Langer R, Vacanti J. Principles of tissue engineering. 34th Ed. Burlington, MA: Academic Press; 2013.

[8] Eguizabal C, Montserrat N, Veiga A, Izpisua Belmonte JC. Dedifferentiation, transdifferentiation, and reprogramming: future directions in regenerative medicine. Semin Reprod Med. 2013; 31(1) 82-94.

[9] Hench LL. Biomaterials: a forecast for the future. Biomaterials 1998; 19(16) 1419-1423.

[10] Wolpert L. Princípios de Biologia do Desenvolvimento, Porto Alegre: Artmed; 2000.

[11] Stevens A, Lowe J. Patologia, 2a ed, Barueri: Editora Manole; 2000.

[12] Santos Jr Ar, Wada MLF, Carvalho HF. Diferenciação Celular. In: Carvalho HF, Recco-Pimentel SM. (ed.) A Célula. 3ed., Barueri: Manole; 2013. p553-570.

[13] Bedi A, Feeley BT, Williams RJ. Management of articular cartilage defects of the knee. J Bone Joint Surg Am 2010; 92(4) 994-1009.

[14] Cotran RS, Kumar Y, Collins T. Patologia celular I: lesão e morte da célula. In: Cotran RS, Kumar Y, Collins T. (ed.) Patologia Estrutural e Funcional, 6ª ed., Rio de Janeiro: Editora Guanabara Koogan; 2000. p.1-25.

[15] Cohen M, Nery C, Peccin MS, Ressio CR, Asaumi ID, Lombello CB. Autologous chondrocyte implantation to treat femoral condyle and talar lesions. Einstein (São Paulo) 2008; 6(1) 37-41.

[16] Nery C, Lombello CB, Ressio CR, Asaumi ID. Implante Autólogo de Condrócitos no Tratamento das Lesões Osteocondrais do Talo. Revista ABTPé 2010; 4(2) 113-123.

[17] Robey PG. Stem cells near the century mark. Journal of Clinical Investigation 2000; 105(11) 1489-1491.

[18] Junqueira LC, Carneiro J. Biologia Celular e Molecular, 12ª Ed., Rio de Janeiro: Guanabara Koogan; 2013.

[19] Lombello CB, Malmonge SM, Wada MLF. PolyHEMA and polyHEMA-poly(MMA-co-AA) as substrates for culturing cells. Journal of Materials Science. Materials in Medicine; 2000 11(9) 541-546.

[20] Strehl R, Schumacher K, de Vries U, Minuth WW. Proliferating cells versus differentiated cells in tissue engineering. Tissue Engineering 2002; 18(1) 37–42.

[21] Mimeault M, Hauke R, Batra SK. Stem cells: a revolution in therapeutics-recent advances in stem cell biology and their therapeutic applications in regenerative medicine and cancer therapies. Clinical Pharmacology and Therapeutics. 2007; 82(3) 252–264.

[22] Jopling C, Boue S, Belmonte JCI. Dedifferentiation, transdifferentiation and reprogramming: three routes to regeneration. Nature Reviews. Molecular Cell Biology 2011; 12(2) 79-89.

[23] Thomson JA, Itskovitz-Eldor J, Shapiro SS, Waknitz, MA, Swiergiel, JJ, Marshal VS, et al. Embryonic stem cell lines derived from hum an blastocysts. Science 1998; 282(5391) 1145-1147.

[24] Yarak S, Okamoto OK. Células-tronco derivadas de tecido adiposo humano: desafios atuais e perspectivas clínicas. The journal Brazilian Annals of Dermatology 2010; 85 (5) 647-656.

[25] Piccin D, Morshead CM Potential and pitfalls of stem cell therapy in old age Disease Models and Mechanisms 2010; 3(7-8) 421-425.

[26] Sharpless NE, DePinho RA. How stem cells age and why this makes us grow old. Nature Reviews Molecular Cell Biology 2007; 8(9) 703-13.

[27] Jones DL, Wagers AJ. No place like home: anatomy and function of stem cell niche. Nature Reviews Molecular Cell Biology 2008; 9(1) 11-21.

[28] Rezza A, Sennett R, Rendl M. Stem cells in development and disease. Current Topics in Developmental Biology 2014; 107 333–372.

[29] Kimelman D, Kirschner M. Synergistic induction of mesoderm by FGF and TGF-beta and the identification of an mRNA coding for FGF in the early Xenopus embryo. Cell 1987; 51(5) 869-77.

[30] Schwank G, Basler K. Regulation of organ growth by morphogen gradients. Cold Spring Harbor Perspectives in Biology 2010; 2(1) 001669.

[31] Gaissmaier C, Koh JL, Weise K. Growth and differentiation factors for cartilage healing and repair. Injury 2008; 39(S1) S88–S96.

[32] Green JC, Smith. Growth factors as morphogens: do gradients and thresholds establish body plan? Trends in Genetics 1991; 7(8) 245–250.

[33] Fortier LA, Barker JU, Strauss EJ, McCarrel TM, Cole BJ. The role of growth factors in cartilage repair. Clinical Orthopaedics and Related Research 2011; 469(10) 2706-2715.

[34] Zhao L, Jiang S, Hantash BM. Transforming growth factor beta1 induces osteogenic differentiation of murine bone marrow stromal cells. Tissue Engineering, Part A. 2010; 16(2) 725-33.

[35] Lodish H, Berk A, Zipursky SL, Matsudaira P, Baltimore D, Darnell J. Molecular Cell Biology, 4th edition. New York: WH Freeman; 2014.

[36] Borges HL, Rehen SK. Capítuo 32. Controle do ciclo celular. In: Carvalho HF, Recco-Pimentel SM. (ed.) A Célula. 3ed., Barueri: Manole; 2013. p519-534.

[37] Agarwal ML, Agarwal A, Taylor WR, Stark GR. p53 controls both the G2/M and the G1 cell cycle checkpoints and mediates reversible growth arrest in human fibroblasts. Proceedings of the National Academy of Sciences of the USA 1995; 92(18): 8493–8497.

[38] Riley T, Sontag E, Chen P, Levine A. Transcriptional control of human p53-regulated genes. Nature Reviews Molecular Cell Biology 2008; 9(5) 402-412.

[39] Paunesku T, Mittal S, Protić M, Oryhon J, Korolev SV, Joachimiak A, Woloschak GE. Proliferating cell nuclear antigen (PCNA): ringmaster of the genome. International Journal of Radiation Biology 2001; 77(10) 1007-1021.

[40] LaBaer J, Garrett MD, Stevenson LF, Slingerland JM, Sandhu C, Chou HS, Fattaey A, Harlow E. New functional activities for the p21 family of CDK inhibitors. Genes and Development. 1997; 11(1) 847-862.

[41] Doran TI, Vidrich A, Sun TT. Intrinsic and extrinsic regulation of the differentiation of skin, corneal and esophageal epithelial cells. Cell. 1980; 22(1Pt 1) 17-25.

[42] Adams J. The complexity of gene expression, protein interaction, and cell differentiation. Nature Education 2008; 1(1) 1101-1104.

[43] Levine M, Tijan, R. Transcription regulation and animal diversity. Nature 2003; 424(6945) 147–151.

[44] Torres-Padilla ME, Parfitt DE, Kouzarides T, Zernicka-Goetz M. Histone arginine methylation regulates pluripotency in the early mouse embryo. Nature 2007; 445(7124) 214–218.

[45] Robertson KD. DNA methylation and chromatin: Unraveling the tangled web. Oncogene 2002; 21(35) 5361–5379.

[46] Egger G, Liang G, Aparicio A, Jones PA. Epigenetics in human disease and prospects for epigenetic therapy. Nature 2004; 429(6990) 457–463.

[47] Riddihough G, Zahn LM. What Is Epigenetics? Science 2010; 330(6004) 611.

[48] Lin CQ, Bissell MJ. Multi-faceted regulation of cell differentiation by extracellular matrix. FASEB Journal 1993; 7(9) 737-43.

[49] Sheppard D. Functions of pulmonary epithelial integrins: from development to disease. Physiology Reviews 2003; 83(3) 673–686.

[50] Madala SK, Korfhagen TR, Schmidt S, Davidson C, Edukulla R, Ikegami M, Violette SM, Weinreb PH, Sheppard D, Hardie WD. Inhibition of the $\alpha v\beta 6$ integrin leads to limited alteration of TGF-α-induced pulmonary fibrosis. American Journal of Physiology: Lung Cell and Molecular Physiology 2014; 306(8) L726-35.

[51] Sarrazin S, Lamanna WC, Esko JD. Heparan sulfate proteoglycans. Cold Spring Harb Perspectives in Biology 2011; 3(7) a004952.

[52] Hay ED. Collagen and embryonic development. In: Hay ED (ed.). Cell Biology of the Extracellular Matrix. Plenum Press, New York; 1981. p 379–409,

[53] Folkman J, Moscona A. Role of cellular shape in growth control. Nature 1978; 273(5661) 345-349.

[54] Lombello CB, Reis Jr GM, Cohen M. Novas perspectivas no tratamento de lesões da cartilagem. Revista Joelho 2002; 2(1) 48-53.

[55] Gao Y, Liu S, Huang J, Guo W, Chen J, Zhang L, Zhao B, Peng J, Wang A, Wang Y, Xu W, Lu S M, Guo Q. The ECM-Cell interaction of cartilage extracellular matrix on chondrocytes. BioMed Research International Volume 2014 (2014), Article ID 648459, 8 pages. http://dx.doi.org/10.1155/2014/648459 (acessed 20 july 2014).

[56] Clause KC, Barker TH. Extracellular matrix signaling in morphogenesis and repair. Current Opinion in Biotechnology 2013; 24(5) 830-833.

[57] Colnot JA, Helms A. molecular analysis of matrix remodeling and angiogenesis during long bone development. Mechanisms of Development 2001; 100(2) 245–250.

[58] Brockes JP, Kumar A. Comparative aspects of animal regeneration. Annual Review of Cell and Developmental Biology 2008; 24 525–549.

[59] Jopling C, Sleep E, Raya M, Martí M, Raya A, Izpisúa Belmonte JC. Zebrafish heart regeneration occurs by cardiomyocyte dedifferentiation and proliferation. Nature 2010; 464(7288) 606-609.

[60] Kami D, Gojo S. Tuning cell fate-From insights to vertebrate regeneration. Organogenesis 2014; 10(2) 1-10.

[61] Hughes DW, Galau GA. Temporally modular gene expression during cotyledon development. Genes and Development 1989; 3(3) 358-369.

[62] Han M, Yang X, Taylor G, Burdsal CA, Anderson RA, Muneoka K. Limb regeneration in higher vertebrates: developing a roadmap. Anatomical Record (Part B: The New Anatomist website) 2005; 287(1)14-24.

[63] Borgens RB. Natural voltage gradients and the generation and regeneration of limbs. In: Sicard RE (ed.) Regulation of Veretebrate Limb Regeneration, New York/Oxford: Oxford Univ. Press, 1985, p6-31.

[64] Adams DS, Masi A, Levin M. H+pump-dependent changes in membrane voltage are an early mechanism necessary and sufficient to induce Xenopus tail regeneration. Development 2007; 134 (7) 1323-35.

[65] Levin M. Large-scale biophysics: ion flows and regeneration. Trends in Cell Biology 2007; 17(6) 261-70.

[66] Stewart R, Rascón CA, Tian S, Nie J, Barry C, Chu LF, Ardalani H, Wagner RJ, Probasco MD, Bolin JM, et al. Comparative RNA-seq analysis in the unsequenced axolotl: the oncogene burst highlights early gene expression in the blastema. PLoS Computational Biology 2013; 9:e1002936.

[67] Carlson BM. Principles of Regenerative Biology. London: Elsevier; 2007.

[68] O'Steen WK. Regeneration of the intestine in adult urodeles. Journal of Morphology 1958; 103(3) 435-78

[69] Donaldson DJ, Mahan JT, Hasty DL, McCarthy JB, Furcht LT. Location of a fibronectin domain involved in newt epidermal cell migration. Journal of Cell Biology 1985; 101(1) 73-78

[70] Oberpriller JO, Oberpriller JC. Response of the adult newt ventricle to injury. Journal of Expimental Zoology 1974; 187(2) 249–53

[71] Tanaka EM. Cell differentiation and cell fate during urodele tail and limb regeneration. Current Opinion in Genetics and Development 2003; 13(5) 497-501.

[72] Imokawa Y, Brockes JP. Selective activation of thrombin is a critical determinant for vertebrate lens regeneration. Current Biology 2003; 13(10) 877-881.

[73] Imokawa Y, Simon A, Brockes JP. A critical role for thrombin in vertebrate lens regeneration. Philos. Trans. R. Soc. London Ser. B 2004; 359(1445) 765-76

[74] Hayashi T, Mizuno N, Takada R, Takada S, Kondoh H. Determinative role of Wnt signals in dorsal iris-derived lens regeneration in newt eye. Mechanisms of Development 2006; 123(11) 793-800

[75] Lesurtel M, Graf R, Aleil B,Walther DJ, Tian Y, Jochum W, Gachet C, Bader M, Clavien PA. Platelet-derived serotonin mediates liver regeneration. Science 2006; 312(5770) 104-107

[76] Lepilina A, Coon AN, Kikuchi K, Holdway JE, Roberts RW, Burns CG, Poss KD. A dynamic epicardial injury response supports progenitor cell activity during zebrafish heart regeneration. Cell 2006; 127(3) 607-19

[77] Campbell LJ, CrewsCM. Molecular and cellular basis of regeneration and tissue repair: wound epidermis formation and function in urodele amphibian limb regeneration. Cellular and Molecular Life Sciences 2008; 65(1) 73-79

[78] Whitehead GG, Makino S, Lien CL, Keating MT. Fgf20 is essential for initiating zebrafish fin regeneration. Science 2005; 310(5756) 1957-60

[79] Stoick-Cooper CL, Weidinger G, Riehle KJ, Hubbert C, Major MB, Fausto N, Moon RT. Distinct Wnt signaling pathways have opposing roles in appendage regeneration. Development 2007; 134(3) 479-89

[80] Brockes JP, Kumar A. Plasticity and reprogramming of diferentiated cells in amphibian regeneration. Nature Reviews. Molecular Cell Biology 2002; 3(8) 566-574.

[81] Candia-Carnevali M. Regeneration in echinoderms: repair, regrowth, cloning. Invertebrate Survival Journal 2006; 3(1) 64-76

[82] Boilly-Marer Y. Role du syst`eme nerveux parapodial dans l′induction de parapodes supernumeraries par greffes h′et′erologues chez Nereis pelagica. Comptes Rendus de l'Académie des Sciences 1971; 272(1) 261-64

[83] Brockes JP, Kumar A. Appendage regeneration in adult vertebrates and implications for regenerative medicine. Science 2005; 310(5756) 1919-1923.

[84] Taub R. Liver regeneration: from myth to mechanism. Nature Reviews Molecular Cell Biology, 2004; 5(10) 836-847.

[85] Endo T, Bryant SV, Gardiner DM. A stepwise model system for limb regeneration Developmental Biology 2004; 270 135-145.

[86] Tsonis PA. Regeneration in vertebrates. Developmental Biology 2000; 221(2) 273-84.

[87] Han M, Yang X, Farrington JE, Muneoka K. Digit regeneration is regulated by Msx1 and BMP4 in fetal mice. Development 2003; 130 5123-5132.

[88] Michalopoulos GK, DeFrances MC. Liver regeneration. Science 1997; 276(5309) 60-66.

[89] Stocum DL, Cameron JA. Looking proximally and distally: 100 years of limb regeneration and beyond. Developmental Dynamics 2011; 240(5) 943-968.

[90] Kumar A, Godwin JW, Gates PB, Garza-Garcia AA, Brockes JP. Molecular basis for the nerve dependence of limb regeneration in an adult vertebrate. Science 2007; 318(5851) 772-77.

[91] Chera S, Kaloulis K, Galliot B. The cAMP response element binding protein (CREB) as an integrative HUB selector in metazoans: clues from the hydra model system. Biosystems 2007; 87(2-3) 191-203.

[92] Knopf F, Hammond C, Chekuru A, Kurth T, Hans S, Weber CW, Mahatma G, Fisher S, Brand M, Schulte-Merker S, et al. Bone regenerates via dedifferentiation of osteoblasts in the zebrafish fin. Developmental Cell 2011; 20(5) 713-24.

[93] Morrison JI, Borg P, Simon A. Plasticity and recovery of skeletal muscle satellite cells during limb regeneration. FASEB J 2010; 24(3) 750-756.

[94] Poss KD,Wilson LG, Keating MT. Heart regeneration in zebrafish. Science 2002; 298(5601) 2188-2190.

[95] Gargioli C, Slack JMW. Cell lineage tracing during Xenopus tail regeneration. Development 2004; 131() 2669-2679.

[96] Calve S, Odelberg SJ, Simon H-G. A transitional extracellular matrix instructs cell behavior during muscle regeneration. Developmental Biology 2010; 344(1) 259-271.

[97] Morrison JI, Loof S, He P, Simon A. Salamander limb regeneration involves the activation of a multipotent skeletal muscle satellite cell population. Journal of Cell Biology 2006; 172(3) 433-440.

[98] Kurosaka H, Takano-Yamamoto T, Yamashiro T, Agata K. Comparison of molecular and cellular events during lower jaw regeneration of newt (Cynops pyrrhogaster) andWest African clawed frog (Xenopus tropicalis). Developmental Dynamics 2008; 237(2) 354-365.

[99] Tsonis PA,Madhavan M, Tancous EE, Del Rio-Tsonis K. A newt's eye view of lens regeneration. International Journal of Developmental Biology 2004; 48(8-9) 975-980.

[100] Grogg MW, Call MK, Okamoto M, Vergara MN, Del Rio-Tsonis K, Tsonis PA. BMP inhibition-driven regulation of six-3 underlies induction of newt lens regeneration. Nature 2005; 438(7069) 858-862.

[101] Bosch TC. Why polyps regenerate and we don't: towards a cellular and molecular framework for hydra regeneration. Developmental Biology 2007; 303(2) 421-433

[102] Galliot B, Miljkovic-Licina M, de Rosa R, Chera S. Hydra, a niche for cell and developmental plasticity. Seminars in Cell Developmental Biology 2006; 17(4) 492-502.

[103] Mansouri L, Xie Y, Rappolee D. A. Adaptive and pathogenic responses to stress by stem cells during development. Cells 2012; 1(4) 1197-1224.

[104] Woodgett JR, Kyriakis JM, Avruch J, Zon LI, Zanke B, Templeton DJ. Reconstitution of novel signalling cascades responding to cellular stresses. Philosophical Transaction of Royal Society Lond B Biological Science 1996; 351(1336) 135-141.

[105] Lindquist S, Craig EA. The heat shock proteins. Annual Review of Genetics 1988; 22, 631-77.

[106] Cullen KE, Sarge KD. Characterization of hypothermia-induced cellular stress response in mouse tissues. Journal of Biological Chemistry 1997; 272(3) 1742-1746.

[107] Zhong W, Xie Y, Wang Y, Lewis J, Trostinskaia A, Wang F, Puscheck E, Eappolee DA. Use of hyperosmolar stress to measure stress-activated protein kinase activation and function in human HTR cells and mouse trophoblast stem cells. Reproductive Sciences 2007; 14(6) 53-547.

[108] Liu J, Xu W, Sun T, Wang F, Puscheck E, Brigstock D, Wang QT, Davis R, Rappolee DA. Hyperosmolar stress induces global mRNA responses in placental trophoblast stem cells that emulate early post-implantation differentiation. Placenta 2009; 30(1) 6-73.

[109] Genari SC, Dolder H, Wada MLF. Scanning and transmission electron microscopy of transformed Vero cells with altered in vitro growth characteristics. Journal of Submicroscopic Cytology and Pathology 1996; 28(4) 565-572.

[110] Genari SC, Gomes L, Wada MLF. Alterations in the growth and adhesion pattern of Vero cells induced by nutritional stress conditions. International Cell Biology 1998; 22(4) 285-294.

[111] Genari SC, Wada MLF. The influence of the nutricional stress conditions on differentiation of epithelial cells in vitro. Brazilian Journal of Morphological Sciences 2003; 20(3) 135-140.

[112] Rappolee DA, Awonuga AO, Puscheck EE, Zhou S, Xie Y. Benzopyrene and experimental stressors cause compensatory differentiation in placental trophoblast stem cells. Systems Biology in Reproductive Medicine 2010; 56(2) 168-183.

[113] Zhou S, Xie Y, Puscheck EE, Rappolee DA. Oxygen levels that optimize TSC culture are identified by maximizing growth rates and minimizing stress. Placenta 2011; 32(6) 475-481.

[114] Morimoto RI. Heat shock: the role of transient inducible responses in cell damage, transformation, and differentiation. Cancer Cells 1991; 3(8) 295-301.

[115] Kregel K. Heat shock proteins: modifying factors in physiological stress responses and acquired thermotolerance. Journal of Applied Physiology 2002; 92(5) 2177-2186.

[116] Karp G. Biologia celular e molecular: conceitos e experimentos, 3a Ed., São Paulo: Manole, 2005.

[117] Bukau B, Horwich A. The Hsp70 and Hsp60 chaperonis machines. Cell 1998; 92(3) 351-366.

[118] Meyer MP, Bukau B. Hsp70 chaperones: cellular functions and molecular mechanism. Cellular and Molecular Life Sciences 2005; 62(6) 670-684.

[119] Almeida MB, Nascimento JLM, Herculano AM, Crespo-Lopez ME. Molecular chaperones: toward new therapeutic tools. Biomedicine and Pharmacotherapy 2011; 65(4) 239-243.

[120] Lanneau D, Thonel A, Maurel S, Didelot C, Garrido C. Apoptosis versus cell differentiation: role of heat shock proteins hsp90, hsp70 and hsp27. Prion 2007; 1(1) 53-60.

[121] Tower J. Stress and stem cells. Wiley Interdisciplinary Reviews: Developmental Biology 2012; 1(6) 1-21.

[122] Zoubeidi A, Gleave M. Small heat shock proteins in cancer therapy and prognosis. International Journal of Biochemistry and Cell Biology 2012; 44(10) 1646-1656.

[123] Whitesell L, Lindquist SL. HSP 90 and the chapering of cancer. Nature Reviews: Cancer 2005; 5(10) 761-72.

[124] Arrigo AP. In search of the molecular mechanism by which small stress proteins counteract apoptosis during cellular differentiation. Journal of Cellular Biochemistry 2005; 94(2) 241-246.

[125] Ciocca D, Calderwood S. Heat shock proteins in cancer: diagnostic, prognostic, predictive, and treatment implications. Cell Stress Chaperones 2005; 10(2) 86–103.

[126] Murphy M. The HSP70 family and cancer. Carcinogenesis 2013; 34(6) 1181-1188.

[127] Goloudina AR, Demidov ON, Garrido C. Inhibition of HSP70: A challenging anticancer strategy. Cancer Letters 2012; 325(2) 117-124.

[128] Jego G, Hazoumé A, Seigneuric R, Garrido C. Targeting heat shock proteins in cancer. Cancer Letters 2013; 332(2) 275-285

Chapter 3

Regulatory Loops Consisting of Transcription Factors and microRNA Species Determine the Mineralizing Characteristics of Cell Phenotypes — Implications for Bone Engineering and Prevention of Soft Tissue Mineralization

Jan O. Gordeladze, Øystein Stakkestad, Sigrid Haugen, Janne E. Reseland, Unni Syversen, Gaute F. Johnsen, Håvard J. Haugen, Ståle P. Lyngstadaas, Mona Møller and Mauro Valtieri

Additional information is available at the end of the chapter

http://dx.doi.org/10.5772/59149

1. Introduction

1.1. Regulation of mineralization in bone and soft tissue

The first stages of cell mediated tissue bio-mineralization involve the activation of various signaling pathways, such as those activated by growth factors and hormones like BMPs, TGFs, PTH, Leptin, Vitamin D and many others.

One major signaling pathway responsible for bio-mineralization is the Wnt-pathway. Wnt-mediated transduction occurs via both β-catenin dependent and independent signaling. In β-catenin dependent signaling, extracellular Wnt ligands bind to the LRP5-Frizzled (Frz) complex, subsequently inhibiting an intracellular cluster of molecules comprised of axin, glycogen synthase kinase 3 (GSK3), and the adenomatosis polyposis coli (APC) protein. This complex then inhibits the cytosolic degradation of β-catenin, which accumulates and subsequently enters the nucleus to heterodimerize with two important transcription factors, LEF and TCF, conferring the Wnt-effect on gene transcription. Via the β-catenin independent signaling pathway, a similar complex forms between Wnt, Frz, and Ror2, which stimulate the synthesis of secondary messengers [1, 2].

Another important pathway is the one mediated via Hedgehog. The Hedgehog (HH) ligand precursor is subjected to a series of modifications before reaching an active, multimeric form. This process releases the transcription factor Smoothened (SMO), which in turn activates the Gli2/3 complex, which goes on to promote gene expression of Gli1, while repressing the transcriptional repressor Gli3 [3].

A third pathway is the NELL-1 signaling pathway. NELL-1 is a secretory osteoinductive protein, binding to the cell surface receptors Integrinβ1&3. A multitude of intracellular signaling pathways are activated due to NELL-1 stimulation. These include MAPK, Hedgehog, and β-catenin dependent Wnt-signaling. Despite the fact that these pathways are still undefined, NELL-1 activation enhances Runx2 transcription and phosphorylation, and development of an osteogenic phenotype [4].

The last stage of bio-mineralization encompasses the precipitation of hydroxyapatite crystals, which spontaneously results in supersaturated or metastable salt solutions. Whether biomineralization will take place, is determined by the genetic programming of precursor cells into a mineralization-competent state (as in physiological bone formation=osteogenesis) or to pathological mineralization (i.e. ectopic mineralization or calcification) [5]. Hence, all tissues which are not meant to minerealize, should be actively "protected" or inhibited from precipitation of mineral. Fetuin-A is a circulating protein produced by the liver, directly inhibiting ectopic mineralization. Monomeric fetuin-A protein binds small clusters of both calcium and phosphate. This interaction results in the formation of prenucleation cluster-laden fetuin-A mono-and polymers, calciprotein monomers, and considerably larger soluble aggregates of protein and mineral calciprotein particles of colloid (soluble) nature [6, 7].

In this manner, fetuin-A serves as a mineral carrier protein and a systemic inhibitor of pathological mineralization, i.e. mineralization not brought about by bone residing osteoblasts, but by fibroblast like cells demonstrating mineralizing phenotypic characteristics.

1.2. Untoward calcification of soft tissues; some model systems

Vascular calcification inevitably afflicts the aging and dysmetabolic population. Modern concepts state that this process has emerged as a highly regulated form of bio-mineralization organized by collagenous and elastin extracellular matrices. Paracrine osteogenic signals, mediated by potent morphogens of the bone morphogenetic protein (BMPs) and wingless-type integration site family member (Wnt) superfamilies, are also active in the programming of arterial osteoprogenitor cells during vascular and heart valve calcification. Inflammatory cytokines, as well as reactive oxygen species, and oxylipids (which are more active within the clinical settings of atherosclerosis, diabetes, and uremia) elicit the ectopic vascular activation of osteogenic morphogens. Specific inhibitors (e.g. MGP, CV2, COMP2, Noggin, Gremlin, Chordin, and Folliastatin) of bone BMP-Wnt signaling have been identified, contributing to the modulation of osteogenic mineralization during development and disease. These inhibitory pathways and their regulators afford therapeutic strategies to prevent and treat valve and vascular sclerosis [1, 3].

In this context, the expression of vascular BMP-2 and BMP-4 (mostly elaborated by endothelial cells) are actively enhancing [Smad-DNA]-binding, thus trans-activating a cascade of factors like Msx2, Runx2, NFATc1 and Osterix (SP7), deemed necessary in the mineralizing process [1, 3]. The downstream pro-osteogenic signaling cascades initiated through heterodimer BMP receptors is held in check by a panoply of extracellular BMP binding proteins and intracellular transcriptional inhibitors (I-Smads, i.e. Smad6 and Smad7) [1, 3].

Calcific aortic valve stenosis (CAVS) is an increasing health problem affecting aging societies. The discovery of osteoblast-like and osteoclast-like cells in the heart have forced a paradigm shift where CAVS is considered to be actively regulated. It has also been shown that valvular fibrosis, as well as calcification, plays an important role in restraining cusp movement. Furthermore, CAVS should probably be regarded as a fibro-calcific ailment [8, 9].

It has been speculated what type of precursor cells are being able to turn into osteoblast-like cells in aortic valves, giving rise to human calcific aortic valve stenosis. Activated myofibroblasts are likely to come from either quiescent valvular interstitial cells (VICs) or from a subpopulation of endothelial cells that undergo an "endothelial to mesenchymal transformation" (EMT) [1, 8, 10]. Lastly, it has been asserted that circulating osteoprogenitor cells (positive for OPN and ALP) may enter the active side of the valve from the circulation [9]. Hence, it should be noted that a lack of "bone-homing" of osteoblast-precursors, as seen in the elderly population, may lead to the development of CAVS. It is also noteworthy that monocytes may enter the valve tissue, transform into osteoclast-like cells and/or inflammatory Th-cells, and thus affect the above mentioned osteoprogenitor cells to start a mineralization process due to their response to TNFα and interleukins [1, 8, 9].

Blood vessels are the first to form in the developing embryo and build extensive networks supplying all cells and organs with nutrients and oxygen. An ageing blood vessel often becomes abnormal in structure and function, thus contributing to a plethora of age-related diseases like ischemic heart&brain-disease, neurodegeneration, and/or cancer. The first regulators to be linked to the aging process were the Forkhead box "O" (FOXO) transcription factors and sirtuin (Sirt) deacetylases [9, 11]. They are now emerging as key regulators of the vascular development and disease. The integration of FOXO-and Sirt-family members into the aspect of vessel maintenance, offers new perspectives on mechanisms of aging, which is the most important risk factor for diseases of vascular system [9, 10].

The FOXO transcription factors control a plethora of cellular responses, encompassing apoptosis, DNA-repair, metabolism, as well as ROS detoxification and cell proliferation. On stimulation of PI3K/AKT-signaling by growth factors, AKT phosphorylates FOXOs on conserved residues with ensuing cytoplasmic sequestration and inactivation [1, 10]. Of specific interest for the precipitation of atherosclerotic plaques is FOXO3A, however, FOXO1 and FOXA3 should also be mentioned, since they has been shown that they hetero-dimerize with the transcription factor SXR, known to bind vitamin K2 [12-14].

The Sirts (especially Sirt1) are involved in the modulation of key genes involved in regulating lifespan and health span, including AMP-activated protein kinase (AMPK), mammalian target of rapamycin (mTOR), and insulin-like growth factor 1 (IGF-1), and their roles modulating cardiovascular health status [15, 16]. It is to be noted that the FOXO class of transcription factors affect the levels of sirtuins (including Sirt1), thus enabling the body's cell-and tissue armamentarium to respond to the energy state (NAD^+/NADH-ratio) in a proper manner to preserve organ function and longevity [10, 15]. In this context, the FOXOs and sirtuins are both instrumental for the heart and vasculature to avoid the numerous detrimental alterations during ageing.

Finally, the present report focusses on altered gene expression leading to proper or disturbed calcification of bone and soft tissues, as a result of perturbed transcription or posttranscriptional control. Particularly, certain transcription factors (TFs) and microRNAs (miRNAs) will be described [17, 18], which are members of so-called regulatory loops [19].

Interestingly, several groups or clusters of miRNAs, targeting several hundred mRNA species encoding TFs, as well as tissue specific or non-specific genes, have been shown to impact conditions like cardiac hypertrophy and cardiac fibrosis [17, 18].

1.3. MicroRNAs in atherosclerogenesis

The dysregulation of cholesterol homeostasis is one of the underlying causes of atherosclerosis. One regulatory factor in cholesterol metabolism includes the sterol regulatory element-binding protein (SREBP) family, which regulates the expression of a plethora of cholesterogenic genes [20]. The microRNA species 33b and 33a, respectively, have been shown to target the ATP-binding cassette A1 cholesterol transporter [21], which mediates intracellular cholesterol efflux from cells to form HDL that is protective against atherosclerosis [22, 23]. Upon injection of antisense miR-33 into western diet-fed mice, serum HDL levels were significantly elevated, and treatment of LDL-receptor deficient animals with anti-miR-33, resulted in augmented levels A1 in both liver and macrophages, with a net increase in blood HDL-levels along with reduced plaque size and inflammation-related gene expression [23].

However, other microRNA species, such as the miR-145/143 cluster, miR-133, and miR-221, exert a more direct role in atherogenesis. The miR-145/miR-143 cluster is enriched in visceral and vascular smooth muscle cells (VSMC) from early embryonic days and throughout adulthood [24-26]. Genomic deletion of this cluster yielded a mild vascular phenotype with no cardiac abnormalities, however, smooth muscle cells were smaller and exhibited an increase in rough endoplasmic reticulum (RER) and a decrease in actin stress fibers, resulting in thinner tunica media [26]. The failure to detect a consistent increase in VSMC proliferation or apoptosis indicates the miRNA cluster is involved in the ability of VSMC to differentiate in response to contractile demands [25].

Krupple-like factor-5, first identified as a transcription factor inducing the gene expression of smooth muscle myosin heavy chain, is involved in smooth muscle proliferation by stimulating cyclin D1 expression [27, 28]. MiR-133 is known to be enriched in the heart and skeletal muscle, but one report asserts that it is also expressed in smooth muscle cells [29]. Over-expression of miR-133 in the carotid artery has been shown to halt VSMC proliferation and prevent neointimial hyperplasia [17]. Interestingly, the level of miR-133 in this tissue exhibits a strong inverse correlation with its targets, one of which is the transcription factor SP1, and it may be asserted that SP1 inhibition leads to a deactivation of Krupple-like factor-5, which in turn prevents down-regulation of Myh11 (myosin heavy chain gene family), and hence VSMC phenotype switching and proliferation [17].

In contrast to miR-145 and miR-133, miR-221, but not the co-clustered miR-222, positively regulates smooth muscle proliferation [30]. It is believed to be induced by platelet-derived growth factor (PDGF), which is known to stimulate VSMC switching and proliferation during angiogenesis and neo-intimal formation [31]. In gain-and-loss-of function experiments, it was demonstrated that miR-221 was required for mediating the effects of PDGF on the suppression of the cell-cycle inhibitors p27Kip1&2, and c-kit [30, 31]. The effects of miR-221 are reinforced by the concurrent up-regulation of miR-21 in neointimal lesions, and miR-21 was shown to inhibit PTEN, which acts as a dual-specificity protein phosphatase, dephosphorylating tyrosine-, serine- and threonine-phosphorylated proteins [32-34], antagonizing the PI3K-AKT/PKB signaling pathway [35]. In this manner, miR-21 inversely modulates apoptotic VSMC death and reduces neo-intimal thickness. Consequently, neo-intimal formation is brought about by a combinatorial effect of changes in the expression of several miRs and their targets, with ensuing regulation of VSMC differentiation, proliferation, and survival.

However, the scope of the present book chapter, is to focus on microRNA species being involved in the process of matrix mineralization, whether or not they are intrinsic miRs residing within osteoblasts or smooth vascular cells or "imported" miRs via exosomes shedded by immune cells (like macrophages and Th-cells) invading the vasculature due to inflammatory processes. One may assert that microRNA species down-regulated in mineralizing osteoblasts should be up-regulated in healthy, non-calcifying soft tissues like blood vessels and heart valves. However, this is a simplification of the issue, since the acquired osteoblastic or mineralizing phenotype is secured by a balanced impact of microRNAs targeting osteogenic markers (or modulators) like BMP-2, SMADs, Runx2, Osterix (SP7), Dkk-1, and RANKL and others, a loss of calcification inhibitors like MGP and Fetuin-A, and finally a loss of smooth muscle cell markers like α-actin and certain MHC-class of antigens, as shown by Goettsch and co-workers [18]. This major compilation of published literature indicates that several microRNAs well known to be down-regulated in osteoblasts (like the miR-species 23a, 24, 27a, 29a, 34c, 133a, 135a, 149, 204, and 328) [36-39] are not necessarily down-regulated in "mineralizing" or "calcifying" vascular cells.

Interestingly, it has been shown that Th17-cells produce exosomes containing high levels of microRNAs (like miR-16, -24, -27a ,-27b, -125b, and -586) known to be low in developing and mature osteoblasts [36, 37], while also carrying high levels of, amongst many; miR-21, -22, -221, -222, and -520 (data not shown). According to Goettsch et al. [18], miR-21 is up-regulated in atherosclerotic arteries, while all the other mentioned miRNAs are down-regulated in arteries, plaques or bicuspid aortic valves. It is therefore necessary to define a cluster of minimal and sufficient microRNAs, which may represent a signature for soft tissue calcification. Hitherto, we do not see that such a cluster has been suggested, nor validated.

It is therefore of major interest to seek help from bioinformatics software to define regulatory loops, consisting of microRNAs, transcription factors and marker genes, known to be sensitive and prone to alterations in precursor cells, but resilient to changes in defined phenotypes. It is therefore exciting to learn that miR-133, already known to affect SP1 [40] seems to be players in regulatory loops, consisting of a selection of microRNA species and transcription factors, where miR-22, -27, -29, -133, -149, and -328, are connected to FOS, ETS1, SP1, SP3, RUNX1, FOSB, and EGR2, in a reciprocal feed-forward and feed-back stabilizing network.

2. The histopathology of vascular calcification

Vascular calcium deposition can be usefully organized into four histoanatomic variants. As outlined, each type of vascular calcification is associated with a characteristic spectrum of vascular disease processes. Moreover, in the setting of the calcified atherosclerotic plaque and senile calcific aortic sclerosis, the initial dystrophic calcification process evolves during vascular injury and remodeling into endochondral, non-endochondral, or mixed ossification mechanisms.

2.1. Atherosclerotic calcification

Atherosclerotic calcification is a dystrophic calcification characterized by cellular necrosis, inflammation, and the presence of lipoprotein and phospholipid complexes [41, 42]. The lipid complexes, originating from cellular membranes, thrombo-fibrinoid complexes, and circulating lipoproteins precipitate calcium in association with atherosclerotic plaques. The ensuing endothelial cell dysfunction provides a thrombogenic surface coated with fibrin and phospholipids, which drive additional lipid deposition. Oxidized lipid products provide several signals that recruit and activate macrophages and T-cells [43]. The calcification appears first in the lipid core of the fibro-calcific plaques, juxtaposed to inflammatory cell infiltrates and necrotic areas. Eventually, calcified cartilage formation follows the degenerative tissue calcification via a vascular remodeling process, leading to the deposition of endochondral bone [43].

2.2. Calcification of cardiac valves

Cardiac valve calcification occurs following mechanical stress and inflammation, leading to dystrophic mineralization and non-endochondral ossification. Degenerative lipid accumulation, fatty expansion of the *valvular fibrosa,* and interstitial calcium deposition are rapidly followed by the invasion of macrophages and T-cells [44, 45]. Hence, valve calcification of the young and elderly, is initiated via overlapping mechanisms, but distinct from calcification observed within atherosclerotic plaques [46-49]. However, during progression of the calcification process, histological and molecular analyses indicate that a secondary phase of active, osteogenic mineral deposition perpetuates the vascular calcium accumulation, but via non-endochondral processes [50, 51].

2.3. Medial artery calcification

Calcification of the medial artery is a non-endochondral ossification process of the arterial *tunica media,* occurring frequently in patients suffering diabetes and end-stage renal disease

[52]. The mineralization pattern resembles intra-membranous calvarial bone formation and odontogenesis [53], driven by the BMP-2/Msx1&2-dependent signaling pathway in the absence of any cartilaginous precursor [54-56]. Many hypotheses of the major molecular determinants of medial calcification vs atherosclerotic calcification with cartilage metaplasia, have been launched, however, these determinants remain elusive. Two model systems may render some clues, though: the characteristics of vascular calcification responses of the ($LDLR^{-/-}$) and ($apoE^{-/-}$) mice [55-57]. Both animal model systems develop atheroma; however, $LDLR^{-/-}$ mice calcify valves and tunica media via a non-endochondral osteogenic process, while $apoE^{-/-}$ mice calcify vessels via cartilage metaplasia.

It has been shown that a myofibroblast cell population, activated by diet-induced diabetes, is diverted to the osteoblast lineage by Msx2-dependent transcriptional programming [54-56]. It was therefore asserted that this migratory myofibroblast population, which responds to vascular smooth muscle cell (VSMC) osteopontin (OPN) production [56, 58, 59], contributes to vascular remodeling and the medial calcification of diabetes. Interestingly, both hyperglycemia [56, 58] and hyperphosphatemia [60] induce OPN expression, which *per se* is a consistent and predicted feature of medial calcification [61-63]. Furthermore, the absence of the osteoprotegerin (OPG) gene expression in mutant mice also results in medial calcification with ensuing vascular T-cell infiltration [64]. Noteworthy is the observation that blood levels of OPG are increased in diabetic humans [65] and diabetic ($LDLR^{-/-}$) mice [55], which may indicate a state of resistance to OPG, namely inhibition of RANK-L signaling. It has been shown that the OPG/RANK-L secretion ratio is enhanced in carotid plaques compared to femoral plaques, which explains why carotid plaques contain less calcium and number of macrophages and T-cells [43].

2.4. Vascular calciphylaxis (soft tissue calcification)

Vascular calciphylaxis occurs when the calcium phosphate solubility threshold is exceeded [66]. When the serum calcium-phosphate product exceeds this threshold, widespread soft tissue deposition of amorphous calcium phosphate will occur. To prevent this happening, an array of mineralization inhibitors have evolved, encompassing fetuin-A, tissue pyrophosphate generating systems, as well as tissue OPN production [67].

Fetuin-A, which is an abundant serum glycoprotein, has been shown to limit organ and soft tissue calcification, including vascular calcium deposition [67]. Pyrophosphate inhibits nucleation and epitaxial calcification and also up-regulates the expression of OPN [68, 69]. The generation of tissue pyrophosphate is obtained via ectonucleotide pyrophosphatases/phosphodiesterases (ENPPs). ENPP1 (or PC-1) is instrumental in limiting calcification in "soft tissues", like blood vessels, skeletal muscle fibers, ligaments and tendons [68, 69].

2.5. Vascular calcification and cartilage metaplasia

In vertebrates, bone formation occurs via both endochondral and non-endochondral mechanisms [53]. In endochondral ossification, avascular cartilage is subjected to vascular invasion, cartilage calcification, remodeling by osteoclasts, and eventually deposition of bone by osteoblastic cells. Endochondral bone formation and neovascularization depend heavily on the expression of Runx2, acting in concert with Sox9 [53, 70]. The array of molecular players in endochondral ossification is also associated with the progression of untoward, degenerative atherosclerotic calcification [71], as shown in models of heritable vascular calcification [72, 73]; matrix Gla protein (MGP) knockout mice and apoE knockout mice, which both develop arterial cartilage metaplasia.

In this context, MGP functions as a noggin-like inhibitor of vascular BMP-2 mediated signaling [74]. Hence, in the absence of MGP, enhanced BMP-2 signaling may promote vascular cartilage metaplasia. Loss of the intracellular BMP-R2 signaling inhibitor Smad6 also provokes arterial cartilage metaplasia and medial endochondral bone formation [75]. Noteworthy is that BMP-2 signaling can drive both chondrogenic and osteogenic differentiation of multipotent mesenchymal precursor cells [76]. However, it is important to emphasize that there is a finely tuned balance between sustained Msx2 expression, which will promote osteogenic differentiation at the expense of adipocyte and chondrocyte development [54, 63] and "intermittent" Msx2 expression. The former will secure functional bone remodeling, while the latter is meant to prohibit mineralization of soft tissues [43]. Unfortunately, this balance may be tilted in both directions, resulting in less mineralized bone tissue and untoward calcification of soft tissues.

2.6. The modulatory effect of vitamin K2

Vitamin K1 and the VCOR activation cycle is apparently of great importance for the status of bone health through the influence of bone derived carboxylated osteocalcin (OC) on several other organ systems, like pancreas (insulin secretion), adipose tissue (adiponectin secretion) and testis (testosterone production), which all form reciprocally interacting organs in a homeostatic organ cross-talk system [77-80]. However, since vitamin K2 (MK-7 and/or MK-4) has been shown to be a ligand for the steroid and xenobiotic receptor SXR also designated pregnane X receptor PXR, orphan nuclear receptor PAR1, and NR1/2 [81], the knowledge of which genes are transcribed with the aid of MK-7 or MK-4 in different organ systems or cells [12-14, 82, 83], is vital to understand how vitamin K2 status affects organ homeostasis in general, and especially bone health and the regulation of detrimental mineralization of soft tissues in particular [1, 8-10, 84].

Apart from the up-regulation of mRNAs for OC, BMP-2, and RANK-L, and alleviation of TNFα-mediated suppression of pro-bone SMAD expression in osteoblasts [13, 14], it has been shown that MK-7/4 in bone acts as a general transcriptional regulator of extracellular matrix-related genes that are involved in the collagen assembly (indicated by gene ontology analyses). Hence, MK-7/4 synergizes with calcitriol (active vitamin D=1,25(OH)$_2$D$_3$) in maintaining bone growth and homeostasis throughout life. However, it seems that MK-7/4 deserve a closer scrutiny, since there are a plethora of additional modulatory hormone-like effects exerted by this "vitamin", as envisaged by its modulatory effect on gene expression in osteoblastic cells. Some interesting features are summarized in the table underneath:

Gene	*Biological effect related to biomineralization processes*
CSNK1A (casein kinase 1) (see Gene Card)	Modulates Wnt and hedgehog (HH) signaling pathways, which play a major role in the differentiation of stem cells to osteoblasts, and contributes to the features of mineralizing osteoblasts
FETUB, fetuin B (see Gene Card)	Enables super-saturation of free Ca-P_i molecules, thus prohibiting precipitation of calcium-hydroxyapatite crystals in soft tissues
MAPK9 (MAP Kinase 9) (see Gene Card)	When activated, promotes β-catenin (CTNNB1) degradation and inhibits the canonical Wnt signaling pathway
SMAD2 (SMAD family member 2) (see Gene Card)	SMAD2 is involved as a transcription factor contributing to the activity of SMAD2/ SMAD3:SMAD4 heterotrimer and down-regulation of TGF-β receptor signaling. Involved in the regulation of osteoblastogenesis and acquisition of mineralizing properties of cells in general
JUN (c-JUN, JUNB, JUND) (see Gene Card)	Part of the dimeric AP1 complex. JUN associates with FOS, thereby affecting differentiation of osteoblasts, osteoclasts, as well as immune cells (e.g. T-cells)
RYK (Receptor-Like Tyrosin Kinase) (see Gene Card)	May be a co-receptor along with FZD8 of Wnt proteins, such as WNT1, WNT3, WNT3A and WNT5A (of which WNT3A&5A are important for osteoblastogenesis and mineralization)
CRSP3 (Mediator complex subunit 23) (see Gene Card)	**The protein encoded by this gene is a subunit of the CRSP complex, necessary for efficient activation by SP1**

Table 1. Genes affected by Vitamin K2 (MK-7 or MK-4), as described by "Gene Card" on the Web. The genes tabulated are involved in biomineralization processes as described by Slatter and co-workers [85].

In short, vitamin K2, without any doubt, heavily affects the phenotypic characteristics of osteoblastic cells, i.e. both differentiation of stem cells to mature osteoblasts, as well as their mineralizing capacity. Furthermore, it seems that vitamin K2 is involved in the steady state level of SP1, the transcription factor, shown to be implicated in reciprocal regulatory loops with several microRNA species known to be discriminators of osteoblast differentiation and function [1, 3, 36-38]. Hence, it may be postulated that vitamin K2 may serve as an important modulator of the process of bio-mineralization, stabilizing cell phenotypes within a narrower span of features. In other words, vitamin K2 may optimize mineral deposition in bone, while blocking untoward mineralization in soft tissues.

3. Putative problems, and future directions

The interaction between hormonal, metabolic, inflammatory, and mechanical stressor molecules, as well as passive mineralization inhibitors determines the ultimate "phenotype" of osteoblasts mediating mineralization in bone, as well as vascular progenitors regulating the deposition of calcium in soft tissues. However, several issues have yet to be resolved concerning the regulation of vascular calcification. Firstly, a detailed comprehension of the origins of the chondro-and osteo-progenitors is still elusive [43]. Secondly, the relative contribution of trans-differentiating VSMCs and migratory myofibroblasts to vascular calcification responses is not fully understood [43]. Additionally, a "homing response" induced by a vascular injury will probably recruit circulating marrow skeletal progenitors [86] and contribute to damage progression, including the development of ectopic vascular bone marrow in association with vascular osteogenesis [43]. And finally, a better (i.e. minimal, but necessary and/or exhaustive) set cell surface markers of the mesenchymal cell lineage seem to be required.

Furthermore, the impact of T-cells and macrophages in vascular calcification is not fully understood, but may explain the reciprocal relationships between bone turnover and vascular calcification. One interesting phenomenon is the shedding of microRNA-containing exosomes from these immune cells [87]. Hence, future efforts in identifying which of these miRNA species and other factors, which may "alter" the phenotype of the fibroblast like cells in vessels and heart valves, are welcomed.

In a clinical setting, once vascular osteogenic tissue has acquired the ability to form mature bone, there is a possibility that anabolic bone-building hormones intended to treat osteoporosis may also augment mineral deposition in soft tissues via coupled matrix turnover processes. It should also be taken into consideration that diseases, like end stage renal disease (ESRD), hyperglycemia, hypertension, hypercholesterolemia, hyperphosphatemia, PTH-resistance, and iatrogenic calcitriol excess may all contribute to the vascular calcium load [88-90].

Hence, it seems that the reciprocally regulated calcification of various body organs/ tissues leaves little "space" for non-overlapping phenomena: calcification in bone with no soft tissue mineralization, and decalcification of soft tissues without losing bone mass. To solve the problematic issue of these tied and reciprocal phenomena, on may resort to apply the fine-tuning ability of bio-molecular "players" like vitamin K2. This vitamin, with its hormonal actions through the transcription factor SXR, seems to enable the body to control both bone mineralization and strength [91-93] and counteract soft tissue calcification in an optimal fashion [81, 94].

4. Materials and methods

In general, the description of materials and methods used is depicted in Figure 1. The results presented in this chapter are based on the following:

Materials: Human mesenchymal stem cells (hMSCs), human fibroblasts/HUVECs were either from in-house stock strains, or obtained from commercial sources. Human MSCs were differentiated to osteoblasts and heart valvular interstitial cells (VICs) according to standard procedures [95], while the fibroblasts/HUVECs were used directly on demand. The cells were exposed to appropriate media and differentiation protocols described elsewhere [96]. Furthermore, bovine calf bone chips were obtained fresh from the slaughterhouse and incubated with or without PBMCs differentiated to osteoclasts, according to standard procedures [97-99].

http://dx.doi.org/10.5772/59149

Methods (incubations, manipulations and analyses): The data presented herein mainly emanate from manipulation of differentiated cells (transfected with vectors expressing various transcription factors, i.e. SP1, ETS1, and RUNX1) in control and mineralizing media (see above) with siRNAs, pre-miRNAs and/or antago-miRNAs (corresponding to miR-149 and miR-328) in the presence or absence of either cytokines (TNFα, IL-1, IL-6, and IL-17A) or vitamin K2 (MK-7). End point analyses of results obtained were performed using mRNA isolation techniques (Quiagen), Q-PCR analyses of pertinent gene transcripts according to the literature [100-102], ELISA-based quantification of secreted cell marker growth factors/cytokines and others (e.g. osteocalcin, IL-10, TGFβ, OPG, and RANK-L). Finally, mineralization surface was measured using the Alizarin red S dye on cells in monolayers [103].

Bioinformatics: Interactions between microRNA-species, transcription factors (TFs) and cell phenotype "specific" marker genes were emulated using the Mir@nt@n algorithm [19]. On the charts, genes and microRNAs are visualized like this: TFs in red boxes; miRNAs in orange rectangles.

Statistics: Mean values were considered significantly different from controls when $p < 0.05$ (non-parametric testing, n=9) (refs). Differences compared to controls are marked with a star *.

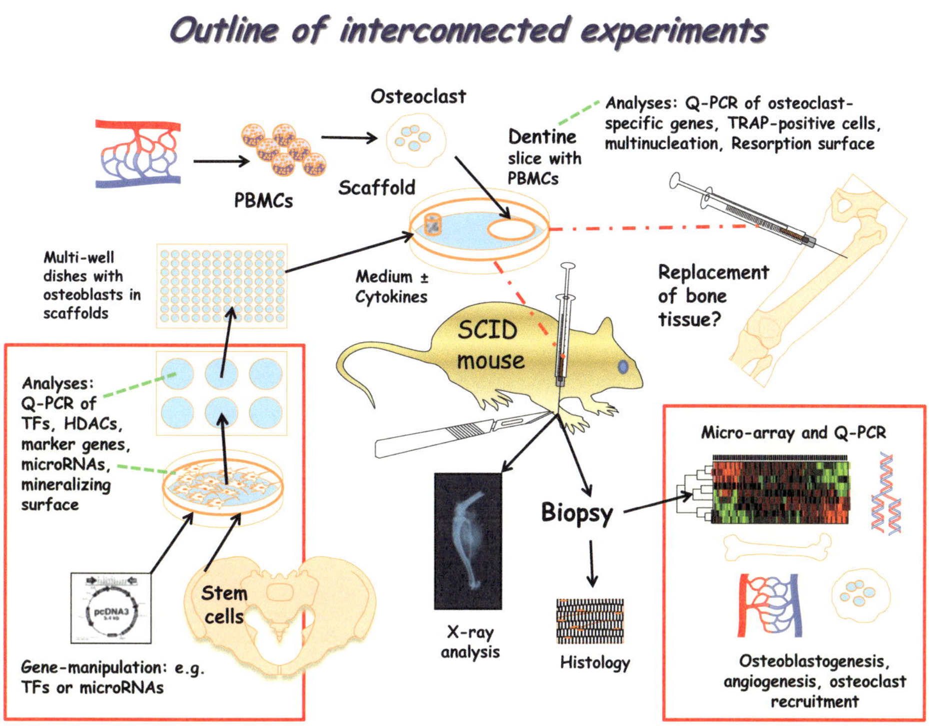

Figure 1. Illustration of interconnected experiments contained within a differentiation scheme of stem cells towards osteoblasts, manipulation of cellular phenotypes in the absence and presence of inflammatory cells or their cytokines, as well as analyses of parameters pertaining to matrix deposition and its calcification.

5. Results

5.1. Bioinformatics analyses using the Mir@nt@n algorithm

Figure 2, which is based on available osteoblast transcriptome data, featuring modulated transcription factor (TF) mRNAs and microRNAs (miRs), shows that ETS1, SP1, and RUNX1 (encircled in green) are involved in feed-forward/feed-back regulatory loops with several miRs included within the Stein [38] and Gordeladze [37] "miR-signatures" for osteoblastic cells. The emulation was performed with the highest possible stringency. Other TFs within the same network system are FOS, FOSB, SP3, and EGR2.

Transcription factors	MicroRNA species « connected»
SP1	Let-7f, 18b, 24, 27a, 27b, 29a, 29c, 31, 96, 135b, 141, 149, 182, 200a, 377, 522, 597
ETS1	19a, 34a, 125a-5p, 133b, 135b, 148b, 206, 222, 328, 377, 492, 522
FOSB	23a, 23b, 27a, 27b, 152, 182, 200a, 204, 220c, 224, 637, 638
RUNX1	17, 20a, 20b, 23a, 27a, 27b, 30a, 91, 93, 106a, 141, 494
JUNB	Let-7a,c,d,f, 15a, 93, 326, 494, 597
EGR2	20b, 23a, 25, 106a, 137
SPI1	34a, 155, 326, 663
FOS	29a, 29c, 149, 597
SP3	133b, 135b, 182, 191
JUN	9, 200b, 522, 637
STAT1	20a, 20b
SOX9	206

Figure 2. Tabulation of transcription factors (TFs) targeted by various microRNA species (either predicted by the Mir@nt@n algorithm or directly shown in experiments with reporter constructs). TFs and microRNAs focused on in this chapter are highlighted in yellow and red (the Gordeladze osteoblast-"signature") or green (the Stein-"signature"), respectively.

Figure 3 shows the result of a Mir@nt@n emulation with low stringency, indicating that SP1, ETS1, FOSB, and RUNX1, are the TFs putatively binding the largest number of microRNA species. It should be emphasized that the "Stein-signature" of miRs (highlighted in green) are involved with more TFs than the "Gordeladze-signature" (highlighted in red), indicating that the former is a better marker and/or predictor of the osteoblast phenotype.

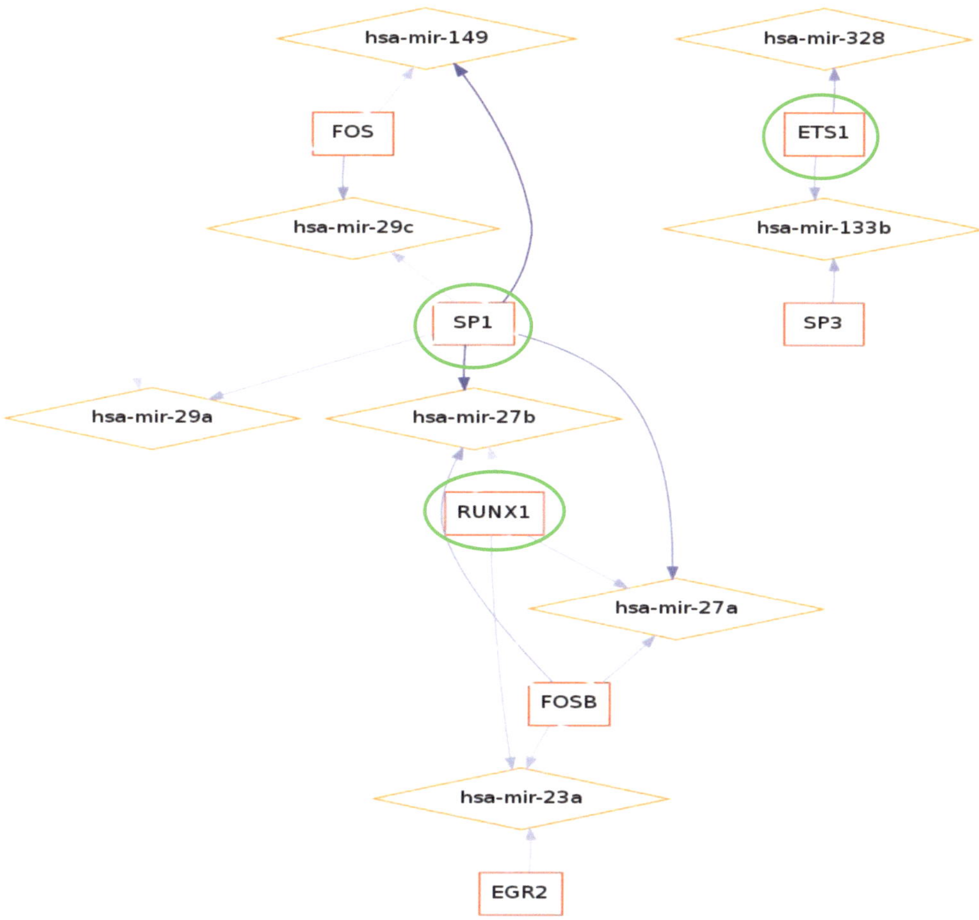

Figure 3. High stringency emulation of feed-forward and feed-back ("reciprocal") regulatory loops encompassing transcription factors and microRNA species using osteoblast transcriptomes and microRNA species known to be involved in the differentiation and/or function of mineralizing osteoblastic cells [19].

Figure 4 shows the network of interacting TFs and miRs of the "Gordeladze-signature" including histone deacetylaces (HDACs) emphasizing the important feed forward/feed-back loops involving SP1 and ETS1 with the microRNA species 149 and 328. Many of the "Gordeladze-signature miRs are interacting with sirtuins (e.g. SIRT1, one HDAC subspecies), which indicates that the regulatory TF-miR system is influenced by the osteoblast's energy status, since it is known that the NAD^{+}/NADH-ratio is influencing the activity of HDACs [10].

Figure 5a&b shows Q-PCR-data on transcription factor, marker gene and microRNA levels, as well as mineralizing surface (%) in osteoblasts exposed to cytokines (cfr. **Methods**) transfected with vectors containing SP1, ETS1 or RUNX1, or TF+cytokines (average of all three TFs from separate experiments). In essence, the two panels indicate that the cytokine mix is detrimental to the osteoblasts, and that over-expression of the transcription factors or antago-

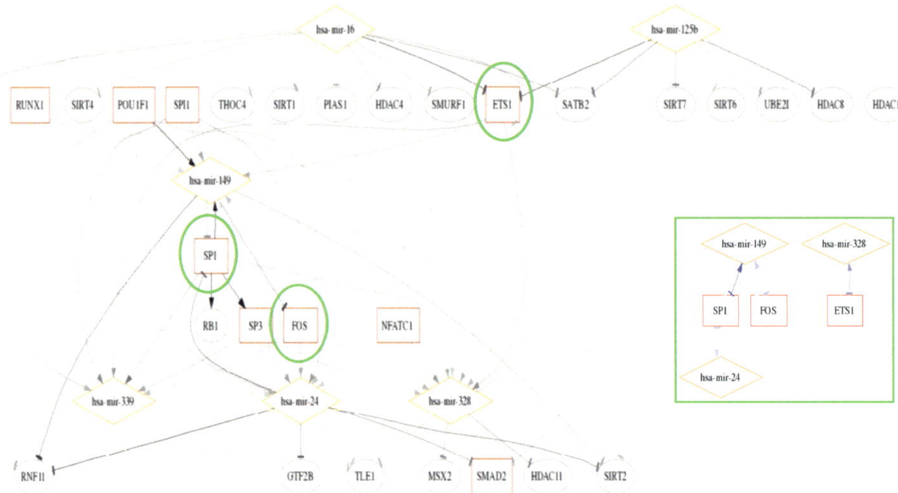

Figure 4. High stringency emulation of microRNA species interacting with transcription factors (TFs), functional marker genes and the complete published "Gordeladze mini-signature" of microRNAs in osteoblasts, including histone deacetylases (HDACs and Sirtuins=Sirts). Insert: Resiprocal regulatory loops involving the miR-24,-149, and-328, and the TFs SP1, FOS, and ETS1.

miRs counteract the negative impact of the cytokines on the osteoblast phenotype. Sections of Alizarin red S colored osteoblasts are depicted underneath the tables.

Figure 6 a&b indicates the same type of analyses/data as Figure 5. Here, Q-PCR-data and per cent mineralizing surface are presented. DKK1, which is an inhibitor of osteoblastogenesis is, inversely regulated within the various treatment groups, compared to the other markers. However, the data set grossly indicates that cytokine exposure "deprives" the osteoblasts of their phenotype markers, while overexpression of either transcription factor (SP1, ETS1 or RUNX1) in the presence of all cytokines "reinstates" the osteoblast phenotype characteristics.

Figure 7 shows one Mir@nt@n-emulation of regulatory network lattices consisting of osteoblast related transcription factors and microRNA species. The insert shows that the vitamin K2 receptor SXR (also known as NR1I2) associates with miR-760, which is tied to JUNB in a reciprocal looping system, connecting vitamin K2 to a network resembling the one featured in Fig. 2. However, vitamin K2 apparently needs miR-597 to connect to this looping system. The significance of miR-597 is discussed in **Section 6.3**.

Figure 8 depicts the effect of the vitamin K2 analogue MK-7 (menaquinone-7) on the secretory function of osteoblasts embedded in bovine bone slices exposed to either, MK-7 (10 ng/ml), SXR-siRNA or pre-miR-760. Clearly, MK-7 exposure of the bone chips for 7 days enhances the secretory profile (osteocalcin, IL-10, TGFβ, OPG and RANKL) measured with ELISA-kits. Furthermore, concomitant incubations with either SXR-siRNA or pre-miR-760 obliterate the effect of MK-7.

Figure 9 shows the impact of the transcription factors FoxA3 and FoxO1 on Runx2 expression in osteoblastic cells (left). Clearly, siRNA directed against either transcription factor obliterates

Osteoblast Parameters	Control = 100%	+ Cytokines	+ TF	+ TF + Cytokines
TF = SP1, ETS1 or RUNX1	100	31*	544*	389*
Runx2	100	17*	432*	324*
Collagen1a1	100	47*	145*	133*
Osteocalcin	100	23*	345*	288*
Osterix (SP7)	100	18*	182*	171*
Mineralizing «surface»	100	22*	234*	198*
OPG/Rank-L ratio	100	435*	27*	31*
MiR-149	100	546*	34*	75
MiR-328	100	465*	37*	58*
MiR-27b	100	388*	31*	36*
MiR-29a	100	481*	28*	30*
MiR-204	100	376*	27*	38*
MiR-211	100	422*	38*	41*

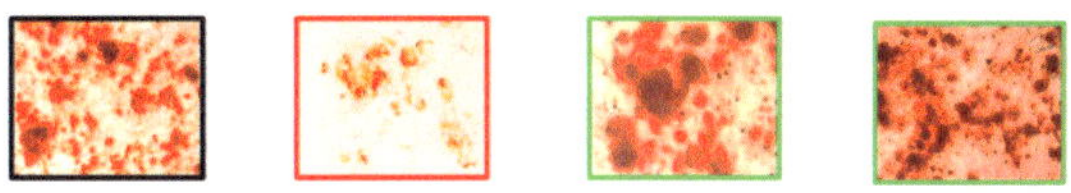

(a)

Osteoblast Parameters	Control = 100%	+ Cytokines	+ Antago-miRs	+ Antago-miRs + Cytokines
TF = SP1, ETS1 or RUNX1	100	39*	573*	366*
Runx2	100	45*	389*	319*
Collagen1a1	100	36*	319*	272*
Osteocalcin	100	27*	321*	251*
Osterix (SP7)	100	31*	301*	199*
Mineralizing «surface»	100	47*	247*	167*
OPG/Rank-L ratio	100	378*	21*	271*
MiR-149	100	449*	17*	69
MiR-328	100	577*	21*	48*
MiR-27b	100	466*	18*	31*
MiR-29a	100	521*	21*	33*
MiR-204	100	488*	33*	18*
MiR-211	100	469*	21*	25*

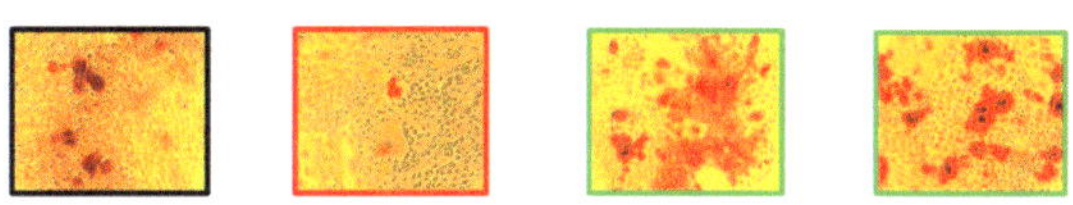

(b)

Figure 5. Tabulation of osteoblast-related parameters (transcription factors=TFs, microRNAs and mineralizing surface) measured in osteoblasts differentiated from hMSCs, in the presence or absence of MK-7, cytokines or both. All values are given as a percentage of controls (=100%). All results are means of separate experiments with the TFs indicated. Figures marked by stars (*) indicate p-values < 0.01.

Osteoblast GENE expression	Control = 100%	+ Cytokines	+ TF	+ TF + Cytokines
SP1, ETS1 or RUNX1	100	41*	17*	11*
MSX2	100	17*	32*	11*
WNT3A	100	21*	15*	23*
DKK1	100	329*	225*	451*
OPN(SPP1)	100	22*	21*	17*
MMP9	100	31*	17*	17*
RUNX2	100	12*	22*	11*
OPG/RANKL-ratio	100	21*	30*	15*
COL1a1	100	28*	29*	18*
PIT-1	100	21*	26*	18*
Mineralizing «surface»	100	18*	27*	15*
MiR-149	100	345*	367*	622*
MiR-29a	100	421*	304*	607*

(a)

Fibr./HUVEC/VIC GENE expression	Control = 100%	+ Cytokines	+ TF	+ TF + Cytokines
SP1, ETS1 or RUNX1	100	431*	21*	89
MSX2	100	288*	23*	78
WNT3A	100	365*	31*	104
DKK1	100	27*	245*	81
OPN(SPP1)	100	433*	12*	67*
MMP9	100	369*	23*	87
RUNX2	100	489*	27*	81
OPG/RANKL-ratio	100	363*	34*	65
COL1a1	100	478*	37*	78
PIT-1	100	387*	31*	88
Mineralizing «surface»	100	416*	68*	135
MiR-149	100	23*	227*	138
MiR-29a	100	18*	238*	97

(b)

Figure 6. Tabulation of osteoblast-related parameters (transcription factors=TFs, microRNAs and mineralizing surface) measured in osteoblasts (fig. a) or fibroblasts/HUVECs/VICs (fig. b) isolated from human tissues or differentiated from hMSCs, in the presence or absence of cytokines and transcription factors (SP1, ETS1 or RUNX1) expressing constructs. All values are given as a percentage of controls (=100%). All results are means of separate experiments with the TFs indicated. Figures marked by stars (*) indicate p-values < 0.01.

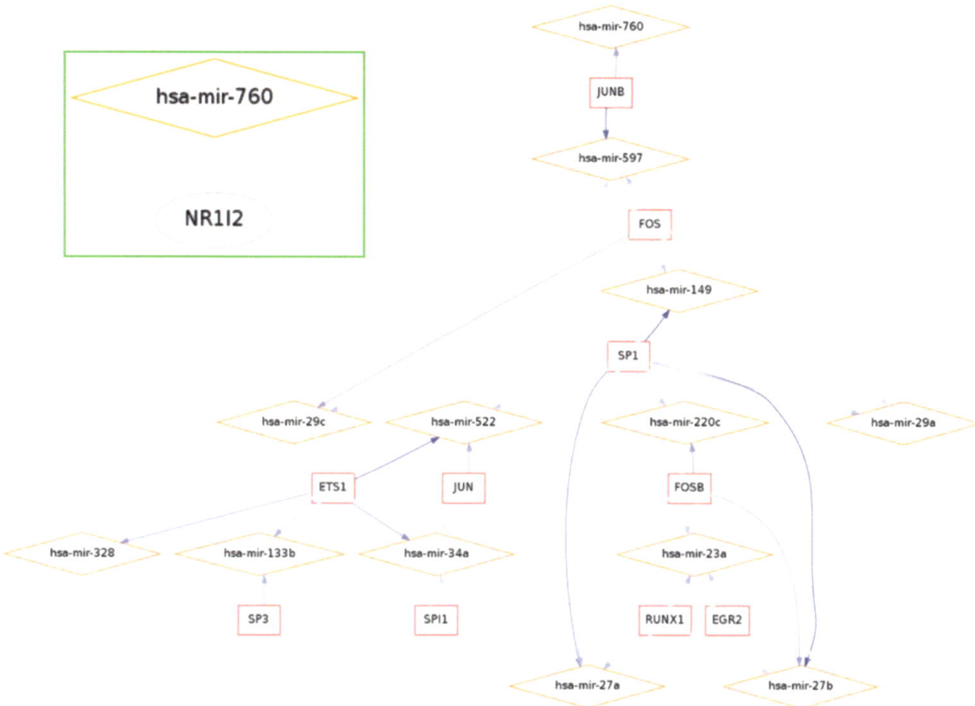

Figure 7. Mir@nt@n emulation [19] of osteoblast transcription factors=TFs and microRNAs, including miR-760, which was shown to putatively couple to NR1I2=SXR with the highest binding characteristics (the Sanger microRNA Database), known to be a receptor for vitamin K2 (i.e. MK-7 and/or the metabolite MK-4.

the stimulatory effect of MK-7 on Runx2 mRNA. It is well known that insulin and various growth factors modulate the activity of both FoxA and FoxO species via the PI3-Kinase/Akt/PKB signaling pathway, indicating that vitamin K2 is able to modulate hormonal or growth factor mediated impact on osteoblasts and possibly also other cell phenotypes.

Figure 10a&b depicts how osteoblasts (a) or fibroblasts/HUVECs/VICs (b) transfected with empty vectors, respond to the exposure of MK-7, a cytokine mixture (see text-Fig 8) or MK-7+cytokines in terms of transcription of genes and microRNAs, modulating mineral deposition in their surrounding matrix.

Figure 11 summarizes the effect ("correction") of transcription factors (TFs), microRNA species and vitamin K2 on the phenotypic characteristics of osteoblasts, fibroblasts, HUVECs and VICs. By manipulating different cell phenotypes with TFs and miRs engaged in reciprocal regulatory loops, in the presence of vitamin K2, it may be possible to stabilize and/or correct cell phenotypes to restrict mineralization of extracellular matrix to organs where mineral deposition is wanted (bone) and inhibit calcification of soft tissues like blood vessels and heart valves. Pay special attention to the reciprocal regulation of Runx2/SP7/BGP vs Matrix GLA protein (MGP) in osteoblasts and fibroblasts/ HUVECs/VICs.

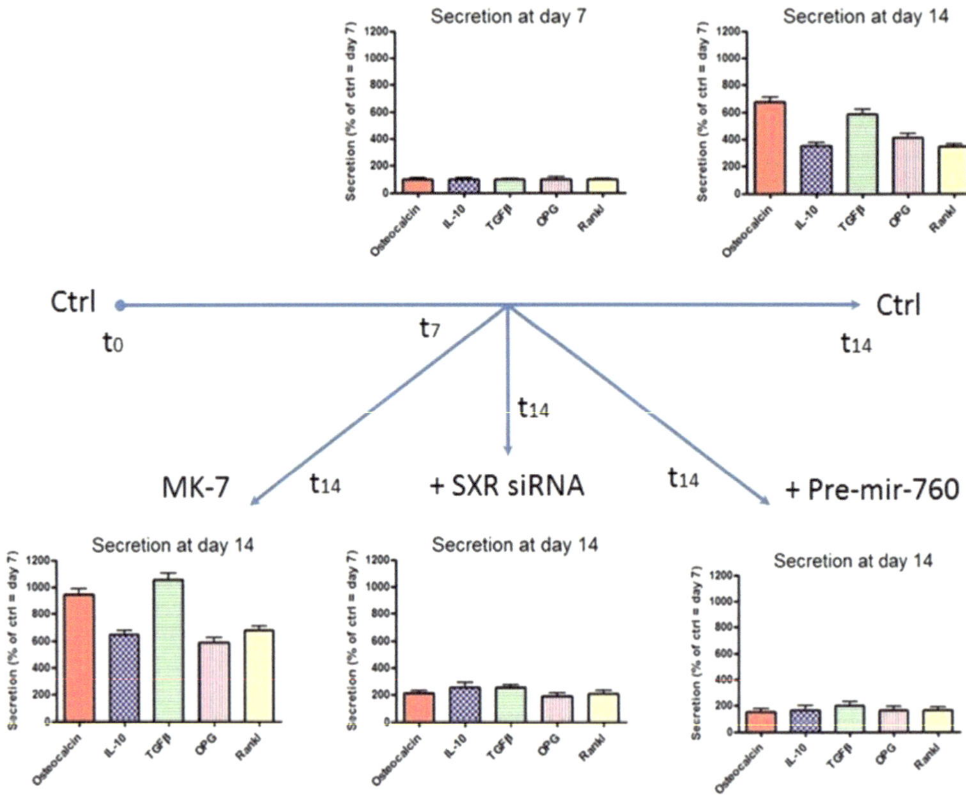

Figure 8. Secretory profile (osteocalcin, IL-10, TGFβ, OPG, and RANK-L) of mature osteoblasts within freshly isolated bovine bone chips for 7 days (days 8-14) after being pre-conditioned for 7 days (day 1-7) with either MK-7 (10 ng/ml), siRNA against SXR, or pre-miR-760.

6. Discussion

6.1. The importance of regulatory loops for the control of cellular phenotype

Feed-forward and feed-back regulatory loops consisting of transcription factors (TFs) and microRNAs are parts of an epigenetically "stabilizing" machinery [104]. Three categories of signals are proposed to operate in establishing a resilient, heritable epigenetic state. An extracellular signal designated the ''epigenator'', conveyed from the environment, triggers the start of the epigenetic pathway. The ''epigenetic initiator'' receives the signal from the ''epigenator'' and determines the precise chromatin location and/or DNA environment for establishing the epigenetic pathway. The ''epigenetic maintainer'' sustains the chromatin environment in both the initial and succeeding generations. Persistence of the chromatin milieu may require cooperation between the initiator and the maintainer. Hence, microRNAs and TFs, once they have attained an interactive "steady state", function as "maintainers" of a given cell phenotype with defined characteristics.

http://dx.doi.org/10.5772/59149

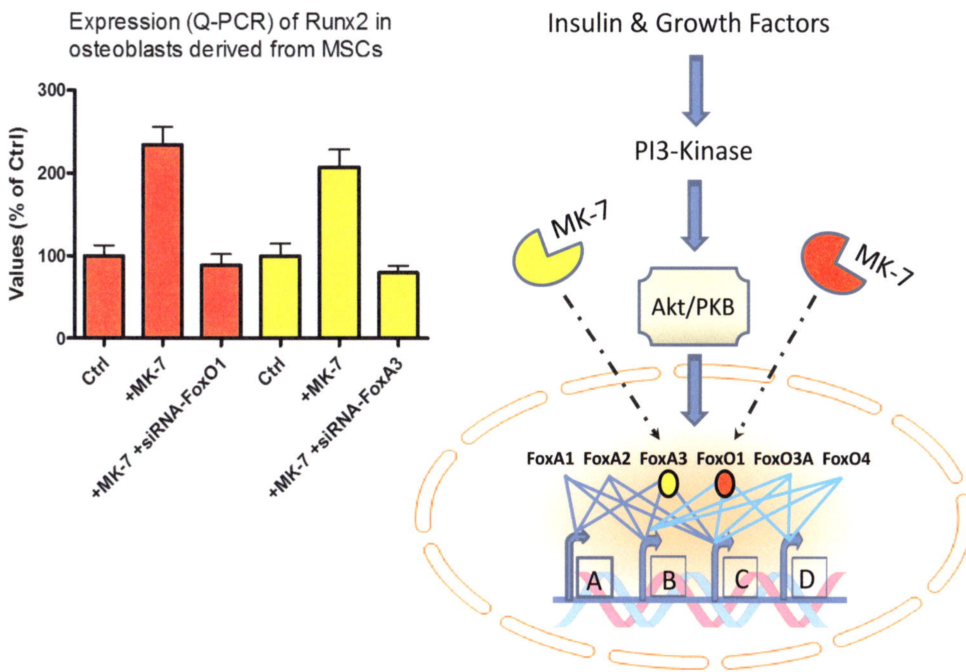

Figure 9. Left: Expression of Runx2 (Q-PCR) in osteoblasts differentiated from hMSCs in the absence (controls) or presence of MK-7 (10 ng/ml), MK-7+either FoxO1 -or FoxOA3-siRNAs (featuring two separate experiments). Right: Schematic representation of the impact of MK-7 on the signaling systems driven by insulin and growth factors (TGFs/BMPs) on the transcription of genes via PI3-kinase and the Akt/PKB-system, as described in the literature [108].

From a literature search (PubMed) on "SP1 transcription factor and osteoblasts", SP1 is somehow interfering with the effect of Runx2, SP7 (Osx), FIAT (inhibitor of ATF4), ETS-like TFs, MZF1 (myeloid zinc finger), JunB, and also directly affecting the transcription of marker genes like Col1α1, Col5α1, Col5α3, Col11α2, Fibromodulin, Osteocalcin, MGP (matrix-gla protein), RANKL, Pit phosphate transporter, Integrin β5, and TGFβ-R1. Furthermore, a similar search on "FosB and osteoblasts" revealed that FosB is interfering with the effect of BMPs and TGFs on the expression of downstream signaling molecules like Smads, TCF/LEF, as well as c-myc, and fra-1, but also directly modulating the transcription of IL-11 (which suppresses DIKK1 & DIKK2, thus enhancing Wnt-signaling), stimulating Pref-mediated dedifferentiation of adipocytes, relaying stretch-mediated osteoblast differentiation, counteracting the negative effect of Notch1 (a decrease in the expression of Col1α1, Osteocalcin, and ALP) by obliterating its negative effect on the Wnt/β-catenin pathway. Finally, the recent literature describes the positioning of ETS1 in the differentiation of osteoblasts in this way: Leptin is a strong inducer of osteoblast differentiation working through Stat3 (ref), and it was shown that ETS1, along with Stat1, Stat3, and VDR were induced by Calcitriol (1,25$(OH)_2D_3$) in UMR-106 osteoblast like cells, and that ETS1 is essential for connective tissue factor (CTGF/CCN2) induction by TGF-β1 in osteoblasts, synergizing with Smad3.

Osteoblast GENE expression	Control = 100%	+ MK-7	+ Cytokines	+ MK-7 + Cytokines
	100	143	47*	121
CSNK1A1	100	433*	47*	121
MSX2	100	356*	65	97
RUNX2	100	476*	33*	88
Osterix (SP7)	100	389*	27*	122
Osteocalcin (BGP)	100	489*	34*	143*
Matr. GLA pr. (MGP)	100	87	47*	112
PIT-1	100	432*	30*	143*
Mineralizing «surface»	100	188*	29*	88
MiR-149	100	17*	188*	78
MiR-328	100	22*	211*	111
Mir-29a	100	11*	251*	123
MiR-133b	100	18*	217*	71*

(a)

Fibr./HUVEC/VIC GENE expression	Control = 100%	+ MK-7	+ Cytokines	+ MK-7 + Cytokines
	100	108	167*	87
CSNK1A1	100	87	321*	122
MSX2	100	54*	188*	91
RUNX2	100	61*	354*	127
Osterix (SP7)	100	31*	449*	86
Osteocalcin (BGP)	100	53*	376*	132
Matr. GLA pr. (MGP)	100	387*	31*	378*
PIT-1	100	78	321*	87
Mineralizing «surface»	100	88	567*	132
MiR-149	100	465*	23*	367*
MiR-328	100	388*	17*	421*
Mir-29a	100	456*	24*	337*
MiR-133b	100	477*	31*	411*

(b)

Figure 10. Tabulation of recently recognized "discriminating" (for comparison, see Figures 6a&b) osteoblast-related parameters (transcription factors=TFs, microRNAs and mineralizing surface) measured in osteoblasts (fig. a) or fibroblasts/HUVECs/VICs (fig. b) isolated from human tissues or differentiated from hMSCs, in the presence or absence of cytokines and transcription factors (SP1, ETS1 or RUNX2) expressing constructs. All values are given as a percentage of controls (=100%). All results are means of separate experiments with the TFs indicated. Figures marked by stars (*) indicate p-values < 0.01.

http://dx.doi.org/10.5772/59149

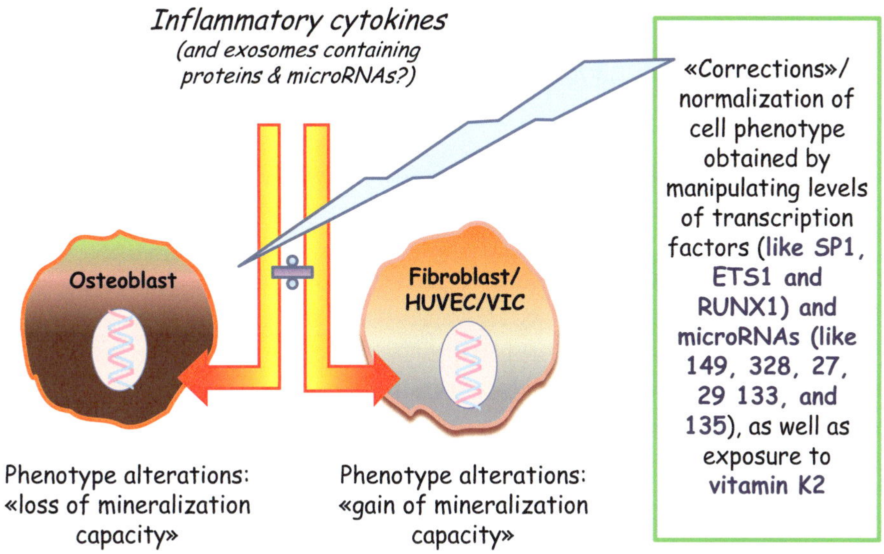

Figure 11. Schematic representation (summary) of the impact of transcription factors (like SP1, ETS1, and RUNX1), microRNA species possibly constituting a minimal and sufficient "osteoblast"-like signature (miR-149,-328,-27,-29,-133, and-135), and vitamin K2 (e.g. MK-7) on the stability of the osteoblast and fibroblast/HUVEC/VIC phenotype (i.e. mineralizing facilitating and mineralization prohibiting) development and/or maintenance.

Furthermore, BMP2 stimulation of pre-osteoblasts is mediated through ETS1, which transactivates Osteopontin, Runx2, PTHrP, and Col1α1 genes. The most compelling evidence for the importance of ETS1 (together with SP1 and SP7=Osx) on conferring the osteoblast phenotype of differentiating cells, is that the Runx2 P1 promoter in mesenchymal cells is co-stimulated at purine-rich DNA sequences (Y-repeats). It has also been shown that mechano-stimulation of MSCs enhances ETS1, Runx2, and ALP transcription and translation in a sequential manner, and that RNF11, also deemed necessary for osteoblastogenesis, in fact, is regulated by ETS1. Finally, the AJ18 gene, which is as a novel KRAB/C82/H(2) gene implicated in the differentiation of osteogenic cells, displays several response elements for proteins like ETS1 and SP1, as well as Runx2, Smads and NFκB.

It may be asserted that the larger number of microRNAs involved in the modulation of the activity of a certain cell's TFs, the more stable the cellular phenotype will become, and the more finely tuned its functions will be. Hence, it may be postulated that the alteration of the steady state levels of players contained within reciprocal regulatory loops in osteoblasts, may also determine whether the cells in other organs than bone may acquire mineralizing characteristics or not. Our results clearly state that this is the case, since it was possible to fortify the mineralizing phenotype of osteoblasts, and weaken the mineralizing properties of fibroblasts, HUVECs and VICs by manipulating the levels of SP1, ETS1, and RUNX1, or the levels of osteoblast-"specific" signature-microRNAs, like miR-149 and miR-328 [36, 37]. Suppression of miR-149 enhances the mRNA level of target proteins associated with the differentiation of osteoblasts (Runx2, APC2, RNF11, and SP1). At the same time, a reduction in miR-149 will enhance the transcription of factors (cEBPA, ATF3, Stat3, and PIAS1) known to counteract chondrocyte development induced by TGFβ-R and Sox9. Enhancing the level of miR-149 (i.e. the corresponding pre-miR) will favor chondrogenesis at the expense of osteoblast differentiation. Hence, miR-149 may be defined as a "switch-miR".

In an excellent review, Goettsch et al, [18] list up-and down-regulated microRNA species in coronary artery disease, aortic stenosis and arteriosclerosis obliterans, of which several, e.g. 29a, 125b, 135b, and 204, having been shown targeting osteogenic markers (like BMP2, DKK1, RANKL, Osterix, Runx2, SMADs, and BMP-2), and 335-5p and 155, known to target calcification inhibitors (like MGP and Fetuin-A). However, this is the first time it has been shown that microRNAs involved in reciprocal regulatory loops with transcription factors are able to "determine" whether a cell type will be mineralizing or not.

6.2. Gene expression profiles as markers for osteoblast and fibroblast/HUVEC/VIC analyses

The tables showing the impact of vitamin K2, and vitamin K2, and cytokines, list various marker gene expression profiles, as well as mineralizing surface obtained with Alizarin red S in 2D-cell cultures. The parameters are selected from a set of articles featuring the characteristics of mineralizing osteoblasts, as well as calcifying soft tissue fibroblast like cells with emphasis on CSNK1A1, WNT3A, DKK1 (signaling molecules), MSX2, Runx2, Osterix=SP7 (transcription factors), collagen1α1, osteocalcin, matrix GLA protein (matrix structural proteins), MMP9, OPN=SPP1 (extracellular bioactive molecules=proteinase, hydroxyapatite binder). PIT-1=POU1F1 (Na^+-phosphate transporter), and RANKL (activator of NFκB) and OPG (osteoclast inhibitory factor), as well as microRNA species contained within various miR-"signatures" [36, 38] of osteoblasts, where they inhibit factors responsible for osteoblastogenesis and expression of biomolecules necessary for mineralization of the cellular matrix. As pointed out elsewhere (under results), the pattern of parameter modulation of osteoblasts and fibroblasts/HUVEC/VIC genes and microRNA expression is compatible with a strengthening of the osteoblast phenotype when the transcription factors SP1, ETS1, or RUNX1 are reinforced, or when microRNAs (e.g. mir-149, -328 (as well as -204, -211,- 27b, and -133b; data not shown in this chapter) are targeted in particular.

6.3. The impact of vitamin K2 on the mineralizing properties of cells

Vitamin K2 (MK-7 (or its metabolite MK-4) was shown to be involved in the regulatory loops consisting of microRNAs and transcription factors (TFs), as stated above. MK-7/MK-4, by binding to the transcription factor SXR/PXR/NR1I2 is able to connect to the looping system via miR-760, JUNB, and miR-597. The microRNA species 597 thus putatively targets (as envisaged by the Mir@nt@n and TargetScan algorithms): APC2, BMP-1,CD44, CTNNA1, CTNNB1, CTNNBL1, FOS, FOXA2, FOXA3, FOXO3, FOXP3, FOXP4, GATA6, GATAD2A, HES1, IL-17D, IRF4, JUNB, MSX1, NFATC4, NOTCH2, RORC, RUNX1, RUNX2, RUNX3, SMAD2, SMAD3, SMAD4, SMAD7, SOCS2, SOX4, SOX6, SOX9, SOX11, SP1, SP2, SP4, SP6, SP8, SPRY1, SPRY3, STAT1, TGFB-2, VLDLR, WNT2, and WNT9B.

These genes are all heavily involved in different signaling pathways determining the development of mineralizing properties, as well as the differentiation of Th-2 and Th-17 cells from precursor immune cells, i.e. T-regulatory cells [105]. It may therefore be asserted that vitamin K2 plays an important modulatory role in: 1) the homeostasis of mineral deposits, i.e. calcification of bone and decalcification of soft tissues, as well as 2) the process of soft tissue infiltration by immune cells, which eventually leads to untoward calcification.

However, when searching various databases for putative interactions between microRNA-species involved in reciprocal regulatory loops with transcription factors (i.e. miR-149, -328, -29, -27, -23, -34a, -133, -220c, -597, -522) for putative interference with the transcription/translation of NFKB1, NFKB2, and/or with their subunits RELA/RELB, known to be important for the differentiation of inflammatory macrophages and Th-cells, no interactions could be predicted. Hence, the well-known modulatory effect of vitamin K2 on osteoclasts and Th-cells related to bone mass and inflammation, respectively, most certainly are not determined directly by the subject reciprocal regulatory loops, as suggested by the coupling of SXR=NR1I2 to FOS, via the miR-7/JUNB/miR-597 axis.

6.4. The overall modulatory effect of reciprocal regulatory loops involving transcription factors and microRNAs, and interference from vitamin K2

From the data presented here, it may be hypothesized that certain transcription factors and microRNA species are heavily involved in determining whether a given cell type will express mineralizing properties or not. In this context, it would be beneficial to exploit this knowledge to directly reinforce mineralization of bone (via osteoblasts) and block the calcification of soft tissues (induced by fibroblasts/HUVECs/VICs in blood vessels and heart valves) due to senescence or active inflammatory processes involving macrophages and T-cells, through a gene therapy program.

However, by optimization of tissue exposure to vitamin K2, one may both directly and indirectly obtain similar results, since vitamin K2 (MK-7 or MK-4 or both) affects the reciprocal regulatory loops, reinforcing both the osteoblast and fibroblast/HUVEC/VIC phenotypes, both in the absence and presence of cytokines derived from active, inflammatory Th-cells.

7. Summary and future perspectives

The present text features the dynamic interaction between important biological player molecules, determining the spectrum of features expressed by different cell types exposed to the same ambient environmental factors (e.g. oxygen, nutrients and hormones/growth factors). Osteoblasts and soft tissue cells, like fibroblasts/HUVECs/VICs, respond to the same bioactive molecules, however, osteoblasts mineralize bone matrix, while the others do not normally calcify soft tissues. We have shown that regulatory looping systems consisting of microRNAs and transcription factors (TFs) may determine whether mineralization is going to take place or not, and that the cell/organ homeostasis is disrupted in the presence of cytokines from inflammatory cells (Th-cells).

By manipulating these loops, containing several members of the epigenetic machinery, one may strengthen wanted cell phenotypes (e.g. osteoblasts in patients with osteoporosis) and weaken unwanted characteristics (e.g. calcifying fibroblast like cells in the elderly or patients suffering from hypertension, diabetes, end-stage renal disease, hyperlipidemia and other diseases). If gene therapy is not warranted or wanted, one may resort to optimal "vitamin K2 therapy", since MK-7 was shown to normalize and strengthen the mineralizing osteoblast phenotype, and weaken the mineralizing "fibroblast" phenotype in the presence of inflammatory cytokines.

We have previously shown that the regulatory looping system also contains histone deacetylaces (HDACS, including the Sirtuin class, the latter responding to the energy state of an organ) (see Fig. 4). Sirtuins may respond to small activator molecules [106, 107], such as plant polyphenols, which confer their activating potential through FOXO3 binding to the Sirt1 mRNA. Sirt1 transcription is activated in mature osteoblasts (where miR-16 levels are diminished), compared to other cell types. Hence, the FoxA/FoxO gene activating system, also driven by vitamin K2, may be further stimulated (synergistically?) by Sirt activators found in natural foods, thus conserving the osteoblast phenotype better during ageing and or inflammatory processes. It should not surprise anyone, if soft tissue cells like fibroblasts/HUVECs/VICs will respond in an opposite manner to the same polyols, reinforcing the anti-calcifying properties of vitamin K2 on soft tissues like blood vessels and heart valves.

Author details

Jan O. Gordeladze[1*], Øystein Stakkestad[2], Sigrid Haugen[2], Janne E. Reseland[2], Unni Syversen[3], Gaute F. Johnsen[2], Håvard J. Haugen[2], Ståle P. Lyngstadaas[2], Mona Møller[4] and Mauro Valtieri[5]

*Address all correspondence to: j.o.gordeladze@medisin.uio.no

1 Institute of Basic Medical Science, Department of Biochemistry, Department of Biomaterials University of Oslo, Norway

2 Institute of Odontology, Department of Biomaterials University of Oslo, Norway

3 Institute of Cancer Research and Molecular Medicine, Norwegian University of Science and Technology, Trondheim, Norway

4 Kappa Bioscience, Norway

5 Istituto Superiore di Sanita, University of Rome, Italy

References

[1] Bostrom, K.I., N.M. Rajamannan, and D.A. Towler, *The regulation of valvular and vascular sclerosis by osteogenic morphogens.* Circ Res, 2011. 109(5): p. 564-77.

[2] Fukuyo, S., et al., *IL-6-accelerated calcification by induction of ROR2 in human adipose tissue-derived mesenchymal stem cells is STAT3 dependent.* Rheumatology (Oxford), 2014. 53(7): p. 1282-90.

[3] Fuchs, S., E. Dohle, and C.J. Kirkpatrick, *Sonic Hedgehog-mediated synergistic effects guiding angiogenesis and osteogenesis.* Vitam Horm, 2012. 88: p. 491-506.

[4] Zhang, X., et al., *The role of NELL-1, a growth factor associated with craniosynostosis, in promoting bone regeneration.* J Dent Res, 2010. 89(9): p. 865-78.

[5] Kirsch, T., *Biomineralization--an active or passive process?* Connect Tissue Res, 2012. 53(6): p. 438-45.

[6] Brylka, L. and W. Jahnen-Dechent, *The role of fetuin-A in physiological and pathological mineralization.* Calcif Tissue Int, 2013. 93(4): p. 355-64.

[7] Jahnen-Dechent, W., et al., *Fetuin-A regulation of calcified matrix metabolism.* Circ Res, 2011. 108(12): p. 1494-509.

[8] Miller, J.D., R.M. Weiss, and D.D. Heistad, *Calcific aortic valve stenosis: methods, models, and mechanisms.* Circ Res, 2011. 108(11): p. 1392-412.

[9] North, B.J. and D.A. Sinclair, *The intersection between aging and cardiovascular disease.* Circ Res, 2012. 110(8): p. 1097-108.

[10] Oellerich, M.F. and M. Potente, *FOXOs and sirtuins in vascular growth, maintenance, and aging.* Circ Res, 2012. 110(9): p. 1238-51.

[11] Novack, D.V., *The FOX(O1) blasts off.* Cell Metab, 2010. 11(3): p. 175-6.

[12] Ichikawa, T., et al., *Steroid and xenobiotic receptor SXR mediates vitamin K2-activated transcription of extracellular matrix-related genes and collagen accumulation in osteoblastic cells.* J Biol Chem, 2006. 281(25): p. 16927-34.

[13] Igarashi, M., et al., *Vitamin K induces osteoblast differentiation through pregnane X receptor-mediated transcriptional control of the Msx2 gene.* Mol Cell Biol, 2007. 27(22): p. 7947-54.

[14] Katsuyama, H., et al., *Menaquinone-7 regulates gene expression in osteoblastic MC3T3E1 cells.* Int J Mol Med, 2007. 19(2): p. 279-84.

[15] Pucci, B., et al., *Sirtuins: the molecular basis of beneficial effects of physical activity.* Intern Emerg Med, 2013. 8 Suppl 1: p. S23-5.

[16] Sanchez-Fidalgo, S., et al., *Sirtuin modulators: mechanisms and potential clinical implications.* Curr Med Chem, 2012. 19(15): p. 2414-41.

[17] Abdellatif, M., *Differential expression of microRNAs in different disease states.* Circ Res, 2012. 110(4): p. 638-50.

[18] Goettsch, C., J.D. Hutcheson, and E. Aikawa, *MicroRNA in cardiovascular calcification: focus on targets and extracellular vesicle delivery mechanisms.* Circ Res, 2013. 112(7): p. 1073-84.

[19] Le Bechec, A., et al., *MIR@NT@N: a framework integrating transcription factors, microRNAs and their targets to identify sub-network motifs in a meta-regulation network model.* BMC Bioinformatics, 2011. 12: p. 67.

[20] Vallett, S.M., et al., *A direct role for sterol regulatory element binding protein in activation of 3-hydroxy-3-methylglutaryl coenzyme A reductase gene.* J Biol Chem, 1996. 271(21): p. 12247-53.

[21] Najafi-Shoushtari, S.H., et al., *MicroRNA-33 and the SREBP host genes cooperate to control cholesterol homeostasis.* Science, 2010. 328(5985): p. 1566-9.

[22] Brooks-Wilson, A., et al., *Mutations in ABC1 in Tangier disease and familial high-density lipoprotein deficiency.* Nat Genet, 1999. 22(4): p. 336-45.

[23] Rayner, K.J., et al., *Antagonism of miR-33 in mice promotes reverse cholesterol transport and regression of atherosclerosis.* J Clin Invest, 2011. 121(7): p. 2921-31.

[24] Xin, M., et al., *MicroRNAs miR-143 and miR-145 modulate cytoskeletal dynamics and responsiveness of smooth muscle cells to injury.* Genes Dev, 2009. 23(18): p. 2166-78.

[25] Boettger, T., et al., *Acquisition of the contractile phenotype by murine arterial smooth muscle cells depends on the Mir143/145 gene cluster.* J Clin Invest, 2009. 119(9): p. 2634-47.

[26] Elia, L., et al., *The knockout of miR-143 and-145 alters smooth muscle cell maintenance and vascular homeostasis in mice: correlates with human disease.* Cell Death Differ, 2009. 16(12): p. 1590-8.

[27] Watanabe, N., et al., *BTEB2, a Kruppel-like transcription factor, regulates expression of the SMemb/Nonmuscle myosin heavy chain B (SMemb/NMHC-B) gene.* Circ Res, 1999. 85(2): p. 182-91.

[28] Suzuki, T., et al., *Kruppel-like factor 5 shows proliferation-specific roles in vascular remodeling, direct stimulation of cell growth, and inhibition of apoptosis.* J Biol Chem, 2009. 284(14): p. 9549-57.

[29] Torella, D., et al., *MicroRNA-133 controls vascular smooth muscle cell phenotypic switch in vitro and vascular remodeling in vivo.* Circ Res, 2011. 109(8): p. 880-93.

[30] Davis, B.N., et al., *Induction of microRNA-221 by platelet-derived growth factor signaling is critical for modulation of vascular smooth muscle phenotype.* J Biol Chem, 2009. 284(6): p. 3728-38.

[31] Liu, X., et al., *A necessary role of miR-221 and miR-222 in vascular smooth muscle cell proliferation and neointimal hyperplasia.* Circ Res, 2009. 104(4): p. 476-87.

[32] Ji, R., et al., *MicroRNA expression signature and antisense-mediated depletion reveal an essential role of MicroRNA in vascular neointimal lesion formation.* Circ Res, 2007. 100(11): p. 1579-88.

[33] Meng, F., et al., *Involvement of human micro-RNA in growth and response to chemotherapy in human cholangiocarcinoma cell lines.* Gastroenterology, 2006. 130(7): p. 2113-29.

[34] Meng, F., et al., *MicroRNA-21 regulates expression of the PTEN tumor suppressor gene in human hepatocellular cancer.* Gastroenterology, 2007. 133(2): p. 647-58.

[35] Martini, M., et al., *PI3K/AKT signaling pathway and cancer: an updated review.* Ann Med, 2014: p. 1-12.

[36] Gordeladze, J.O., et al., *Concerted stimuli regulating osteo-chondral differentiation from stem cells: phenotype acquisition regulated by microRNAs.* Acta Pharmacol Sin, 2009. 30(10): p. 1369-84.

[37] Gordeladze, J.O., et al., *From stem cells to bone: phenotype acquisition, stabilization, and tissue engineering in animal models.* ILAR J, 2009. 51(1): p. 42-61.

[38] Lian, J.B., et al., *MicroRNA control of bone formation and homeostasis.* Nat Rev Endocrinol, 2012. 8(4): p. 212-27.

[39] van Wijnen, A.J., et al., *MicroRNA functions in osteogenesis and dysfunctions in osteoporosis.* Curr Osteoporos Rep, 2013. 11(2): p. 72-82.

[40] Qiu, T., et al., *MiR-145, miR-133a and miR-133b inhibit proliferation, migration, invasion and cell cycle progression via targeting transcription factor Sp1 in gastric cancer.* FEBS Lett, 2014. 588(7): p. 1168-77.

[41] Bostrom, K., et al., *Bone morphogenetic protein expression in human atherosclerotic lesions.* J Clin Invest, 1993. 91(4): p. 1800-9.

[42] Demer, L.L., *Vascular calcification and osteoporosis: inflammatory responses to oxidized lipids.* Int J Epidemiol, 2002. 31(4): p. 737-41.

[43] Vattikuti, R. and D.A. Towler, *Osteogenic regulation of vascular calcification: an early perspective.* Am J Physiol Endocrinol Metab, 2004. 286(5): p. E686-96.

[44] Otto, C.M., et al., *Characterization of the early lesion of 'degenerative' valvular aortic stenosis. Histological and immunohistochemical studies.* Circulation, 1994. 90(2): p. 844-53.

[45] Wallby, L., et al., *T lymphocyte infiltration in non-rheumatic aortic stenosis: a comparative descriptive study between tricuspid and bicuspid aortic valves.* Heart, 2002. 88(4): p. 348-51.

[46] Hisar, I., et al., *Aortic valve calcification: its significance and limitation as a marker for coronary artery disease.* Angiology, 2002. 53(2): p. 165-9.

[47] Ortlepp, J.R., et al., *Cardiovascular risk factors in patients with aortic stenosis predict prevalence of coronary artery disease but not of aortic stenosis: an angiographic pair matched case-control study.* Heart, 2003. 89(9): p. 1019-22.

[48] Stewart, B.F., et al., *Clinical factors associated with calcific aortic valve disease. Cardiovascular Health Study.* J Am Coll Cardiol, 1997. 29(3): p. 630-4.

[49] Yamamoto, H., et al., *Valvular and thoracic aortic calcium as a marker of the extent and severity of angiographic coronary artery disease.* Am Heart J, 2003. 146(1): p. 153-9.

[50] Mohler, E.R., 3rd, et al., *Bone formation and inflammation in cardiac valves.* Circulation, 2001. 103(11): p. 1522-8.

[51] Rajamannan, N.M., et al., *Human aortic valve calcification is associated with an osteoblast phenotype.* Circulation, 2003. 107(17): p. 2181-4.

[52] Chen, N.X. and S.M. Moe, *Arterial calcification in diabetes.* Curr Diab Rep, 2003. 3(1): p. 28-32.

[53] Karsenty, G., *Minireview: transcriptional control of osteoblast differentiation.* Endocrinology, 2001. 142(7): p. 2731-3.

[54] Cheng, S.L., et al., *MSX2 promotes osteogenesis and suppresses adipogenic differentiation of multipotent mesenchymal progenitors.* J Biol Chem, 2003. 278(46): p. 45969-77.

[55] Shao, J.S., et al., *Teriparatide (human parathyroid hormone (1-34)) inhibits osteogenic vascular calcification in diabetic low density lipoprotein receptor-deficient mice.* J Biol Chem, 2003. 278(50): p. 50195-202.

[56] Towler, D.A., et al., *Diet-induced diabetes activates an osteogenic gene regulatory program in the aortas of low density lipoprotein receptor-deficient mice.* J Biol Chem, 1998. 273(46): p. 30427-34.

[57] Davies, M.R., R.J. Lund, and K.A. Hruska, *BMP-7 is an efficacious treatment of vascular calcification in a murine model of atherosclerosis and chronic renal failure.* J Am Soc Nephrol, 2003. 14(6): p. 1559-67.

[58] Bidder, M., et al., *Osteopontin transcription in aortic vascular smooth muscle cells is controlled by glucose-regulated upstream stimulatory factor and activator protein-1 activities.* J Biol Chem, 2002. 277(46): p. 44485-96.

[59] Li, G., et al., *Estrogen attenuates integrin-beta(3)-dependent adventitial fibroblast migration after inhibition of osteopontin production in vascular smooth muscle cells.* Circulation, 2000. 101(25): p. 2949-55.

[60] Chen, N.X., et al., *Phosphorus and uremic serum up-regulate osteopontin expression in vascular smooth muscle cells.* Kidney Int, 2002. 62(5): p. 1724-31.

[61] Moe, S.M., et al., *Medial artery calcification in ESRD patients is associated with deposition of bone matrix proteins.* Kidney Int, 2002. 61(2): p. 638-47.

[62] Shanahan, C.M., et al., *Medial localization of mineralization-regulating proteins in association with Monckeberg's sclerosis: evidence for smooth muscle cell-mediated vascular calcification.* Circulation, 1999. 100(21): p. 2168-76.

[63] Takemoto, M., et al., *Enhanced expression of osteopontin in human diabetic artery and analysis of its functional role in accelerated atherogenesis.* Arterioscler Thromb Vasc Biol, 2000. 20(3): p. 624-8.

[64] Bucay, N., et al., *osteoprotegerin-deficient mice develop early onset osteoporosis and arterial calcification.* Genes Dev, 1998. 12(9): p. 1260-8.

[65] Browner, W.S., L.Y. Lui, and S.R. Cummings, *Associations of serum osteoprotegerin levels with diabetes, stroke, bone density, fractures, and mortality in elderly women.* J Clin Endocrinol Metab, 2001. 86(2): p. 631-7.

[66] Qunibi, W.Y., C.A. Nolan, and J.C. Ayus, *Cardiovascular calcification in patients with end-stage renal disease: a century-old phenomenon.* Kidney Int Suppl, 2002(82): p. S73-80.

[67] Schafer, C., et al., *The serum protein alpha 2-Heremans-Schmid glycoprotein/fetuin-A is a systemically acting inhibitor of ectopic calcification.* J Clin Invest, 2003. 112(3): p. 357-66.

[68] Johnson, K., et al., *Linked deficiencies in extracellular PP(i) and osteopontin mediate pathologic calcification associated with defective PC-1 and ANK expression.* J Bone Miner Res, 2003. 18(6): p. 994-1004.

[69] Rutsch, F., et al., *PC-1 nucleoside triphosphate pyrophosphohydrolase deficiency in idiopathic infantile arterial calcification.* Am J Pathol, 2001. 158(2): p. 543-54.

[70] Zelzer, E., et al., *Tissue specific regulation of VEGF expression during bone development requires Cbfa1/Runx2.* Mech Dev, 2001. 106(1-2): p. 97-106.

[71] Tyson, K.L., et al., *Osteo/chondrocytic transcription factors and their target genes exhibit distinct patterns of expression in human arterial calcification.* Arterioscler Thromb Vasc Biol, 2003. 23(3): p. 489-94.

[72] Luo, G., et al., *Spontaneous calcification of arteries and cartilage in mice lacking matrix GLA protein.* Nature, 1997. 386(6620): p. 78-81.

[73] Matsui, Y., et al., *Osteopontin deficiency attenuates atherosclerosis in female apolipoprotein E-deficient mice.* Arterioscler Thromb Vasc Biol, 2003. 23(6): p. 1029-34.

[74] Zebboudj, A.F., M. Imura, and K. Bostrom, *Matrix GLA protein, a regulatory protein for bone morphogenetic protein-2.* J Biol Chem, 2002. 277(6): p. 4388-94.

[75] Schneider, M.D., V. Gaussin, and K.M. Lyons, *Tempting fate: BMP signals for cardiac morphogenesis.* Cytokine Growth Factor Rev, 2003. 14(1): p. 1-4.

[76] Galvin, K.M., et al., *A role for smad6 in development and homeostasis of the cardiovascular system.* Nat Genet, 2000. 24(2): p. 171-4.

[77] Hamidi, M.S., O. Gajic-Veljanoski, and A.M. Cheung, *Vitamin K and bone health.* J Clin Densitom, 2013. 16(4): p. 409-13.

[78] Feigh, M., et al., *Oral salmon calcitonin enhances insulin action and glucose metabolism in diet-induced obese streptozotocin-diabetic rats.* Eur J Pharmacol, 2014. 737: p. 91-6.

[79] Oury, F., et al., *Osteocalcin regulates murine and human fertility through a pancreas-bone-testis axis.* J Clin Invest, 2013. 123(6): p. 2421-33.

[80] Riddle, R.C. and T.L. Clemens, *Insulin, osteoblasts, and energy metabolism: why bone counts calories.* J Clin Invest, 2014. 124(4): p. 1465-7.

[81] Suhara, Y., et al., *[Structure-activity relationship of novel vitamin K analogues as steroid and xenobiotic receptor (SXR) agonists].* Yakugaku Zasshi, 2012. 132(8): p. 881-6.

[82] Tabb, M.M., et al., *Vitamin K2 regulation of bone homeostasis is mediated by the steroid and xenobiotic receptor SXR.* J Biol Chem, 2003. 278(45): p. 43919-27.

[83] Yamaguchi, M. and M.N. Weitzmann, *Vitamin K2 stimulates osteoblastogenesis and suppresses osteoclastogenesis by suppressing NF-kappaB activation.* Int J Mol Med, 2011. 27(1): p. 3-14.

[84] Zhou, C., S. Verma, and B. Blumberg, *The steroid and xenobiotic receptor (SXR), beyond xenobiotic metabolism.* Nucl Recept Signal, 2009. 7: p. e001.

[85] Slatter, J.G., et al., *Microarray-based compendium of hepatic gene expression profiles for prototypical ADME gene-inducing compounds in rats and mice in vivo.* Xenobiotica, 2006. 36(10-11): p. 902-37.

[86] Kuznetsov, S.A., et al., *Circulating skeletal stem cells.* J Cell Biol, 2001. 153(5): p. 1133-40.

[87] de Candia, P., et al., *Serum microRNAs as Biomarkers of Human Lymphocyte Activation in Health and Disease.* Front Immunol, 2014. 5: p. 43.

[88] Ishimura, E., et al., *Different risk factors for peripheral vascular calcification between diabetic and non-diabetic haemodialysis patients--importance of glycaemic control.* Diabetologia, 2002. 45(10): p. 1446-8.

[89] Goodman, W.G., et al., *Coronary-artery calcification in young adults with end-stage renal disease who are undergoing dialysis.* N Engl J Med, 2000. 342(20): p. 1478-83.

[90] El Asmar, M.S., J.J. Naoum, and E.J. Arbid, *Vitamin k dependent proteins and the role of vitamin k2 in the modulation of vascular calcification: a review.* Oman Med J, 2014. 29(3): p. 172-7.

[91] Iwamoto, J., et al., *Bone quality and vitamin K2 in type 2 diabetes: review of preclinical and clinical studies.* Nutr Rev, 2011. 69(3): p. 162-7.

[92] Iwamoto, J., T. Takeda, and Y. Sato, *Menatetrenone (vitamin K2) and bone quality in the treatment of postmenopausal osteoporosis.* Nutr Rev, 2006. 64(12): p. 509-17.

[93] Tanana, I. and H. Oshima, *[Vitamin K2 as a potential therapeutic agent for glucocorticoid-induced osteoporosis].* Clin Calcium, 2006. 16(11): p. 1851-7.

[94] Horie-Inoue, K. and S. Inoue, *Steroid and xenobiotic receptor mediates a novel vitamin K2 signaling pathway in osteoblastic cells.* J Bone Miner Metab, 2008. 26(1): p. 9-12.

[95] .Huang, W., et al., *Fn14 promotes differentiation of human mesenchymal stem cells into heart valvular interstitial cells by phenotypic characterization.* J Cell Physiol, 2014. 229(5): p. 580-7.

[96] Campillo-Fernandez, A.J., et al., *Analysis of the biological response of endothelial and fibroblast cells cultured on synthetic scaffolds with various hydrophilic/hydrophobic ratios: influence of fibronectin adsorption and conformation.* Tissue Eng Part A, 2009. 15(6): p. 1331-41.

[97] Davies, C.M., et al., *Mechanically loaded ex vivo bone culture system 'Zetos': systems and culture preparation.* Eur Cell Mater, 2006. 11: p. 57-75; discussion 75.

[98] David, V., et al., *Ex Vivo bone formation in bovine trabecular bone cultured in a dynamic 3D bioreactor is enhanced by compressive mechanical strain.* Tissue Eng Part A, 2008. 14(1): p. 117-26.

[99] Fritz, V., et al., *Bone-metastatic prostate carcinoma favors mesenchymal stem cell differentiation toward osteoblasts and reduces their osteoclastogenic potential.* J Cell Biochem, 2011. 112(11): p. 3234-45.

[100] James, A.W., *Review of Signaling Pathways Governing MSC Osteogenic and Adipogenic Differentiation.* Scientifica (Cairo), 2013. 2013: p. 684736.

[101] Persy, V. and P. D'Haese, *Vascular calcification and bone disease: the calcification paradox.* Trends Mol Med, 2009. 15(9): p. 405-16.

[102] Jensen, E.D., R. Gopalakrishnan, and J.J. Westendorf, *Regulation of gene expression in osteoblasts.* Biofactors, 2010. 36(1): p. 25-32.

[103] Gordeladze, J.O., et al., *Transient down-regulation of cbfa1/Runx2 by RNA interference in murine C3H10T1/2 mesenchymal stromal cells delays in vitro and in vivo osteogenesis, but does not overtly affect chondrogenesis.* Exp Cell Res, 2008. 314(7): p. 1495-506.

[104] Vrtacnik, P., J. Marc, and B. Ostanek, *Epigenetic mechanisms in bone.* Clin Chem Lab Med, 2014. 52(5): p. 589-608.

[105] Pateras, I., et al., *NF-kappaB signaling at the crossroads of inflammation and atherogenesis: searching for new therapeutic links.* Expert Opin Ther Targets, 2014. 18(9): p. 1089-101.

[106] Cho, S.Y., et al., *Identification of a small molecule activator of SIRT1 gene expression.* Aging (Albany NY), 2013. 5(3): p. 174-82.

[107] Allard, J.S., et al., Dietary activators of Sirt1. Mol Cell Endocrinol, 2009. 299(1): p. 58-63.

[108] Czech, M.P., Insulin's expanding control of forkheads. Proc Natl Acad Sci U S A, 2003. 100(20): p. 11198-200.

Chapter 4

Adult Mesenchymal Stem Cells in Current Tissue Engineering Concepts

Susanne Jung and Johannes Kleinheinz

Additional information is available at the end of the chapter

http://dx.doi.org/10.5772/59219

1. Introduction

In contemporary regenerative concepts the combination of multipotent cells and individualized scaffolds holds great promise. Adult stem cells with their regenerative and proliferative characteristics are a very adaptable cell resource for tissue regeneration.

From the support of neuronal healing processes after traumatic lesion to clinical application in plastic surgery especially adult mesenchymal stem cells have already proven their operational capability with good experimental as well as clinical success.

Human adipose derived stromal cells represent a group of mesenchymal stromal cells that come with certain advantages in terms of regenerative medicine: the cells are easily and-if necessary-repeatedly harvested, i.e. by liposuction, without leaving disfiguring scars and the cells occur in abundance, thus a relevant amount of cells can be acquired quickly. The term 'plasticity' describes the potency of adipose derived cells, to cross-differentiate across their original germ layer into cell lineages of non-mesenchymal origin. A feature that possibly opens new vistas in regenerative tissue design.

Their special charm lies not only in their differentiating ability but in their secretory properties also, that seem to be able to further support angiogenesis, inflammation and its regulation or matrix remodelling processes.

This survey aims to present the current state of adult stem cell application in regenerative concepts. With special regard to adipose-derived cells the chances and pitfalls of the appliance of these cells are illuminated.

2. Stem cells in regenerative medicine — Established strategies

When talking about stem cell biology one has to define and to confirm the characteristics of this special cell population in each setting. Genuine stem cells distinguish themselves by typical biological features.

Stem cells are able to renew themselves by symmetrical division; this cell division produces two identical copies (daughter cells) from one original cell. During asymmetrical division one of the originated cells becomes the aforementioned identical daughter cell, whereas the second cell alters its cytological profile and starts irretrievably the differentiation towards a certain specialized cell type.

Figure 1 illustrates symmetrical and asymmetrical stem cell division.

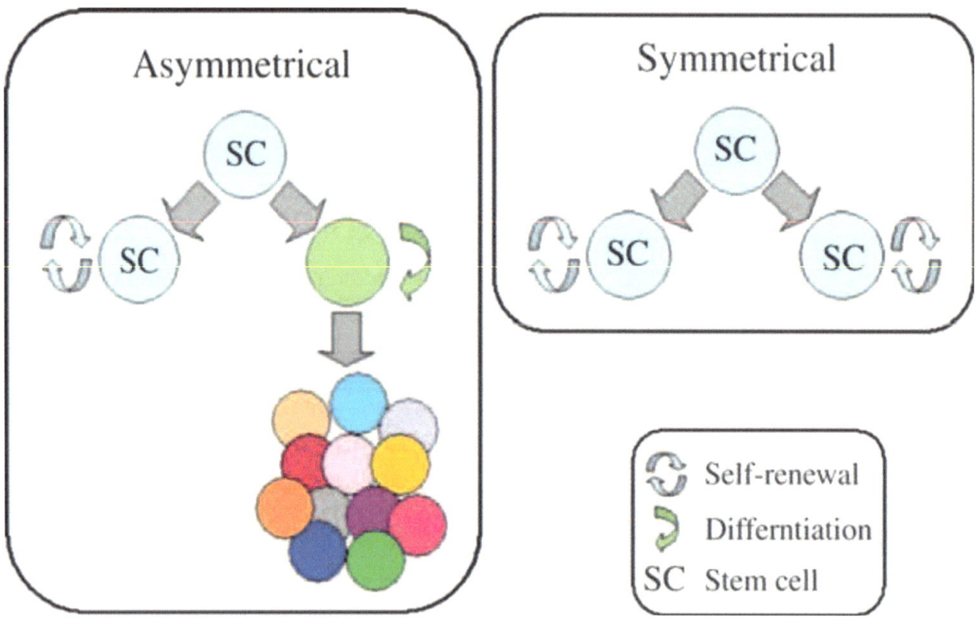

Figure 1. Symmetrical and asymmetrical stem cell division [1]

Stem cells provide a supply of progenitor cells, which are able to give rise to tissues from their original dermal layer or, especially when talking about embryonic stem cells (ESC), to tissues from all three different dermal layers. The ability to differentiate beyond the original cellular derivation means true pluripotency. In contrast, adult stem cells, harvested for example from bone marrow or the liver, are capable to generate specialized cells and tissues of their original germ layer; this ability to bring forth cells similar to their embryonic origin is called multipotency. The plasticity of adult stem cells, the potential to differentiate beyond their original germ layer in a multilineage development is discussed controversially in current literature for they have not been successful in producing veritable offspring after injection into a tetraploid blastocyst in a mouse model, so far [2].

http://dx.doi.org/10.5772/59219

A well defined panel of surface markers characterizes stem cells as multi-or pluripotent cells and separates them from other cell entities. Morphologically, for example multipotent adult cells appear as spindle-shaped, fibroblast-like cells with fine tails. In the presence of other cell populations the cells become more round, similar to cobblestones.

Yet, a morphological description alone is not able to identify the cells due to their similarity to other cells, i.e. to fibroblasts.

The immunohistochemical evidence or the typical lack of characteristic surface markers defines the cells under scrutiny as multi-or pluripotent cells. It is important to acknowledge that the expression profile of the respective cell type is dynamic and might alter during under cultivation and passaging.

Prominent defining surface markers of different human stem cells are summarized in Table 1.

ESC	MSC	ADSC
positive SEA-3, SSEA-4, (), TRA-1-60, TRA-1-81, alkaline phosphatase, GTCM-2 **negative** SSEA-1	**positive** CD105, CD90, CD 271, CD73 **negative** CD34, CD45, CD14 or CD11b, CD79a or CD19, MHC-II molecules, mainly HLA-DR	**positive** CD13, CD29, CD44, CD49, CD90, CD105, CD106, CD166, Fibronectin, aSMA, Vimentin, Collagen-1 **negative** CD 31, CD45, CD 117, CD133, HLA-DR, STRO-1, Lin, HLA II

Table 1. Surface markers of different adult stem cells [3,4]

To define stem cell criteria in a cell population in short the cells in question have to show the capacity of self-renewal, they have to able to survive and proliferate long-term and the unfractioned plastic-adherent cell population has to show multipotency [5].

In many scientific constellations different groups of pluri-and mulitpotent cells have proven their effectiveness. There is a great variety of disorders in which stem cell therapy in different applications might emerge as promising therapeutic strategy.

Table 2 summarizes cell lines and corresponding diseases where stem cell therapy is under investigation.

Bai et al. investigated the effectiveness of adipose tissue derived pluripotent cells on cardiac function after acute myocardial infarction in an in vivo design and were able to show that in severe combined immunodeficient mice not only the cultivated cells but also the freshly harvested and transplanted cells showed significant positive effect on myocardial regeneration and cardiac function after local injection to the peri-infarct region [6].

Another prominent example is the application of bone marrow derived mesenchymal stem cells in the therapy of retinal degeneration. In a rat model of retinal dystrophy the subretinal or intravitreal transplantation of human bone marrow derived stem cells led to a significant amelioration of the retinal function, whereas the effect in the subretinal injection group was more permanent [7]. Both examples underline the potential of these cells, even if many issues remain unaddressed so far and a clinical application needs much more supportive data.

Tissue type	Cell lineage	Disease application
Endocrine	Islet beta cells	Diabetes
Blood	Lymphoid cells Erythrocytes	Immune deficiencies Sickle cell anaemia
Liver	Hepatocyte	Liver failure
Renal	Podocytes and tubular cells	Acute and chronic renal failure
Neural	Oligodendrocyte Dopamine neuron Motor neuron	Spinal cord injury Parkinson's disease Motor neuron disease
Muscle	Skeletal muscle	Muscular dystrophy
Eye	Retinal epithelium	Retinal degeneration
Cardiovascular	Cardiomyocyte Endothelial	Myocardial infarction Stroke Ischaemic limb disease

Table 2. Tissues, cell lines and diseases

Different populations of pluripotent cells could exert their therapeutic efficacy in animal models of meniscus and lung injury as well as in clinical studies of osteogenesis imperfect [8, 9, 10].

3. Adult stem cells: Their pros and cons

Stem cells derived from two different origins are investigated and applied successfully in many scientific settings and some clinical strategies: one source, the most prominent and versatile group of differentiating cells are human embryonic cells. The advantages of these cells are their ability proliferate stably over many passages in in-vitro settings and their potential to differentiate into many different cell types derived from each one of the three embryonic germ layers: the endo-, meso-and ectoderm layer [11].

This true pluripotency defines the unique attractiveness of these cells especially in regenerative concepts that aim to replace complex anatomical structures. The harvesting of embryonic cells nevertheless remains problematic and has been the reason for controversial ethical and forensic debate: human embryonic stem cells are derived from cells of the blastocyst stage of embryos; the embryos in question generally are derived from in vitro fertilization treatment and not used for implantation for different motives; to date over 1000 immortal hESC cell lines are reported.

The process of harvesting and isolating embryonic cells is strictly regulated and under religious control. This reasonable and necessary limitation of the accessibility to the pluripotent cells complicates their use in clinical research. In addition to that, transplanted hESCs often cause strong immunorejection and as a result to their proliferative activity tend to cause neoplasia i.e. teratoma [12]. Despite the controversies, in few clinical trials dealing with spinal cord injuries, ECs are applied successfully: transplanted embryonic stem cell-derived glia cells, which are capable of producing myelin, were able to redevelop damaged spinal cord tissues in rodents with consecutive improved motoric functionality; these cells are utilised and studied in patients with complete subacute spinal cord injury to date (A Phase 1 Safety Study of GRNOPC1 in Patients With Neurologically Complete, Subacute, Spinal Cord Injury [clinical trial]. ClinicalTrials.gov registration number: NCT01217008. Accessed at http://clinicaltrials.gov/ct2/show /NCT01217008 on 11 June 2011.). The safety and tolerability of ECS derived retinal cells are investigated in patients with advanced macular degeneration, a degenerative disease that threatens the eyesight of millions of patients worldwide (A Phase I/II, Open-Label, Multi-Center, Prospective Study to Determine the Safety and Tolerability of Sub-retinal Transplantation of Human Embryonic Stem Cell Derived Retinal Pigmented Epithelial (MA09-hRPE) Cells in Patients With Advanced Dry AMD [clinical trial]. ClinicalTrials.gov registration number: NCT01344993. Accessed at www.clinicaltrials.gov/ct2 /show/NCT01344993 on 11 June 2011.).

Somatic or adult stem cells, the other source of differentiating cells, can be harvested as not-differentiated cells between surrounding specialized cells in the mature organism from various tissues or fluids without or with only little donor site morbidity. The difficulty of embryonic cell harvesting or handling does not occur.

In therapeutic concepts the use of autologous cells provides the base of personalized therapy. Their function in the embryo as well as in the adult organism is to replace and repair damaged or aged cells or tissues: this concept becomes imminent regarding the regenerative potential of the epidermis with its typical cell-based renewal cycle or the regenerative properties of the liver after damage or resection.

The advantage of these mature cells is their capacity to differentiate into various specialized cells and tissue types: mesenchymal stem cells from the bone marrow are able to develop into different cell types within the limits of their original blastodermic layer, i.e. osteoblasts or chondrocytes. Multilineage developmental or plasticity describes the stem cells' potential to overcome the histological limitations of their initial germ layer and their capacity to differentiate into progenitor cells and ultimately into cell types of another origin. This promising attribute of adult stem cells is discussed controversially in current literature:

Krampera et al. compared adult stem cells from different origins for their study: in an in vitro setting the authors were able to induce an ectodermic differentiation of MSCs from bone marrow, fat, spleen and thymus into neuron-like cells [13]. Ning et al. described in 2009 an in vivo endothelial differentiation of adipose tissue-derived stem cells under the influence of Fibroblast Growth Factor 2 (FGF2) and other growth and proliferation factors in an animal model. In the presence of FGF2 the observed ADSCs proliferated faster, expressed endothelial surface markers as CD31 or von Willebrand Factor (vWF) and formed tube-like formations in matrigel [14]. Following these findings a cross germinal differentiation of adult stem cells – with the support of growth of differentiation factors – seems feasible.

There are several similar investigations to support these findings; many other authors come to the conclusion that this cross-germinal specialization cannot be observed reproducibly. Especially the biological functionality of the differentiating cells is questioned critically.

In a clinical context, the capacity of adult stem cells to cross their germinal borderline remains unclear so far; there is particularly no sufficient and finally convincing in-vivo data to support the hypothesis.

Adult stem cells are detected in different tissues:

Mesenchymal stromal or stem cells (MSCs) are fibroblast-like shaped cells and appear ubiquitously in the human organism; they seem to originate from perivascular or adventitial tissue. They have the potential of multilineage differentiation and seem to exert numerous paracrine effects: by secreting different growth and signalling factors they seem to be able to modulate angiogenesis, (anti) inflammation or apoptosis.

The first MSCs described were bone marrow derived cells [15]: they are harvested from bone marrow aspirates. Their cultivation, functional characterization and differentiation behaviour has been investigated since the 1970s. Under standard culture conditions they have to be able to specialize into osteoblasts and chondrocytes.

Adipose derived stromal cells (ADSCs) come with many advantages: their harvest is much easier and safer than the isolation of bone marrow. One is able to produce the cells in larger amounts and the cells can be isolated repeatedly from the same donor region. The isolation and characterization is performed without difficulty and has often been described in literature.

Figure 2 gives a survey of the isolation of hADSCs.

Another interesting source of adult stem cells is the dental pulp stem cells (DPSCs). The probably most attractive feature of these cells that develop from the neural crest is their easy approachability from removed (wisdom) teeth as a by-product of the necessary surgery [16].

Due to their cytological similarity to bone marrow derived stem cells, these dental derived cells are under special investigation concerning not only repair or replacement of dental structures but in musculoskeletal regeneration also. During their characterisation in vitro it could be shown that DPSCs – compared to bone marrow derived cells for – show striking similarity to neuroglia cells and differentiated into neural and endothelial cells reproductively [17].

It has been possible to reprogramme dental pulp cells after third molar removal to induced pluripotent cells (iPS) that differentiated in vitro and in-vivo expressing cell and tissue types from all three germ layers. Even the typical formation of teratoma could be observed [18].

In mouse models of myocardial infarction and hind limb ischemia the therapeutic potential of endothelial and smooth muscle cells derived from DPSC induced pluripotent cells has been demonstrated [19]. These investigations encourage further scientific work with dental pulp derived cells, which are gained in many cases as a by-product of a necessary tooth removal;

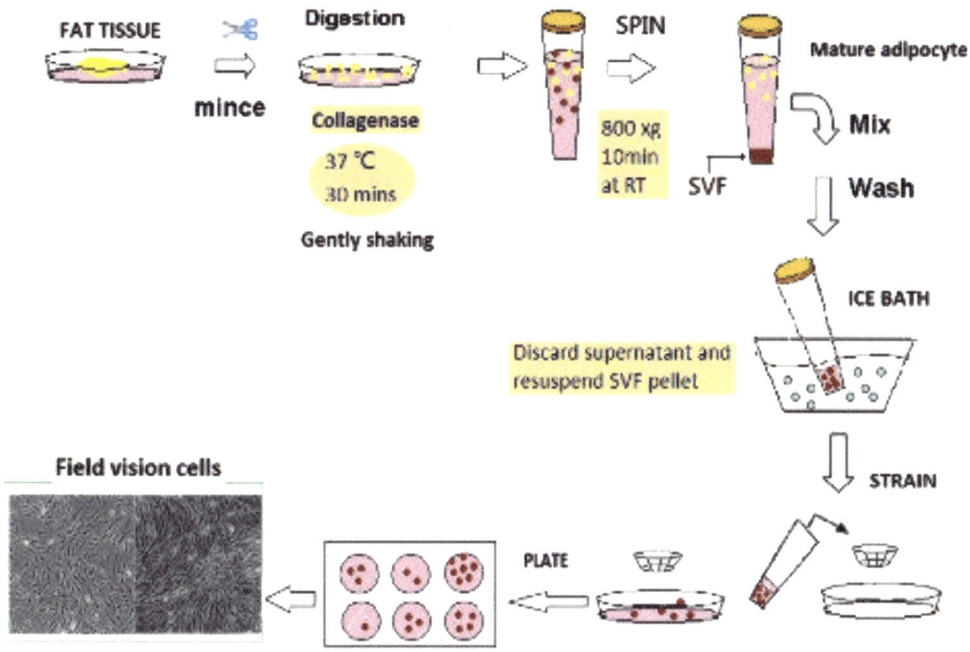

Figure 2. The preparation and isolation process, RT=room temperature, SVF=stromal-vascular cell fraction [3]

in future concepts the dental pulp cells from third molars might be preserved and stored for later use in cell based autologous tissue regeneration.

To combine the good accessibility of adult cells with the excellent differentiating potential another cell source was to be established. Induced pluripotent cells (iPS) surmount the problem of limited access: the cells, for example skin fibroblasts, can be harvested (repeatedly) after informed consent in the mature organism from various tissues as adult cells with differentiating capacity. Therefore they prevail the difficulty of immunoreactivity: they can be harvested autologously. By inducing certain surface/ interaction markers retrovirally, among others Sox2, c-Myc, the so-called Yamanaka factors, these cells forget their initial programming and regress to an earlier precursor stage with enhanced, ES cell-like differentiating properties. Similar to ECS they are capable of unlimited self-renewal in vitro.

The iPS cells show very good proliferative activity; this feature comes with the induction of neoplasia after transplantation, similar to embryonic stem cells.

Their potential and special charm lies in the fact, that with the help of this technique it is generally feasible to isolate an individual cell line from almost every donor. This cell line might be the basis of cell pool, from which many different functional cell and tissue type can be generated autologously for individualized regenerative concepts. In addition iPS cells are used for the construction of disease specific cell and tissue models. With these the potential of therapeutic strategies can be explored.

Figure 3 illustrates the experimental process.

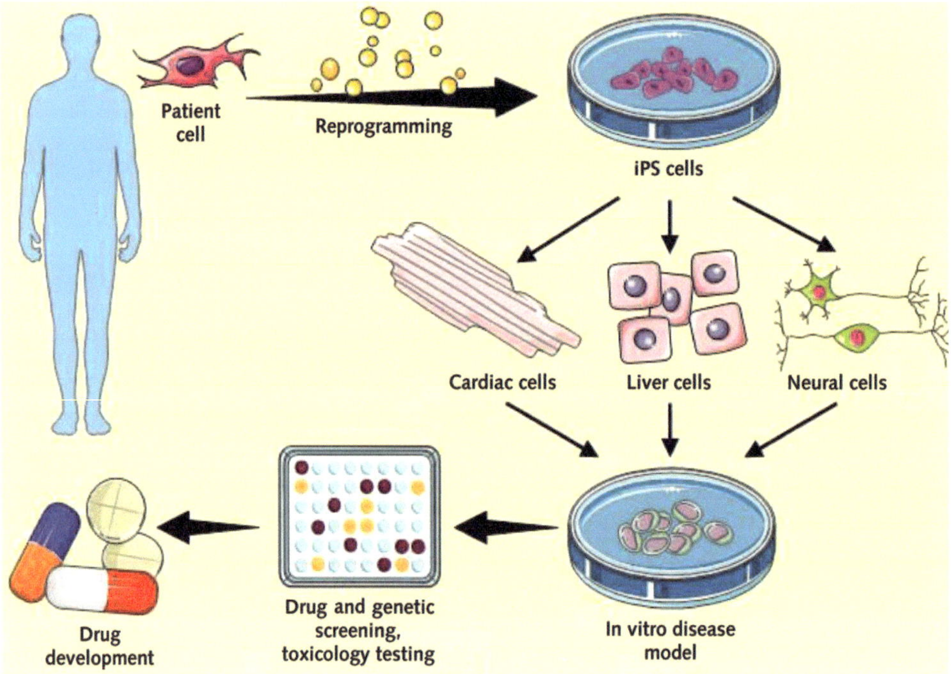

Figure 3. iPS cells as models of different diseases [20, 21]

The harvest, isolation and application of somatic or adult stem cells provide the potential for personalized regenerative concepts. The application especially of autologous cells comes with the advantage of avoiding immunorejection and has verified its prospective in different clinical concepts.

4. Adipose-derived cells vs. bone marrow-derived cells

Human adipose derived stromal cells (hADSCs) are localized in different fatty tissues as subcutaneous, visceral and omental fat. Morphologically these cells appear typically as spindle-shaped. Their harvesting and isolation can be easily performed with minimal invasive approach under local anaesthesia with minimal tissue damage and without excessive scarring, for example by liposuction or –aspiration, if necessary the collect can be repeated without difficulty. The cells can be harvested in amounts that up to 500 times higher than mesenchymal stromal cells from other localizations. This feature comes with the advantage that the cells could be implanted at a therapeutic amount without a preparing culturing and expansion phase; thus, adverse culture effects that might impair their biological function and effectiveness can be avoided [22].

In many investigations it was possible to show that the adipose derived cells are able to differentiate within their original embryonic germ layer to chondrocytes or myoblasts. In addition some authors state that a capacity for differentiation beyond the cytological restrictions of the mesoderm to neuronal and endothelial cells is possible. Rehman et al. among others described the potential of adipose derived cells for endothelial and macrophage differentiation [23]. These traits make adipose-derived cells an ideal cell source for personalized regenerative concepts including bone or chondral tissue. In addition to that, adipose-derived tissue represents an enormous reservoir of potentially angiogenic cells. Their prospective in therapeutic (neo-)vascularization is not yet fully elucidated.

Table 3 gives a survey of modulating factors that initiate the respective differentiation of ADSCs.

Adipogenic specialization	Insulin, IBMX, dexamethasone, rosiglitazone, indomethacin
Osteogenic specialization	1,25-dihydroxy-cholecalciferol, beta-glycerophosphate, ascorbic acid
Neurogenic specialization	EGF, FGF, 5-azacytidine

Table 3. Selection of growth factors that support the various differentiations of hADSCs, EGF: epidermal growth factor; FGF: fibroblast growth factor, BMX: 3-isobutyl-1-meyhylxanthine; IGF: insulin-like growth factor, [24]

In in-vitro settings the culture with certain supplements und defined conditions led to the assumption, that a differentiation of ADSCs to hepatocytes of beta cells of the langerhans islets might be feasible, hinting to a probable autologous cell based approach to chronic liver failure or diabetes mellitus. However, these interesting attempts are not ready for therapeutic application yet [25, 26].

Bone marrow derived mesenchymal stromal cells (BM-MSCs) were the first pluripotent adult cells to be detected and investigated and therefore are characterized adequately. Bone marrow derived MSCs are gained from bone marrow aspirates, a painful clinical procedure that comes with potential donor site morbidity as an impaired sensitivity of the lateral thigh or pelvis fracture. The output of pluripotent cells is relatively small; only 0.001%-0.01% of the nucleated bone marrow cells can be expected to have differentiating potential, with decreasing percentage in elder donors [22]. BM-MSCs have been applied in various in vitro settings, in animal models and in some clinical trials.

When comparing adipose-and bone marrow derived cells, one has to regard the immunophenotype. The expression of surface markers on both cell types is widely identical. Whereas adipose derived cells tend to express CD34 especially in early stages, bone marrow derived cells and MSCs from other sources do not express this surface marker [27].

Furthermore there is some hint that bone marrow derived MSCs show a significantly reduced proliferation rate compared to other adult pluripotent cells as adipose derived cells or umbilical cord cells [28]. This question is discussed controversially though; harvesting, isolation and culture protocols or conditions as media or various supplements might be responsible for enhanced or reduced proliferation rates – independent from the cell source.

The question of superior differentiation capacity is debated controversially also. There are many investigations supporting the thesis that the one or the other cell type might be superior in differentiating to an adipogenic or osteogenic direction or they detect no difference at all. In this case, too, the impact of special culture conditions as cause of the different findings has not been ruled out fully.

Finally, the bone marrow and adipose derived cells seem to exert different immunomodulatory effects on their host tissue. In different in vitro studies ADSCs seem to have a more pronounced suppressive effect on the immune response than BM-MSCs. Adipose derived cells were more efficient to inhibit the maturation of dendritic cells and in stimulating the secretion of the immunomodulatory cytokine IL 10 [29, 30].

The analysis of the respective gene expression profile of ADSCs and BM-MSCs indicated that a differential expression of about 13% of the studied genes differentiates both cell populations on gene level. It was not possible to define certain key markers that clearly identify their origin. Different cell passages or varying culture conditions, too, influence the expression profile [31].

5. Current concepts of stem cell supported tissue engineering

The first reported success of the clinical use of autologous differentiating cells in regenerative medicine was the successful treatment of traumatic calvarian defects with a combination of autologous iliac bone, adipose derived stromal cells and fibrin in a degradable scaffold leading to bone formation and good clinical regeneration of the impaired scull [32].

Clinical trials with small patient groups report, that during the treatment of chronic perianal fistula of Crohn patients, a local therapy of the affected area with autologous ADSCs, in combination with fibrin glue, seems to effect a higher healing rate than many other therapeutic strategies [33, 34].

Especially in plastic surgery clinical concepts based on the distinctive properties of adipose derived stem cells are established; actually the cells' potential has been discovered by plastic surgeons [35].

One clinical challenge in plastic surgery is the surgical approach to chronic wounds as diabetic ulcers or skin defects after radiation due to the impaired vascularisation of the affected area. Allogeneic dermis or collagen scaffolds are cultivated with adipose derived pluripotent cells that lead to good clinical results, avoiding the accompanying scarring and donor site morbidity of local flap surgery. The procedure can easily be repeated in larger and deeper defects [36].

Injection of autologous adipose derived stromal cells is effective in the treatment of arterial clotting after filler injection in the face for augmentation purposes. When this complication happens, the area in question becomes sensitive and painful due to the hypoxia of the affected tissue. The arterial obstruction has to be overcome within 4-5 days to avoid tissue necrosis. In case reports it has been described that the injection of autologous ADCs results in reduced ischemia and immediate neo-vascularisation. In Europe the first clinical trial with ADSCs in the treatment of cardiovascular diseases as acute myocardial infarction or chronic heart failure have started.

Dental pulp derived stem cells have been studied and applied in regenerative concepts concerning the tooth structures particularly. In an animal model Iohara et al. demonstrated the entire regeneration of the histologically complex pulp tissue including neuro-and vasculogenesis after pulpectomy supported by autologous PDSCs transplantation [37].

According to their original repair and renewal function autologous DPSCs have been eingesetzt to replace dentin. In a porcine model direct pulp capping has been performed using autologous DPSCs together with a synthetic restaurative dental material (beta-tricalcium phosphate); in the follow up control the effective dentin regeneration was demonstrated [38]. The aim to completely replace a missed tooth requires a complex communication, interaction and vice-versa induction of mesenchymal cells and epithelial cells from the oral cavity; different investigative settings aim to orchestrate this complex interaction. Ikeda et al. were able to complete the bioengineering of a primordial tooth that has been inserted in the oral cavity and developed roots and really erupted into the oral cavity [39]. Beyond the dental structures DPSCs have shown their capacity for osteogenic differentiation in vitro as well as in in-vitro trials. Dental pulp derived cells were used among others combined with beta tricalcium phosphate scaffolds in critical size defects in rat calvaria. Annibaldi et al. reported in their preliminary data a trend to increased bone density and mineralisation in the presence of the DPSCs [40].

Beyond their potential concerning hard tissue regeneration, dental pulp derived cells were under investigation in terms of neurological and endocrinological issues also. The hopeful results report an improved and accelerated functional regeneration after focal cerebral ischemia and therapeutic local transplantation of DPSCs; especially the impact of secreted mediators on the repair mechanisms seems to cause the observed effects [41]. Not only a differentiation to hepatocyte-like cells with the ability to store glycogen and to produce urea but also islet-like cells being able to release insulin glucose-dependently has been working out in-vitro and in mice models [42, 43].

To date the most encouraging operational area for adult, especially adipose derived stem cells are concepts of bone repair and regeneration. Especially the active physiological behaviour of healthy bone tissue with its permanent remodelling to adapt to mechanical loading or its lack or to accommodate the current nutrient supply makes this tissue an ideal point of application for cell based intervention. Additionally, the therapy of bony critical size defects often comes with clinical problems of tedious distraction, insufficient long-term osteosynthesis or donor site morbidity after autologous bone grafting.

Current research therefore deals with interventional concepts that combine the differentiating precursor cells from adipose tissue with the enormous paracrine effect of these cells, secreting a plethora of proliferation, growth and differentiation factors and thus activating not only the regenerative potential of the resident tissue but also attracting circulating precursor cells.

Structural base of the most regenerative approaches is still a 3D scaffold to support the homing of the proliferating cells. Not only are the mechanical properties of these scaffolds of importance, their composition affects the biology of the cells also: Marino et al. were able to depict, that the osteogenic differentiation of human adipose derived stem cells onto a porous beta-tricalcium phosphate (beta-TCP) matrix was triggered by the scaffold alone without the use of differentiative media [44]. This effect might be enhanced by modern composite scaffolds: akermanite scaffolds, consisting of calcium-, magnesium-, and silicon-containing ceramic show improved stability and better degradation. In addition they seem to positively influence the functionality of differentiating cells by inducing their calcium deposition [45, 46].

6. Future prospect

In different experimental settings the transplanted pluripotent cells do not exert their regenerative potential by actually differentiating and replacing damages tissue but exert their effects by secreting growth, proliferation and differentiation factors and thus in a paracrine way or by cell-cell interaction.

In terms of stem cell science many biological and technical questions remain unanswered so far and years of intense research lay ahead before autologous pluripotent cells become a standard tool in clinical routine.

The current restrictions of embryonic stem cell research are necessary to avoid abuse but the promising, steadily growing comprehension about adult stem cells and induced pluripotent cells with their ability to regain their original differentiating potential seem to hint to the fact that these cells might be the target of interest in future regenerative concepts in personalized medicine.

The challenge of the coming decade of stem cell research will have to deal with safety issues, the problem to reliably avoid oncogenic growth and to develop reproducible, secure and payable concepts of cell isolation, culturing, (re-) programming and storage. There have to be protocols for the autologous transplantation to the region of interest and techniques to improve cell homing.

Despite many unanswered questions the future of many regenerative concepts will be supported by the potential of differentiating cells.

Author details

Susanne Jung[1,2*] and Johannes Kleinheinz[1,2]

*Address all correspondence to: Susanne.Jung@ukmuenster.de

1 Department of Cranio-Maxillofacial Surgery, University Hospital Muenster, Germany

2 Research Unit Vascular Biology of Oral Structures (VABOS), Germany

References

[1] Mountford JC. Human embryonic stem cells: origins, characteristics and potential for regenerative therapy, Transfus Med. 2008;18(1) 1-12.

[2] Eggan K, Jaenisch R. Differentiation of F1 embryonic stem cells into viable male and female mice by tetraploid embryo complementation, Methods Enzymol 2003;365 25–39.

[3] Cheng H. Human adipose-derived stem cells: Isolation, characterization and current application in regeneration medicine. Genomiv Medicine, Biomarkers and Health Sciences. 2011;3 53-62.

[4] Jarocha D, Lukasiewich E, Majka M. Advantage of mesenchymal stem cells (MSC) expansion directly from purifiedbone marrow CD105+and CD271+cells,Folia Histochem Cytobiol 2008;46 307–314.

[5] Bianco P, Robey PG, Simmons PJ. Mesenchymal stem cells: revisiting history, concepts and assays. Cell Stem Cell. 2008;2(4) 313-9.

[6] Bai X, Yan Y, Song YH, Seidensticker M, Rabinovich B, Metzele R, Bankson JA, Vykoukal D, Alt E. Both cultured and freshly isolated adipose tissue-derived stem cells enhance cardiac function after acute myocardial infarction. Eur Heart J. 2010;31(4) 489-501.

[7] Tzameret A, Sher I, Belkin M, Treves AJ, Meir A, Nagler A, Levkovitch-Verbin H, Barshack I, Rosner M, Rotenstreich Y. Transplantation of human bone marrow mesenchymal stem cells as a thin subretinal layer ameliorates retinal degeneration in a rat model of retinal dystrophy, Exp Eye Res. 2014;118 135-44. doi: 10.1016/j.exer.2013.10.023.

[8] Horwitz EM, Gordon PL, Koo WK, Marx JC, Neel MD, McNall RY, Muul L, Hofmann T. Isolated allogeneic bone marrow-derived mesenchymal cells engraft and stimulate growth in children with osteogenesis imperfecta: Implications for cell therapy of bone, Proc Natl Acad Sci (U S A). 2002;99 8932–8937.

[9] Ortiz LA, Gambelli F, McBride C, Gaupp D, Baddoo M, Kaminski N, Phinney DG. Mesenchymal stem cell engraftment in lung is enhanced in response to bleomycin exposure and ameliorates its fibrotic effects, Proc Natl Acad Sci (U S A) 2003;100 8407–8411.

[10] Phinney DG, Isakova I. Plasticity and therapeutic potential of mesenchymal stem cells in the nervous system, Curr Pharm Des 2005;11 1255–1265.

[11] Stojkovic M, Lako M, Strachan T et al. Derivation, growth and applications of human embryonic stem cells, Reproduction. 2004;128 259–67.

[12] Przyborski SA. Differentiation of human embryonic stem cells after transplantation in immune-deficient mice, Stem Cells. 2005;23(9) 1242–1250.

[13] Krampera M, Marconi S, Pasini A, Galie M, Rigotti G, Mosna F, Tinelli M, Lovato L, Anghileri E et al. Induction of neural-like differentiation in human mesenchymal stem cells derived from bone marrow, fat, spleen and thymus, Bone 2007;40 382–490.

[14] Ning H, Liu G, Lin G, Yang R, Lue TF, Lin CS. Fibroblast growth factor 2 promotes endothelial differentiation of adipose tissue-derived stem cells, J Sex Med. 2009;6(4) 967-79.

[15] Friedenstein AJ, Petrakova KV, Kurolesova AI, Frolova GP. Heterotopic of bone marrow. Analysis of precursor cells for osteogenic and hematopoietic tissues, Transplantation 1968;6 230–247.

[16] Gronthos S, Mankani M, Brahim J et al. Postnatal human dental pulp stem cells (DPSCs) in vitro and in vivo, Proc Natl Acad Sci 2000;97 13625–13630.

[17] Karaoz E, Demircan PC, Saglam Ö et al. Human dental pulp stem cells demonstrate better neural and epithelial stem cell properties than bone marrow-derived mesenchymal stem cells, Histochem Cell Biol 2011;136 455–473.

[18] Yan X, Qin H, Qu C et al. iPS cells reprogrammed from human mesenchymal-like stem/progenitor cells of dental tissue origin, Stem Cells Dev 2010;19 469–480.

[19] Yoo CH, Na HJ, Lee DS et al. Endothelial progenitor cells from human dental pulp-derived iPS cells as a therapeutic target for ischemic vascular diseases, Biomaterials 2013;34 8149–8160.

[20] Power C, Rasko JE. Promises and challenges of stem cell research for regenerative medicine, Ann Intern Med. 2011;155(10) 706–713.

[21] Fong CY, Gauthaman K, Bongso A. Teratomas from pluripotent stem cells: A clinical hurdle, J Cell Biochem. 2010;111(4) 769–781.

[22] Bernardo ME, Locatelli F, Fibbe WE. Mesenchymal stromal cells. Ann N Y Acad Sci. 2009;1176 101-17.

[23] Rehman J, Traktuev D, Li J, et al. Secretion of angiogenic and antiapoptotic factors by human adipose stromal cells, Circulation. 2004;109 1292e1298.

[24] Huang SJ, Fu RH, Shyu WC, Liu SP, Jong GP, Chiu YW, Wu HS, Tsou YA, Cheng CW, Lin SZ. Adipose-derived stem cells: isolation, characterization, and differentiation potential, Cell Transplant.2013;22(4) 701-9.

[25] Lue J, Lin G, Ning H, et al. Transdifferentiation of adipose-derived stem cells into hepatocytes: a new approach, Liver Int. 2010;30 913e922.

[26] Kajiyama H, Hamazaki TS, Tokuhara M, et al. Pdx1-transfected adipose tissue-derived stem cells differentiate into insulin-producing cells in vivo and reduce hyperglycemia in diabetic mice, Int J Dev Biol. 2010;54 699e705.

[27] Gronthos S, DM Franklin, HA Leddy, PG Robey, RW Storms and JM Gimble. Surface protein characterization of human adipose tissue-derived stromal cells, J Cell Physiol 2001;189 54–63.

[28] Kern S, Eichler H, Stoeve J, Kluter K, Bieback K. Comparative analysis of mesenchymal stem cells from bone marrow, umbilical cord blood, or adipose tissue, Stem Cells 2006;24 1294–1301.

[29] Noel D, Caton D, Roche S, Bony C, Lehmann S, Casteilla L, Jorgensen C, Cousin B. Cell specific differences between human adipose-derived and mesenchymal-stromal cells despite similar differentiation potentials, Exp Cell Res 2008;314 1575–1584.

[30] Ivanova-Todorova E, Bochev I, Mourdjeva M, Dimitrov R, Bukarev D, Kyurkchiev S, Tivchev P, Altunkova I, Kyurkchiev DS. Adipose tissue-derived mesenchymal stem cells are more potent suppressors of dendritic cells differentiation compared to bone marrow-derived mesenchymal stem cells, Immunol Lett 2009;126 37–42.

[31] Strioga M, Viswanathan S, Darinskas A, Slaby O, Michalek J. Same or Not the Same? Comparison of Adipose Tissue-Derived Versus Bone Marrow-Derived Mesenchymal Stem and Stromal Cells, Stem Cells Dev. 2012;21(14) 2724-52.

[32] Lendeckel S, Jodicke A, Christophis P, et al. Autologous stem cells (adipose) and fibrin glue used to treat widespread traumatic calvarial defects: case report, J Craniomaxillofac Surg. 2004;32 370e373.

[33] Garcia-Olmo D, Garcia-Arranz M, Herreros D, et al. A phase I clinical trial of the treatment of Crohn's fistula by adipose mesenchymal stem cell transplantation, Dis Colon Rectum. 2005;48 1416e1423.

[34] Garcia-Olmo D, Herreros D, Pascual I, et al. Expanded adipose-derived stem cells for the treatment of complex perianal fistula: a phase II clinical trial, Dis Colon Rectum. 2009;52 79e86.

[35] Zuk PA, Zhu M, Mizuno H, Huang J, Futrell JW, Katz AJ, Benhaim P, Lorenz HP, Hedrick MH. Multilineage cells from human adipose tissue: implications for cell-based therapies, Tissue Eng. 2001;7 211-28.

[36] Yun MJ, Park JU, Kwon ST. Surgical options for malignant skin tumors of the hand, Arch Plast Surg. 2013;40 238-43.

[37] Iohara K, Zheng L, Ito M, Ishizaka R, Nakamura H, Into T, Matsushita K, Nakashima M. Regeneration of dental pulp after pulpotomy by transplantation of CD31(-)/CD146(-) side population cells from a canine tooth, Regen Med. 2009;4(3) 377-85.

[38] Zheng Y, Wang XY, Wang YM et al. Dentin regeneration using deciduous pulp stem/progenitor cells, J Dent Res 2012;91 676–682.

[39] Ikeda E, Morita R, Nakao K et al. Fully functional bioengineered tooth replacement as an organ replacement therapy, Proc Natl Acad Sci U S A 2009;106 13475–13480.

[40] Annibali S, Bellavia D, Ottolenghi L, Cicconetti A, Cristalli MP, Quaranta R, Pilloni A. Micro-CT and PET analysis of bone regeneration induced by biodegradable scaffolds as carriers for dental pulp stem cells in a rat model of calvarial "critical size" defect: Preliminary data. J Biomed Mater Res B Appl Biomater. 2014;102(4) 815-25.

[41] Leong WK, Henshall TL, Arthur A, et al. Human adult dental pulp stem cells enhance poststroke functional recovery through non-neural peplacement mechanisms, Stem Cells Transl Med 2012;1 177–187.

[42] Patil R, Kumar BM, Lee WJ et al. Multilineage potential and proteomic profiling of human dental stem cells derived from a single donor, Exp Cell Res 2013;20 92–107.

[43] Kanafi MM, Rajeshwari YB, Gupta S et al. Transplantation of islet-like cell clusters derived from human dental pulp stem cells restores normoglycemia in diabetic mice, Cytotherapy 2013;15 1228–1236.

[44] Marino G, Rosso F, Cafiero G, Tortora C, Moraci M, Barbarisi M, Barbarisi A. Beta-tricalcium phosphate 3D scaffold promote alone osteogenic differentiation of human adipose stem cells: in vitro study, J Mater Sci Mater Med. 2010;21(1) 353-63.

[45] Barba M, Cicione C, Bernardini C, Michetti F, Lattanzi W. Adipose-derived mesenchymal cells for bone regereneration: state of the art. Biomed Res Int. 2013;2013 416391.

[46] Zanetti A, McCandless G, Chan J, Gimble J, Hayes D. Characterization of novel akermanite:poly-ε-caprolactone scaffolds for human adipose-derived stem cells bone tissue engineering, J Tissue Eng Regen Med. 2012 doi: 10.1002/term.1646.

Chapter 5

Mass Production of Mesenchymal Stem Cells — Impact of Bioreactor Design and Flow Conditions on Proliferation and Differentiation

Valentin Jossen, Ralf Pörtner, Stephan C. Kaiser, Matthias Kraume, Dieter Eibl and Regine Eibl

Additional information is available at the end of the chapter

http://dx.doi.org/10.5772/59385

1. Introduction

Stem cells have enormous potential in health and medical research due to their ability to differentiate into specialized cells and to self-renew. Applications can be seen in cell-based therapies, e.g. for the treatment of Parkinson's disease, type I diabetes, arthritis, burn victims, and cardiovascular diseases, as well as in tissue engineering of artificial organs, development and testing of drugs, and *in vitro* toxicity tests [1-9].

In general, stem cells are cells generating identical daughter and progenitor cells, which both have the ability to develop terminally differentiated functional cells after going through a defined differentiation process [10,11]. There are two types of stem cells, varying in their ability to differentiate, methods of isolation and propagation and, thus, their application potential: embryonic stem cells and adult stem cells. Whereas embryonic stem cells represent pluripotent cells which are able to grow into all derivatives of the three germ layers, adult stem cells are only able to develop into a limited number of cells. Adult stem cells, to which mesenchymal stem cells (also referred to as mesenchymal stromal cells) belong, are of postnatal origin, are more tissue-specific, and regarded as multipotent. In this work the focus is laid on mesenchymal stem cells (MSCs) as they have become increasingly important in the field of regenerative medicine in recent years. Numerous clinical studies have confirmed the safety of allogenic and autologous MSCs for treatment of human diseases [12,13]. Currently, MSCs are being investigated in terms of their therapeutic potential for inflammatory, autoimmune and degenerative conditions in preclinical and clinical studies [14,15].

Between the late 1960s and the 1980s, Friedenstein et al. [16-18] and Owen et al. [19,20] described tissue-specific progenitor cells capable of differentiation to cells of mature tissues. They worked with subpopulations of bone marrow cells characterized by osteogenic potential. Interestingly, the term MSC was introduced in 1991 for the first time. Caplan [21] proposed classifying adult stem cells as MSCs, if they were able to differentiate to all cells of mesodermal lineage, although their multilineage potential was only demonstrated later by Pittinger et al. [22] (for review see [23]). Between the 1990s and 2000s numerous studies were conducted with cell populations that were, strictly speaking, not MSCs according to Caplan`s proposed definition. Finally, as it became impossible to compare results, a revision of the MSC definition was introduced by the International Society for Cellular Therapy (ISCT) in 2006 [24]. The ISCT definition is based on two cellular properties of MSCs which are morphologically similar to fibroblasts: (1) their adherence to plastic and (2) their trilineage differentiation to adipogenic, chondrogenic and osteogenic cells [25]. Additional requirements include positive expression of CD105 (endoglin marker), CD73 (ecto-5`-nucleotidase marker) and CD90 (Thy1 marker), and negative expression of CD45 (leukocyte marker), CD34 (hematopoietic stem cell marker), CD14 or CD11b (monocyte and macrophage markers), CD79 or CD19 (B cell marker), and human leukocyte antigen class II.

As things stand, bone marrow aspirate represents the most often used source [22] of MSCs. In addition, MSCs are obtained from adipose tissue [26], placenta as well as umbilical tissue [15] and blood [27,28], and peripheral blood [29] (Figure 1). Other sources, such as periosteum, trabecular bone, synovia, skeletal muscle, deciduous teeth, fetal pancreas, lung, liver and amniotic fluid, have also been reported [15]. But independent of the source, the low frequency of MSCs makes their direct collection for the majority of MSC applications unfeasible. A therapeutic dose has been reported to range between 1 and $1000 \cdot 10^6$ cells per kg body weight [30-33], whereby the exact dosage is determined by the type of disorder. However, when expanding MSCs for particular therapies, the type of therapy (that is, either an allogeneic or autologous one) also needs to be taken into account. In doing so, the ability of MSCs (which display heterogeneous cell populations that vary from donor to donor and depend on donor age and tissue) [34,35] to rapidly proliferate *in vitro* is exploited.

In order to fulfill the potential of MSCs as therapeutic agents for a wide range of applications, culture conditions need to be optimized to obtain clinical grade MSCs with defined safety standards at large scale [14,15,31,41-46]. To achieve this, strategies for preparation of the required quantities of cells, have to be implemented, which allow the undifferentiated state to be preserved or direct stem cell differentiation into the desired lineage. Scalable culture systems with optimized culture strategies and controlled culture environment are prerequisites for the successful transfer of stem cell-based therapy concepts into clinical routine. In the following, we describe the main characteristics of MSCs with respect to cultivation, appropriate cultivation methods for mass production, and a new scale-up-approach for expanding human mesenchymal stem cells (hMSCs) derived from adipose tissue (hADSCs) and bone marrow (hBM-MSCs).

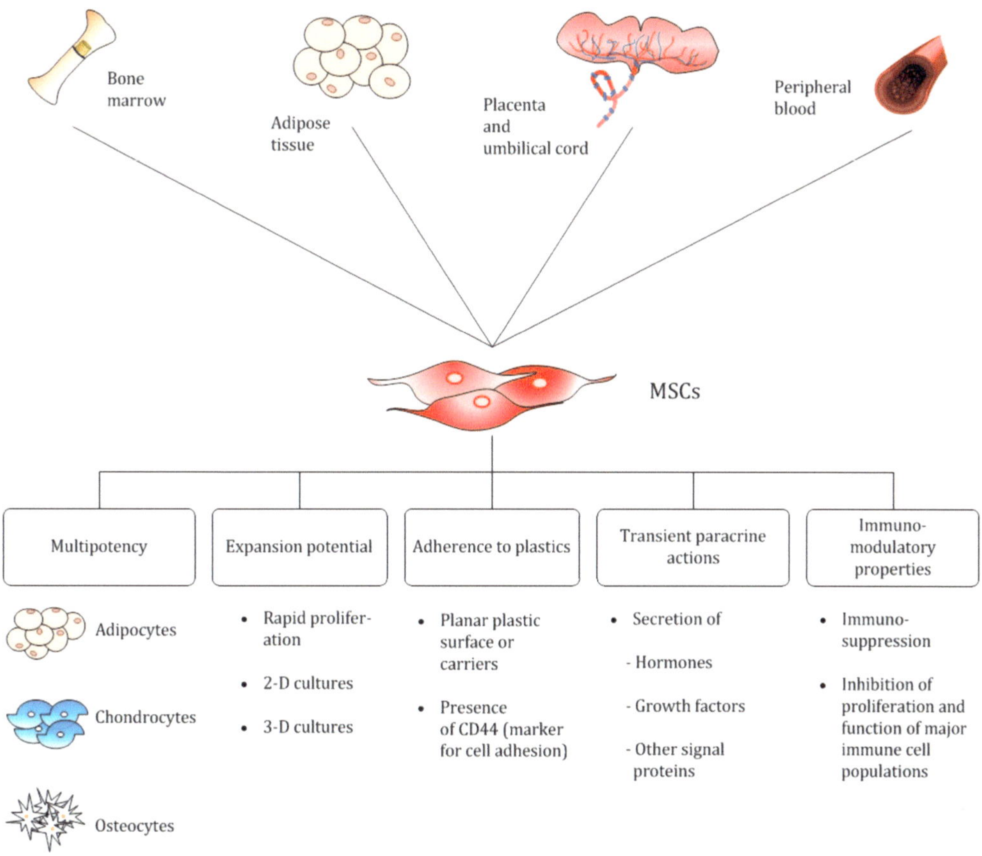

Figure 1. Main sources and properties of MSCs. Today`s high pre-and clinical interest in MSCs is ascribed not only to their multipotency, but also to their paracrine secretion of angiogenic factors, cytokines and immunomodulatory substances, which is documented by a body of literature (e.g. [36-40]).

2. Characteristics of MSCs and the influence of the engineering parameters on the cultivation

2.1. Main factors influencing MSC cultivation

For the successful expansion of stem cells appropriate culture conditions are essential. When working with MSCs *in vitro*, the problems are non-infinite growth and non-complete directed differentiation [47]. A critical step is the isolation of specific stem cells, as they often occur in small quantities and only in a tissue formed by other cells. Therefore, special protocols for their isolation and preservation are required [48]. Furthermore, *in vivo* stem cells live in a highly specialized microenvironment. This microenvironment interacts with a number of characteristics such as the extracellular matrix surrounding the cells and mediators on growth and differentiation. These factors have to be modeled for *in vitro* expansion by appropriate selection of the matrix for cell adhesion, the culture medium including growth factors, and the physical environment, i.e. temperature, pH, oxygen and shear effects.

Self-renewal and the differentiation potential of stem cells are influenced to a large extent by donor age and passage number [49,50]. In general, MSCs can be expanded by several orders of magnitude in a few weeks [51] due to their high expansion potential. But with increasing passage number they can lose their multipotency or their specific cell surface markers might be downregulated, resulting in a loss of the therapeutic properties of these cells [52]. After 30 to 40 doublings MSCs become senescent, accompanied by inhibited growth and induction of apoptosis [53]. It is assumed that this effect is induced to some extent by the harsh enzymatic treatment during subcultivation, e.g. by trypsinization. Consequently, techniques without subcultivation have been investigated to prolong the proliferation phase and to increase the expansion factor [54]. Kretlow et al. [50] observed that both increasing age and the number of passages have lineage dependent effects on MSC differentiation potential. In addition, there seemed to be an interplay between donor age and cell passage number. These effects are of high relevance for clinical therapies, because they might have strong impact on the cultivation protocol.

As mentioned above, the extracellular microenvironment plays a significant role in controlling cellular behavior. In recent years, different biomaterials have been studied to find a microenvironment that is conducive to stem cell growth and differentiation, and that mimics the *in vivo* situation, at least to some extent (reviewed by [55]). Appropriate biomaterials can support the cellular attachment, proliferation, and lineage-specific differentiation of stem cells. In the case of MSC differentiation growth factors or factors known to induce lineage commitment of stem cells were incorporated into cultures with scaffolding materials, quite often for regulation of osteogenic, chondrogenic or adipogenic differentiation [56-59]. Because such scaffolding materials are not normally used for MSC expansion, they are not further discussed here.

For expansion with the final goal of harvesting large quantities of MSCs, they are typically adhered to a surface with optimized surface chemistry and topography, which supports cell attachment, spreading and proliferation (Figure 2). Today, routine cultivation of MSCs is performed in planar plastic plates and flasks ensuring two-dimensional (2-D) growth [60]. Reusable glass systems are very seldom utilized at this time. As plates and flasks allow for a limited expansion only and subcultivations exceeding 6 passages (see above) should be avoided (high risk of cell differentiation), scalable matrices have been established. These matrices cover hollow fiber membranes, macrocarriers [61,62] and, in particular, microcarriers (see also Section 3).

Microcarriers are small spherical particles differing in core material, density, diameter and surface charge, and are intended for use in suspension bioreactors. Table 1 gives an overview of commercially available microcarriers that are suitable in MSC expansions. Depending on the porosity, the microcarriers are classified into non-porous, micro-porous (pore sizes < 1 μm) and macro-porous types (pore sizes between 10 and 50 μm). Whereas in non-and microporous microcarrier cultures the cells attach to the surface of the beads, in macro-porous microcarrier cultures they grow in the pores. However, cell harvest is more complicated, which is a general challenge in microcarrier-based cultivations. While taking care to avoid differentiation or even

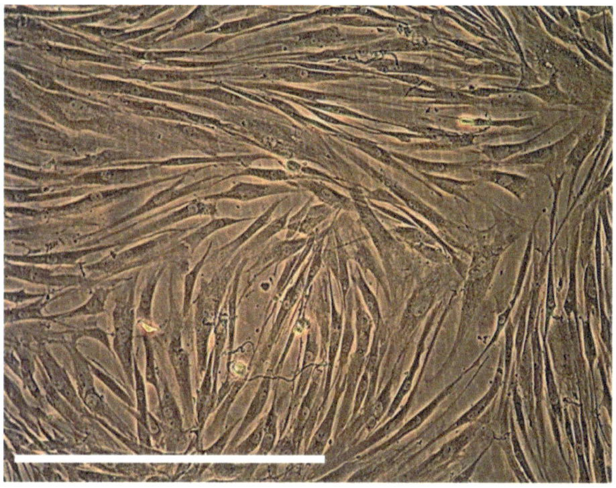

Figure 2. Adherent glandular stem cells grown in a T-Flask (bar: 500 μm).

damage, the cells are normally harvested after chemical treatment [63-65]. Examples of stem cell proliferation on microcarriers can be found in [66-69].

Type	Diameter [μm]	Density [kg·m³]	Surface area [cm²·g⁻¹]	Charge [+/-]	Core material
CultiSpher-G	130-380	1020	-	none	Gelatin
CultiSpher-S	130-380	1020	-	none	Gelatin
Cytodex I	147- 248	1030	4400	+	Cross-linked dextran
Cytodex III	141-211	1040	2700	none	Cross-linked dextran
Collagen	90-150 / 125-212	1020	480 / 360	none	Cross-linked polystyrene
FACT III	90-150 / 125-212	1020 / 1040	480 / 360	+	Cross-linked polystyrene
Hillex II	160-180	1110	515	+	Modified polystyrene
Hillex-CT	160-180	1110	515	+	Modified polystyrene
Plastic	90-150 / 125-212	1020 / 1040	480 / 360	none	Cross-linked polystyrene
Plastic Plus	90-150 / 125-212	1020 / 1040	480 / 360	+	Cross-linked polystyrene
ProNectin F	90-150 / 125-212	1020 / 1040	480 / 360	none	Cross-linked polystyrene
Glass	90-150 / 125-212	1020 / 1040	480 / 360	none	Cross-linked polystyrene

Table 1. Overview of commercially available microcarriers for the cultivation of MSCs. Non-porous microcarriers=Hillex II, Hillex-CT, Plastic, Plastic Plus, ProNectin F, Glass; micro-porous microcarriers=Cytodex I, Cytodex III; macro-porous microcarriers=CultiSpher-G, CultiSpher-S.

An alternative is carrier-free cultivation in 3-D (three-dimensional) aggregates or spheroids, or encapsulation, e.g., in alginate [70] or in hydrogels [57,71-79]. These techniques are often used to induce a lineage specific differentiation. With respect to the expansion of MSCs, there is conflicting evidence in the literature with regard to the usefulness of these techniques. Detrimental effects such as altered proliferation, stem cell marker expression, cell shape and modified differentiation potential have been reported [80,81]. A general problem in 3-D aggregates can be seen in the inhomogeneous supply of oxygen and nutrients. Therefore, it is unlikely that carrier-free cultivation of MSCs is an appropriate approach for the standardized production of large quantities of cells with consistent product quality.

A further strong impact on MSC proliferation and differentiation is exerted by the culture medium through its composition (content of nutrients such as glucose or glutamine, growth factors, type and concentration of serum), culture conditions such as pH or oxygen concentration, and the density, in which the cells were plated/seeded [47].

Due to the importance of hyperglycemia *in vivo* and the hypothesis, that high glucose could be deleterious to stem cell therapy, the impact of glucose on proliferation, potential for differentiation, and other physiological effects have been studied intensively. Glucose concentration in basal media has been shown to have both positive and negative impacts on MSC growth [47,82-93]. It is beyond the scope of this chapter to discuss the various effects in detail but certain findings are presented below to highlight the complexity of this topic:

- Evidence of retention of MSC characteristics from different sources with regard to surface marker profiling, proliferation, differentiation and karyotyping when cultured extensively in DMEM-HG medium containing a high glucose concentration of 25 mM [83].
- Increase in peak cell density by 40 % at low glucose concentration (5.5 mM) when compared with an expansion at high glucose concentration (25 mM) [84].
- High glucose concentration in cell culture medium did not acutely affect hMSC growth factor or proliferation [88].
- High glucose (25mM) enhanced telomerase-immortalized human mesenchymal stem cell (hMSC-TERT) proliferation in long-term studies in contrast to hMSCs, where proliferation was unchanged. Thioredoxin-interacting protein, which is involved in apoptosis regulation, was stimulated by glucose in hMSC-TERT. However, in both cell types apoptosis was not influenced by high glucose levels [90].
- Culture in high glucose-containing medium had a negative effect on colony formation and differentiation for rat non-adherent bone marrow MSCs [91].
- Glucose reduction prevents replicative senescence and increases mitochondrial respiration in hMSCs [94].
- High glucose conditions suppress the function of bone marrow-derived endothelial progenitor cells via inhibition of the eNOS-caveolin-1 complex [82].
- High glucose regulates cyclin D1/E of hMSCs through TGF-beta1 expression via Ca2 /PKC/MAPKs and PI3K/Akt/mTOR signal pathways [92].

It seems that MSCs are affected by high glucose concentrations. However, in respect of proliferation and differentiation, no real advantage of low glucose concentrations has been found to date. Our own data (not published) indicate that the effect of high glucose concentrations on MSC-fate depends to a large extent on the medium formulation, e.g. the type and concentration of serum or growth factor concentration (see also below). For practical reasons, cultivation at high glucose levels has some advantages, as in low glucose media more frequent medium-feeding or exchange is required than at higher cell densities.

The impact of glutamine has been studied less extensively than that of glucose. Ferrari et al. observed an increase in peak cell density by approx 25 % when adding of 2 to 6 mM of glutamine [84]. Schop et al. concluded from their studies that glutamine has no importance as an energy source for hMSCs [93].

In the cultivation and differentiation of stem cells, growth factors play an important role. They influence self-renewal, senescence, aging, embryonic development and differentiation [95]. Thus, they create an appropriate microenvironment and provide important messenger signals for cells. In effect, growth factors can induce or inhibit signaling cascades in the cells via receptors on the cell surface which affect cell physiology [96]. In many types of stem cells, regulation of self-renewal is done by proteins of the fibroblast growth factor (FGF)-family [95]. For example, for FGF-2 an improved expansion was demonstrated [97]. The absence of aforementioned growth factors may lead to differentiation of stem cells, which is unwanted in the pure propagation of the cells [98]. Again, the appropriate concentration of growth factors needs to be determined depending on the cells, medium composition etc. [99].

A further important medium compound with respect to stem cell physiology is serum, in particular its type and concentration. At present, most studies are performed using fetal calf serum (FCS) or fetal bovine serum (FBS) in concentrations between 10 % and 20 % [100]. In Section 5 an example with reduced FBS concentration is discussed. Clinical studies require a GMP- compliant medium [43,101] due to the drawbacks of serum (non-human origin, possible contamination etc.) [102]. Although human serum has positive effects compared to FCS or FBS, extended use for large scale processes is doubtful due to limited availability. Replacement of serum containing medium by serum-free medium formulation has been addressed by a number of publications during the last few years [103-113]. The first serum-free culture media (for example StemPro® MSC SFM XF) for MSCs are now commercially available. However, the cultivation results (cell yields) have not proved satisfactory and the matrix is generally being pre-coated with serum [114,115].

Hypoxic conditions are regarded as advantageous for stem cell proliferation and differentiation. MSCs grown *in vitro* with reduced oxygen content in the gassing air were characterized by increased cell proliferation. It has been shown that at 1% and 5% oxygen in the gassing air the cells enter the exponential growth phase earlier and produce less inhibitor substances [114,116]. Our own studies (data not published) underline this, but indicate that with an optimized medium the positive effect of low oxygen concentration seems to be smaller. Further culture parameters such as osmolality, pH and temperature might influence stem cell fate, but these parameters have not yet been studied intensively so far.

For large scale expansion of MSCs, the cell density required for seeding has a strong impact on the expansion capacity. Due to restricted availability of the cell material and limited population doubling or passage number, the lowest possible seeding density of cells should be applied [51,117]. Fortunately, the literature data suggest that seeding MSCs at low densities (2.5 instead of 2500 cells cm^{-2}) can increase cell proliferation and cell density while maintaining marker profile [118]. Colter et al. seeded 1.5 and 3.0 cells cm^{-2} and successfully generated single cell-derived colonies [51]. Hewitt et al. [119] found the best conditions for cell expansion on microcarriers to be 3000 microcarriers mL^{-1} (ca. 1 g dry weight L^{-1}) in flasks. They determined a seeding density of 5 cells per microcarrier. Higher growth rates of MSCs seeded at lower densities are explained by a longer exponential growth phase, more population doublings, and more availability of nutrients per cell [120].

2.2. Shear stress and the influence of engineering parameters on MSC proliferation and differentiation

2.2.1. General basics

There is strong evidence that growth and differentiation of stem cells are effected by several types of mechanical forces including stretch, strain, compression, and shear stress [121]. Due to the importance of shear stress, the fundamentals are given in this section before shear effects on expansion of MSCs are discussed.

Cells growing adherent to a solid matrix are exposed to shear forces from the moving fluid. By definition shear in a fluid system has two components, shear stress τ and shear rate γ. Shear stress is a force per unit area acting on and parallel to a surface. Shear rate is a measure of a velocity gradient (velocity/length). The two quantities are therefore related in laminar Newtonian fluid - and cell culture media are regarded as Newtonian - by

$$\tau = \eta_{fl} \cdot \gamma \tag{1}$$

where η_{fl} is the viscosity of the fluid. In model systems such as laminar flow between two parallel plates, cone-and-plate viscometer or a coaxial cylinder Searle viscometer, the shear rate and corresponding shear stress can be calculated by simple mathematical equations. Shear effects are often investigated in flow chamber bioreactors consisting of parallel plates [122]. In this apparatus, a laminar flow causes a defined wall shear stress τ_w on the bottom plate where the cells grow adherently (see Figure 3). The shear stress for a Newtonian fluid at a surface element parallel to a flat plate at the point y is given by

$$\tau = \eta_{fl} \cdot \left(\frac{dU}{dy} \right) \tag{2}$$

where U is the velocity of the fluid along the boundary and y is the height above the boundary. Specifically, wall shear stress is defined as:

http://dx.doi.org/10.5772/59385

$$\tau_w = \eta_{fl} \cdot \left(\frac{dU}{dy} \right) \quad y{=}0 \tag{3}$$

Flow velocity $U(y)$ can be calculated for given values of the flow rate F, the height h and the width b of the flow chamber (Figure 3):

$$U(y) = \frac{6F}{b \cdot h^3} \cdot \left(h \cdot y - y^2 \right) \tag{4}$$

Wall shear stress τ_w acting on the cells is given by:

$$\tau_w = \eta_{fl} \cdot \left(\frac{dU}{dy} \right) \quad y{=}0 = \frac{6F\eta_{fl}}{b\,h^2} \tag{5}$$

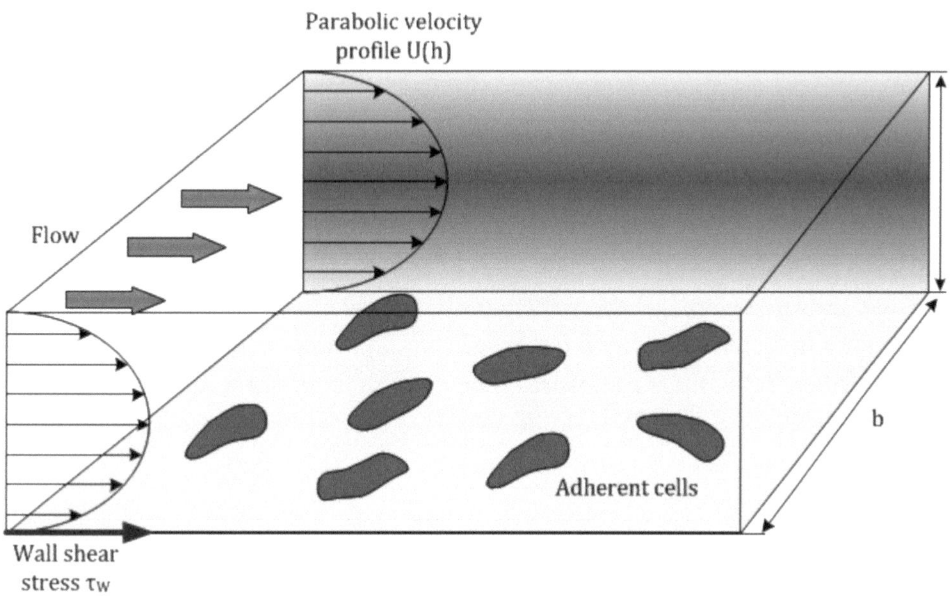

Figure 3. Wall shear stress acting on adherent cells in a flow chamber with parallel plates. For abbreviations see text (modified from [122], with kind permission from Springer Science and Business Media).

In the complex, mostly turbulent environment in a culture system (bioreactor), the local shear rate varies within the vessel and it therefore is more difficult to associate cellular effects (cell damage, differentiation etc.) with the magnitude of the prevailing shear rate or the associated shear stress. An example is given in Figure 4 for cells grown on microcarriers in suspension, e.g. a spinner flask, a stirred bioreactor or a wave-mixed bioreactor (see Section 3).

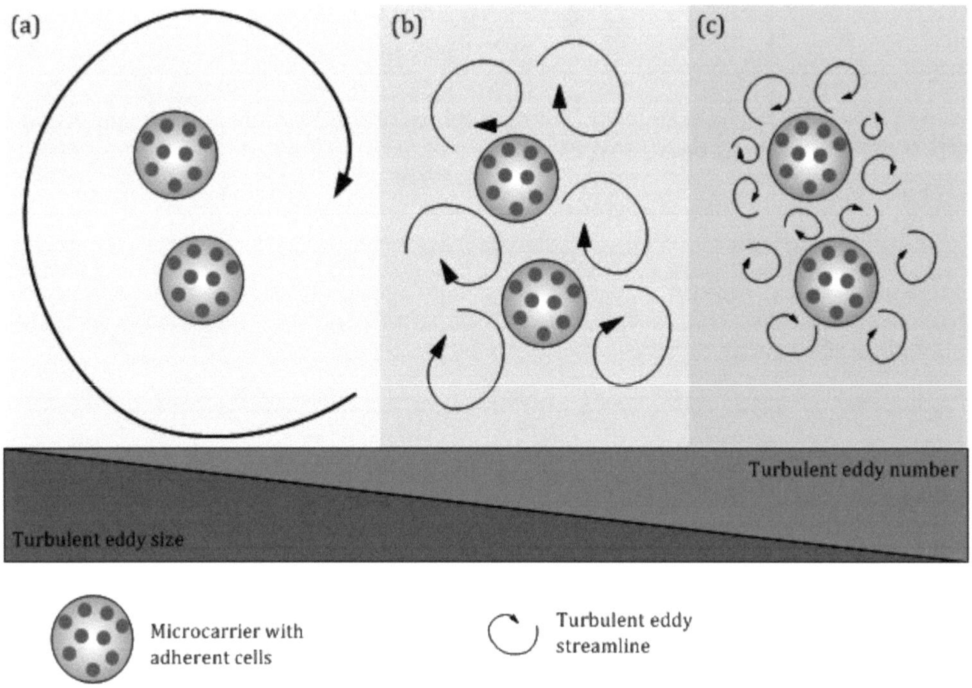

Figure 4. Shear forces on microcarriers in a turbulent flow. Microcarrier-eddy interactions: (a) eddies much larger than beads, (b) multiple eddies the same size as beads, (c) eddies the size same as interbead spacing.

Several concepts to describe the complex shear effects on cells in a turbulent flow were described and evaluated with available data in respect of the impact of shear forces on microcarrier cultures in bioreactor systems (reviewed in [122]). Among others, the concept of an "Integrated Shear Factor" ISF – a measure of the strength of the shear field between the impeller and the spinner flask walls – was developed to describe shear damage to continuous mammalian cell lines [124]. For a stirred bioreactor, the ISF is given by

$$ISF = \frac{2\,\pi\,n_R\,d_R}{D_R - d_R} \tag{6}$$

with rotational speed n_R, vessel diameter D_R and impeller diameter d_R. According to Cherry und Papoutsakis [123] the largest shear stress to which cells grown on microcarriers are exposed occurs in turbulent eddies which are the same size as the microcarriers. The energy of the eddies is transferred to the surface of the microcarriers, resulting in high local velocity gradients between the microcarriers and the fluid, and the highest shear rates on the cells. The microcarriers are caused to rotate within these eddies. Using Kolmogorov´s theory the length scale l_{Kol} of the smallest eddies are in the order of

http://dx.doi.org/10.5772/59385

$$l_{Kol} = \left(\frac{\nu^3}{\varepsilon}\right)^{1/4} \tag{7}$$

where ν is the kinematic viscosity and ε the energy dissipation rate per unit mass

$$\varepsilon = \frac{P}{\rho_{fl} V_L} \tag{8}$$

with power input P and liquid volume V_L. The Kolmogorov eddy length scale corresponds to the diameter of the smallest eddy generated in the bioreactor. In a turbulent environment, eddies break down to form smaller eddies. On the Kolmogorov length scale, viscosity dominates and the turbulent kinetic energy is dissipated into heat. When the Kolmogorov eddy length scale becomes equivalent to the diameter of the microcarrier, movement of the flow lines can shear the cells. The Kolmogorov eddy length scale is affected by stirrer speed, liquid properties and impeller design. Croughan et al. [125] used this concept to describe cell damage in respect of cells grown on microcarriers. Damage became significant when the microscale was about two-thirds the size of the microcarriers, or smaller (discussed by [119]).

Besides the above-mentioned concept for estimation of shear stress in bioreactors, several numbers are used to characterize culture systems (bioreactors) such as (1) geometric dimensions, (2) volumetric power input (P/V), (3) power number (Newton number, Ne), which is defined as

$$Ne = \frac{P}{n_R^3 \, d_R^5 \, \rho_{fl}} \tag{9}$$

for stirred bioreactors (with rotational speed n_R and fluid density ρ), (4) volumetric mass transfer coefficient (k_La) for oxygen or CO_2, (5) mixing time, (6) Reynolds number, which is given by

$$\mathrm{Re} = \frac{\rho_{fl} \, n_R \, d_R^{\,2}}{\eta_{fl}} \tag{10}$$

for stirred bioreactors, (7) impeller tip speed (u_{tip}) for stirred bioreactors, and (8) volumetric gas flow, among others. These numbers can be used to compare process parameters in different bioreactor systems. Recommendations for determination of these parameters have been published by [126] and [127], for example.

The numbers introduced above are not usually suitable for a more in-depth description of fluid flow in a bioreactor. This can be accomplished by computational fluid dynamic (CFD) simulations [128-130]. CFD uses numerical methods and algorithms to solve and analyze problems that involve fluid flows. This is especially important for tissue engineering constructs, in which a set of complex phenomena such as hydrodynamics, nutrient transfer, cell growth, and matrix deposition have to be taken into account [131-133]. Parameters such as fluid velocity, oxygen tension, stress, and strain, which are difficult to determine experimentally, can be derived from CFD simulations and related to cellular parameters [132]. This is advantageous for complex culture systems, where it is impractical or almost impossible to install probes to determine fluid specific parameters (see Section 5).

2.2.2. *Shear stress in MSC cultivations*

Flow-induced shear stress can be stimulating or detrimental to the behavior of MSCs. In the following, firstly a brief review of basic studies is given. These studies were mostly intended to induce a lineage-specific differentiation of stem cells in which the main goal is to engineer a microenvironment for controlled stem cell differentiation [134-136]. Secondly, the impact of shear on stem cells grown in bioreactors for mass cell production is discussed.

2.2.2.1. *Effects of shear stress on stem cell fate*

Shear stress can affect stem cells in different ways [137,138], e.g. with respect to morphology, proliferation capacity, gene expression, cell cycle arrest and apoptosis, proteomic profiling, and differentiation [139-147]. In some cases opposing conclusions have been reported. Adamo and Garcia-Cardena [139] as well as Chang and Wang [140] observed changes in the cell morphology of MSCs due to shear stress (e.g., differences in morphology between static and dynamic culture systems). Changes in morphology can have an influence on cell proliferation potential, as shown in earlier studies by Prockop and coworkers [142]. They reported that smaller agranular cells divided faster than large granular cells, while the doubling time and the differentiation potential remained the same. These results are contrary to those of Luo et al. [143] and Maul et al. [144], who found that shear stress does not support cell proliferation, but leads to cell cycle arrest or a decrease in cell number.

Numerous reports have addressed the induced differentiation of MSCs by shear stress [137], mostly osteogenic [138,148-154] and chondrogenic [155,156]. Meanwhile, a suitable bandwidth for shear effects on MSCs has been determined. In general, increasing shear stress seems to promote osteogenesis and mineralization [137], but some of the findings are again contradictory. This is explainable to some extent through the different cultivation systems or bioreactors used. Studies on the effect of shear forces were performed in either 2-D cultures (flow chambers) with defined shear stress or in perfused macroporous carriers [157]. While in two-dimensional flow chambers (cell growth as monolayer) shear stress levels of 0.5-2 Pa are required for stimulation of osteoblasts [151], which is in the order of the postulated Weinstein shear forces of 0.8-3 Pa in the Haversian channels [158], the results are not readily transferable to 3-D systems. 3-D bioreactors with perfusion or rotational motion (see Section 3) also show the influence of flow on the osteogenic differentiation of osteoblastic cell lines, primary osteoblasts and osteoblast precursor cells [159-163]. However, the calculated shear stresses were in part up to 2-5 orders of magnitude below the values of the 2-D flow chambers. For greater understanding of effects in scaffolds, computer-based simulations are increasingly used to detect the magnitude of shear forces [164,165]. The shear stress acting in the scaffolds is usually in the range of a few mPa. However, this is often not caused by the local shear forces acting on the cells, which are ultimately the important ones, but by those in the bulk flow. Furthermore, it is rarely taken into account that in the various perfusion bioreactors not only the perfusion rate for the development of local flow and shear forces is important, but also matrix properties such as porosity, pore size, interconnectivity of the pores, and elasticity and extensibility of the relevant material [162,166,167]. Moreover, with respect to tissue engineering, fluid dynamic calculations are rarely coupled with mass transfer effects. In principle, it cannot therefore be excluded that the effects observed are not due to a mechano-stimulation of the cells, but to improved mass transport. Furthermore, different culture conditions such as medium composition, type of serum, oxygen concentration etc. might also play a role (see above).

2.2.2.2. Effect of shear stress on stem cells grown in bioreactors for cell expansion

For high level expansion of stem cells, bioreactor cultivations are required that support the required number of cells. This involves minimal variations in lineage specific differentiation while the genetic and epigenetic stability of the cells needs to be kept under control [132]. In order to ensure this, the environment acting on the cells in a bioreactor needs to be characterized and kept similar during upscaling of the process. To date, microcarrier-based suspension cultures have been studied mostly for expansion of stem cells (see above) as they provide a sufficient surface area for attachment and expansion [119,168]. Microcarriers are preferably suspended in stirred bioreactors (see Section 3.2), where shear-influenced stem cell differentiation and shear-induced cell damage can be distinguished. The damaging effects of flow stress in microcarrier cultures were studied in the 1980s and 1990s mostly in relation to permanent or established cell lines [124]. Recent publications show that stem cells are affected by shear stress below damaging levels (reviewed by [132]). With respect to cell expansion, preferably without significant unwanted differentiation, a narrow band of shear stress levels is essential [169]. Turbulent flow in stirred bioreactors, however, represents a scenario that is not characterized by a narrow band of shear stress levels. The effect of broadband stresses associated with turbulence in stirred bioreactors on preservation of the differentiation potential of stem cells and minimization of lineage commitments is still poorly understood [119,132,170-172]. Furthermore, controversial observations on the shear stress tolerance of the cells might be due to protective substances contained in the medium, such as serum or differences in the type of surface or cell density. Thus, for example, cells grown in the pores of macroporous carriers are better protected against shear than cells grown on non-porous carriers [63,67,173].

To summarize, on the one hand shear stress can be deliberately used for targeted differentiation of stem cells. On the other hand, with respect to stem cell expansion, shear effects need to be minimized to prevent an unwanted differentiation. Therefore, in order to optimize process design, the specific task of the cultivation or the target product should always be kept in mind. During production of cell therapy products, cell expansion and harvest are in the foreground, and cell differentiation is generally undesirable. In the discussion that follows, the focus is placed on epansion of MSCs and suitable bioreactors for cell therapy. Based on a classification of the power input type and operating principles, the most common bioreactors are presented and cultivation results are discussed.

3. Bioreactor systems for cell therapeutics

3.1. Classification

Since the early 2000s successful MSC expansions have been reported. In addition to static bioreactors (Figure 5 A), different types of dynamic bioreactor versions (Figure 5 B) have proven themselves. 2-D cultivations performed in planar systems such as petri dishes, T-flasks, and stacked plate systems are still predominant in both the development and production of cell therapeutics. Stacked plate systems (e.g., CellSTACKs®, CellFactories) with 10-or 4-layer vessels are typically made of polystyrene and are non-instrumented, and have become the system of choice in semi-commercial and commercial production processes [31,174]. Their mass and energy transfer is exclusively caused by conduction and reaction processes within the planar system and by interaction with an environment that is typically temperature and humidity controlled. In other words, gas exchange only takes place at liquid-gas interphase and concentration gradients can appear during cultivations. Not surprisingly, culture broth inhomogeneities, which influence cell yield and quality, increase as the number of layers in the vessel rises. Moreover, manipulations become more complicated and risk of contamination also increases. In spite of additional attempts to apply robotics and bioreactor control to static bioreactors [175,176], dynamic bioreactors ensuring 2-D or 3-D cell growth remain superior.

In dynamic bioreactors the power input generated is responsible for mass and energy transfer. To date, a high number of different dynamic bioreactors have been used to expand MSCs. According to the type of power input, they can be divided into mechanically driven, pneumatically (air-) driven and hydraulically driven systems. Mechanically driven bioreactors include stirred systems (Section 3.2), wave-mixed systems (Section 3.3), and a recently introduced rotating bed bioreactor (Section 3.4) [177]. In actual fact, rotating bed (or wall vessel) bioreactors [177,178] are important for tissue engineering applications aimed at cell differentiation (and not at cell expansion). Pneumatically driven versions such as the 3 L Air-Wheel bioreactor from PBS Biotech® [179] are excluded, although a poster presentation on Coronado Island recently showed that this bioreactor shows even slightly higher cell expansion rates than a stirred system in MSC expansion runs. More often used are hydraulically driven representatives, which include parallel plate bioreactors (Section 3.5), hollow fiber bioreactors (Section 3.6), and fixed bed bioreactors (Section 3.7), for which power input is generated by pumps.

In the case of the mechanically driven bioreactors, microcarriers (Section 2.1) or perfused plates have so far provided the growth surface for the MSCs. In hydraulically driven bioreactors the cells are grown either on perfused plastic surfaces (Section 3.5), hollow fibers (Section 3.6) or packed particles (Section 3.7). It is worth mentioning that the majority of dynamic bioreactor types were originally developed for production processes aimed antibodies and vaccines. For this, genetically modified, continuous mammalian cell lines such as Chinese hamster ovary (CHO) cells are generally grown [180-182]. With the exception of spinner flasks (as simple stirred bioreactors), mechanically driven bioreactors are generally instrumented with sensors to enable monitoring and control of temperature, pH value, DO and carbon dioxide concentration, gas and liquid flow rates, impeller or rocking speed.

A second classification trial is based on the number of uses of the bioreactor`s cultivation vessel and distinguishes between reusable and single-use bioreactors. Whereas the vessel of a reusable bioreactor is made of glass or stainless steel, single-use bioreactors have a plastic cultivation vessel. This cultivation vessel is pre-assembled, beta-or gamma-irradiated and, then delivered as ready-to-use. After one use, as the vessel is a rigid polystyrene or polycarbonate container or a flexible multilayer bag with contact layers made of polyethylene or ethylene vinyl acetate, it is discarded [183]. For this reason single-use bioreactors are also often referred to as disposable bioreactors. If single-use bioreactors are operated correctly for the production of high value products at small and medium volume scale (as is the case for productions of cell therapeutics), they contribute to savings in time and costs. Furthermore, process flexibility and safety can be increased, which is an advantage in processes where the cells produced are the target product that is directly given to the patient.

Different authors have described adverse effects on CHO cells grown in polyethylene bags [184-188]. They found poor cell growth arising from leachables such as bis(2.4-di-tert-butylphenyl)phosphate. Such leachables are substances that can migrate from bag layers during processing and leach out into the culture broth in concentrations that are deleterious to cell growth. This phenomenon has not been reported for MSC expansions to date.

There is no doubt that advantages currently prevail if an appropriate single-use bioreactor (regardless of whether it represents a static or dynamic system) is applied for productions of cell therapeutics. This, together with the suitability of single-use bioreactors for commercial production processes with Good Manufacturing Practice (GMP) and regulatory demands, explains their increasing usage. In particular, dynamic systems including stirred, wave-mixed, parallel plate, hollow fiber and fixed bed single-use bioreactors are currently the focus of interest for producers of cell therapeutics.

3.2. Stirred single-use bioreactors in microcarrier-based MSC expansions

In stirred single-use bioreactors, mixing is induced by the mechanical agitation of one or more impellers. For research purposes disposable spinner flasks operating with magnetically driven, slowly moving paddle impellers (60 rpm) are most often used [189]. The cells are propagated on microcarriers until working volumes of 250 mL. In contrast to scalable, automated stirred bioreactor versions, spinner flasks are only surface-aerated. Furthermore, spinner cultivations entail availability of incubators for temperature as well as humidity control. Nevertheless, peak viable cell densities of between $6 \cdot 10^5$ and $1 \cdot 10^6$ cells mL^{-1} and maximum cell expansion factors between 30 and 50 are achievable in spinner flasks when the culture medium contains serum and is cyclically perfused (by manual partial and periodical medium exchange). This was

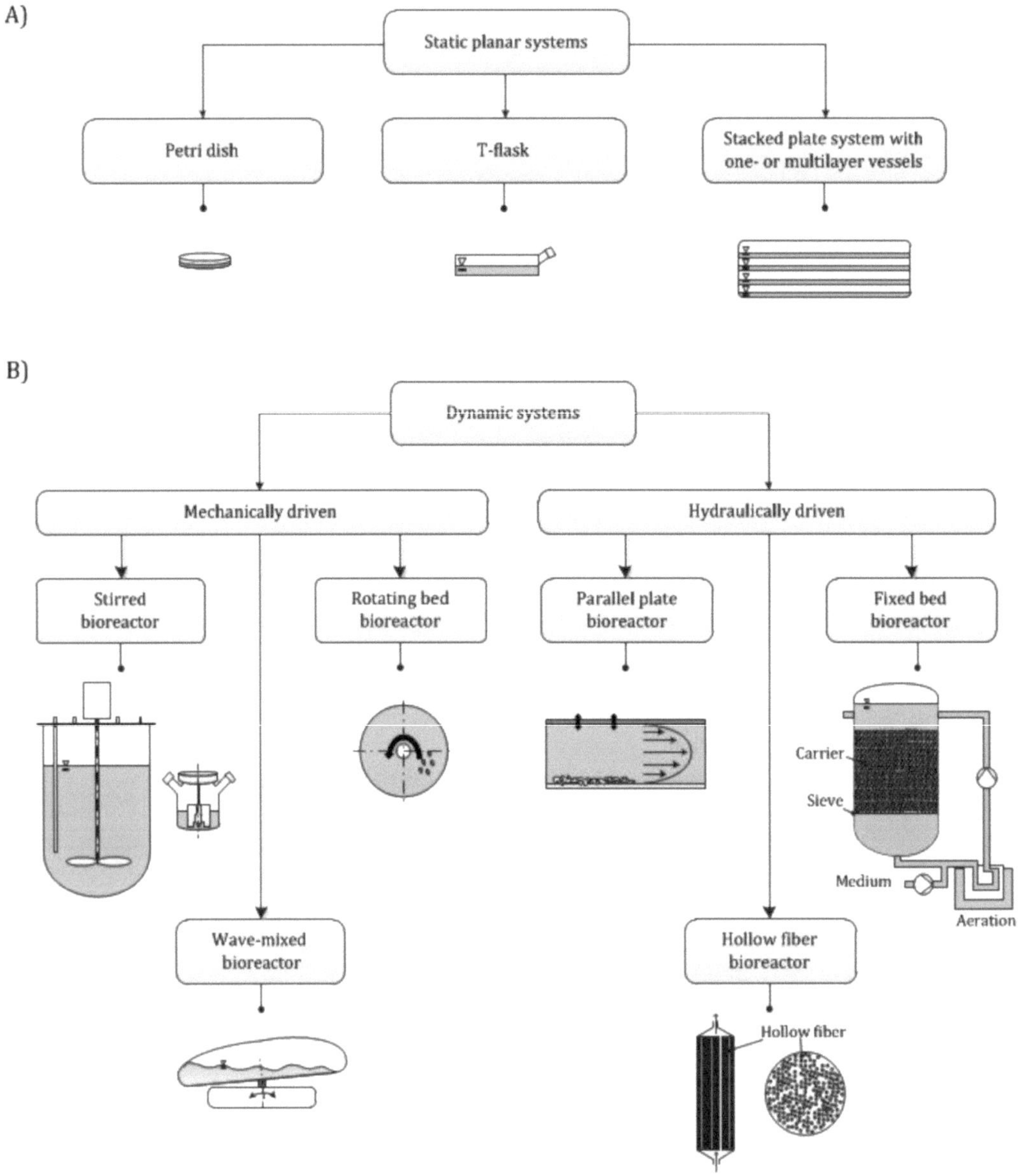

Figure 5. Suitable bioreactors for MSC expansion. Presented classification bases on the power input type.

demonstrated for hADSCs and hBM-MSCs expanded on polystyrene-and gelatin matrix based microcarriers in our lab over a cultivation period of 7 days.

At benchtop scale MSC expansions were successfully carried out in the Mobius® Cell Ready 3L bioreactor (Merck Millipore) and the BIOSTAT® UniVessel 2L SU (Sartorius Stedim Biotech). The key element of both systems is the rigid polycarbonate vessel, contains one or two rotating impellers and an aeration device (sparger). The user can choose between plastic vessels having reusable or single-use temperature, DO and pH probes.

The Mobius® CellReady bioreactor (maximum working volume of 2.4 L) is equipped with a single top-driven marine impeller and an open pipe or micro-sparger. According to its configuration, the main bioengineering parameters were investigated and summarized by Kaiser et al. 2011 [190]. The published results of the bioengineering studies with the Mobius® CellReady 3L show that this system is also usable for MSC expansion. Suitable impeller speeds are adjustable, which enables culture homogeneity while avoiding microcarrier sedimentation, cell differentiation and cell damage that may result from too high shear stress. For hBM-MSCs the accuracy of this hypothesis was demonstrated by the studies of Cierpka et al. [191] and Jing et al. [192]. In cultivations running in serum-supplemented medium (10 % FBS) for between 12 and 14 days, peak viable cell densities between $2.5 \cdot 10^5$ and $2.7 \cdot 10^5$ cells mL^{-1} were obtained. Moreover, Stadler [193] and Ott [189] achieved a serum-reduced expansion of hADSCs within 6 days, which was comparable to a 250 mL reference spinner flask (100 mL working volume). In spite of reduced cultivation time and serum content, the peak viable cell density was more than double ($5.5 \cdot 10^5$ hADSCs mL^{-1}) when compared with the results of Cierpka et al. [191] and Jing et al. [192].

The top-driven UniVessel® SU 2L bioreactor (maximum culture volume of 2 L) has two 3-segment blade impellers and an L-shaped macro-sparger. Similarly to the Mobius® Cell Ready it is hydrodynamically well-characterized [194,195]. Peak viable cell densities between $1.8 \cdot 10^5$ hBM-MSCs mL^{-1} and $3 \cdot 10^5$ hADSCs mL^{-1} were determined in serum-supported production processes on cultivation day 9 [189,196]. In order to increase MSC expansion efficiency in the UniVessel SU 2L, Jossen et al. [197] improved the fluid flow while reducing fluid shear stress for cultivations with higher microcarrier amounts. This was accomplished by modifying the standard vessel design (increasing impeller blade angle from 30° to 45° and reduction of off-bottom clearance from 0.41 to 0.26). As a result, the maximum specific power input was reduced by a factor of 2 and the peak viable cell density was more than 3 times higher than that of the hBM-MSCs, for which a cell expansion factor of 35 was reached.

Recently, the propagation of hADSCs on microcarriers in stirred bag bioreactors exceeding benchtop scale was carried out for the first time. Schirmaier and coauthors [196] worked with Sartorius Stedim`s top-driven BIOSTAT® CultiBag STR 50L, for which there is comprehensive bioengineering characterization [195]. They used the 3-D bag version shown in Figure 7 E (Section 5), in which two 3-segment blade impellers and a micro-sparger have been implemented. Cultivation was executed with 35 L working volume while realizing a partial medium exchange on day 4. Growth on polystyrene microcarriers at serum-reduced conditions (in this case, 5%) allowed $3 \cdot 10^8$ hADSCs to be harvested on cultivation day 9. This provided the basis for the scale-up approach explicated in Section 5.

3.3. Potential of wave-mixed bioreactors for microcarrier-based MSC expansions

MSC expansions have already been carried out in wave-mixed bioreactors, which have a pillow-like culture bag with single-use sensors for temperature, pH and DO control. The bag is fixed on the rocker platform, which makes a one-dimensional (1-D) oscillatory movement.

By rocking the platform a wave is induced in the bag, which contains medium and cells. In this way, the medium surface is continuously renewed while bubble-free oxygen is introduced into the culture broth from the headspace of the bag. Wave generation and propagation, and thus mass and energy transfer, in this bioreactor type are dependent on the rocking rate, rocking angle, aeration rate, filling level (50% maximum) and culture broth viscosity. These parameters affect the fluid flow, mixing time, oxygen mass transfer, shear stress acting on cells, and finally the cultivation result. As discussed by Eibl et al. [198] and Werner et al. [199], shear stress is highest at the lowest filling level together with the highest rocking rate and rocking angle. Indeed, energy dissipation and shear stress pattern were more homogeneous in wave-mixed bioreactors with 1-D motion than in stirred cell culture bioreactors with a paddle impeller. For this reason, these wave-mixed bioreactors are well-suited for productions with shear sensitive cells. A further advantage of wave-mixed bioreactors is their negligible foaming, meaning that there is no need for antifoam agents to be added, which also simplifies downstream processing of the product. In addition, protocols for microcarrier-based productions with different continuous mammalian cell lines are available as well as for primary cells known to be very sensitive to shear and difficult to culture (such as T-cells) [200-204].

The first attempts published demonstrate the feasibility of expanding MSCs in 2 L wave-mixed bags on Cytodex 3-and Cultispher-S-microcarriers. Akerström [205] propagated MSCs over 18 days while feeding carriers on day 11 and 13 into a Wave Bioreactor 2/10 (GE Healthcare). The expansion factors determined after trypsinization were around 6. Timmins et al. [206] achieved expansion factors around 16 on cultivation day 7 when producing MSCs derived from the placenta at reduced O_2 levels (5%). He published a procedure which was realizable in wave-mixed bioreactors from both GE Healthcare and Sartorius Stedim Biotech.

3.4. Single-use rotatory bed bioreactors

An interesting new approach for a perfused dynamic bioreactor represents Zellwerk´s rotating bed bioreactor system ZRP [177]. As shown in Figure 6, a cylindrical culture vessel has a rotating bed of polycarbonate plates. Bed rotation is caused by a non-contact magnetic drive coupled to the culture vessel. The culture vessel is equipped with sampling ports and a measuring device with ports for pH and DO sensors. The external media circulation combined with the rotation of the polycarbonate bed ensures mixing of the culture medium. To provide an oxygen supply, the overlay atmosphere of the headspace of the vessel is aerated with CO_2, N_2 and air. All cultivation parameters such as pH, DO, temperature and bed rotation are monitored online and regulated by a control unit. Reichardt et al. demonstrated a 39 fold expansion of cells derived from human umbilical cord arteries in 9 days [177].

3.5. Single-use parallel plate bioreactors

Whereas stirred and wave-mixed bioreactors used for the expansion of MSCs are operated with microcarriers in fed batch mode (feeding or partial medium exchange), parallel plate bioreactors run carrier-free and in continuous perfusion mode. Single-use parallel plate bioreactors have a modular multiplate design. The plates are typically made from polystyrene, whereby each plate consists of two compartments separated by a gas-permeable membrane

Figure 6. The rotating bed bioreactor system. The bioreactor system is composed of a cylindrical culture vessel and an integrated bed of polycarbonate plates with a cell culture surface area of up to 6000 cm^2 (with courtesy of Zellwerk Gmbh; Oberkrämer, Germany).

(that is impermeable to liquids) [207]. Whereas the upper compartment is filled with air, the bottom compartment contains the cells. The cells grow on the plastic surface of the bottom compartment as monolayer (2-D growth) while being continuously supplied with culture medium.

In general, plate bioreactors have been well-investigated in terms of shear stress, which is regarded to be low in their case. In 1996 Peng and Palsson [208] studied the influence of their geometry on fluid flow and the resulting growth and differentiation of bone marrow-derived stem cells. They found a superiority of radial flow and ascribed this finding to the uniform environment caused by the hyperbolic velocity and tube-like shear stress contribution. Due to the absence of walls in the flow path and the location of the boundary walls parallel to the flow direction, slow flowing regions were obtained on the growth surface of radial flow-type parallel plate bioreactors.

Prominent representatives of single-use parallel plate bioreactors are Corning`s E-Cube™ System and Pall`s Integrity™ Xpansion™ Multiplate Bioreactor. The E-Cube, which is available with 10-or 25-stack modules requires an incubator for operation. It offers a maximum growth surface of 21,250 cm^2. The Integrity™ Xpansion Multiplate Bioreactor is a self-contained bioreactor system that is obtainable with 10 up to 200 plates and provides a maximum growth surface of 122,400 cm^2. Its gas exchange is realized in a central column with channels along the plates through which the medium circulates. CFD studies realized with the Integrity™ Xpansion™ Multiplate Bioreactor revealed the occurrence of gentle laminar flow. Maximum wall shear stress did not exceed 10 mPA, which was up to 1000 times lower than in stirred bioreactors (personal communication W. Kuhlmann, ATMI Life Sciences, June 2013).

Bone marrow-derived MSCs have been successfully expanded without differentiation in this single-use bioreactor type, which generally allows production of up to a billion cells per batch. By using holographic microscopy in cultivations with the Integrity™ Xpansion™ Bioreactor, cell morphology can be monitored on the 10 top plates. Normally, the lack of opportunity to monitor cells is regarded as a drawback of parallel plate bioreactors [209].

3.6. Hollow fiber bioreactors in MSC expansion procedures

Similarly to parallel plate bioreactors, hollow fiber bioreactors provide a low shear stress environment [210] and cell propagation in continuous perfusion mode. Hollow fiber bioreactors are characterized by a high surface area-to-volume ratio (100-200 cm^2 mL^{-1}) and create 3-D environment for cells. They consist of a bundle of parallel hollow fibers constructed from cellulosic, polysulfone, polypropylene or polyethylene materials, which are encased in a cylindrical polycarbonate cartridge with ports for flow around the fibers. The pore size of the semi-permeable hollow fiber membrane determines which molecular species are rejected.

The beginning of hollow fiber technology dates back to the early 1970s [211], when Knazek and his team used tubular membranes upon which cells were able to expand to high cell densities (10^7 to 10^8 cells mL^{-1}) and which were impermeable to the cells. Analogically, the fiber matrix was permeable to gases, nutrients and metabolic waste products. Cells suspended in the culture medium settled on the outer surfaces of hollow fibers while oxygenated culture medium flowed continuously through the fibers. Medium nutrients diffused through the fiber wall into the cells and metabolic waste products diffused from the cells through the fiber wall into the perfused liquid.

Since then, hollow fiber bioreactors have been improved and used extensively, for example to expand lymphocytes [212,213], to produce glycoproteins (in particular, antibodies) and viruses [214-221], to cultivate hepatocytes and as extracorporeal assist devices [222-224]. Modules were designed which distributed the main fluid flow equally to each hollow fiber in the bundle [225]. This generated a parabolic velocity profile in the fiber lumen and resulted in uniform shear stress for the attached cells, which was directly proportional to the intracapillary flow rate [211,212]. Cell adhesion was enhanced by pre-coating the hollow fibers with one or even more extracellular matrix proteins such as fibronectin or collagen [226].

The use of hollow fiber bioreactors to produce clinically relevant numbers of MSCs (total cell yields between 10^8 and 10^9 cells) has been reported in numerous studies [227]. These cell amounts are sufficient for autologous and selected allogeneic therapies. Exemplarily, we would like to refer to serum-supplemented productions recently performed with the FiberCell System (FiberCell Systems, max. growth surface of 2.5 m^2) and the Quantum Cell Expansion System (Caridian BCT, maximum growth surface of 2.1 m^2). The articles describe expansion of MSCs isolated from human bone marrow and placenta over a few weeks [210,228-230]. Nonetheless, hollow fiber bioreactors and fixed bed bioreactors subsequently described are restricted in scalability. They both have limitations in that their nutrient and oxygen gradients in culture restrict the length of the fiber and height of the fixed bed. The biggest challenge, however, for both reactor types is posed by cell harvest.

3.7. Single-use versions of fixed bed bioreactors and MSC expansions

Initial work with fixed bed bioreactors (also referred to as packed bed bioreactors) and mammalian cells started in the 1950s. In fixed bed bioreactors cells are immobilized on or encapsulated in a stationary matrix consisting of particles densely packed in a cylindrical vessel. Several types of particles have been used to date: Macro-porous microcarriers, porous ceramic beads, porous glass beads, glass fibers, polyester discs, alginate beads and hydrogels [62,231-236]. The culture medium is perfused through the bed of the bioreactor and supplies cells with nutrients while removing undesired metabolites. For this purpose, either an external medium reservoir is used or the culture medium is circulated in an internal loop. When fixed bed bioreactors are run in continuous perfusion mode, cell densities of around 510^8 cells mL^{-1} can be achieved. Knowledge about configurations of fixed bed bioreactors and their bioengineering characteristics was summarized by Warnrock et al. [232].

Fixed bed bioreactors have been successfully employed in different investigations with primary cells, such as bioartifical liver support systems and stem cell expansions. It is a well-known fact that the company Pluristem Therapeutics, a leading developer of placenta-based cell therapeutics, uses reusable fixed bed bioreactors (PluriX 3-D bioreactors) in combination with Fibra-Cel® disks. Fibra-Cel® disk carriers are characterized by high porosity and composed of two layers of non-woven polyester and polypropylene [236,237]. These fiber carriers also form the fixed bed in Eppendorf`s single-use version of the BioBLU® bioreactor, which is obtainable up to a vessel volume of 5L. When pre-loaded with 150 g of Fibra-Cel® disks, 0.12 $m^2$ of effective surface area per gram of disks is provided. The largest single-use fixed bed bioreactor is Pall`s Integrity™ iCELLis™ bioreactor, which has a maximum growth surface of 500 m^2 and a fixed bed made from polyethylene terephthalate microfibers [238]. But as in the case of the BioBLU® SU bioreactor, no reviewed literature about MSC cultivation in the Integrity™ iCELLis™ was found at the time of writing, with the exception of an mL-scale application in which a single-use fixed bed bioreactor was used. This concerned a syringe-like fixed bed bioreactor for the cultivation of implantable immortalized hMSCs expanded on alginate beads and applied in cell therapy trials for stroke treatment [1,239].

4. Scale up of MSC cultivations: General considerations

In planar cultures, the growth of the adherent cells has to date been limited by the available growth surface area, which is determined by the geometry of the cultivation vessel. Scaling up of planar static cultures is therefore realized by increasing the growth surface (linear scale-up), either by working with several bioreactors in parallel and/or using multilayer systems. However, as already pointed out in Section 3, with an increasing number of layers and/or bioreactors the risk of contamination and vessel-to-vessel variance rises. It is assumed that lot size of planar culture systems are capped at 1-4 10^{11} cells, whereby lot sizes of > 1 10^{11} cells are not readily achievable without massive automation and parallel processing [174].

The available growth surface area also limits the lot sizes in dynamic parallel plate, hollow fiber bioreactors and fixed bed bioreactors, whereas concentration and temperature gradients are less likely in these systems because of the convective transport. Due to the limited engineering data available for these systems, no systematic scale-up studies have been published so far and scaling-up is mainly realized using *trial-and-error* methods. In contrast, several scale-up rules have now been accepted for stirred bioreactors which is why the main focus of the following is on this bioreactor type. Since it is not possible to simultaneously maintain identical operational characteristics such as mixing time, power input, impeller speed, carbon dioxide removal and oxygen mass transfer over the different scales, the most important factor(s) has (have) to be identified.

In biopharmaceutical production processes, the impeller tip speed (u_{tip}) is a frequently used scale-up criterion [240,241], which is directly proportional to the impeller diameter and the rotational speed. Typically, the impeller tip speed correlates well with the maximum fluid velocities, and consequently, the maximum shear stresses, as long as low aeration rates are used. Otherwise, local shear stresses from rising bubbles and/or bursting bubbles at the liquid surface may exceed impeller-induced shear by several orders of magnitude [242]. Based on model devices, non-lethal responses of hMSCs on shear stresses, including changes in gene expression, signaling pathways and morphology as well as cell differentiation, have been reported over a wide range of shear from 0.01 to 5 Pa [138,145,146,152,243], depending on the cell source and donor as well as the experimental setup (used device, exposure time, medium etc.).

However, the impeller tip speed does not account for the actual impeller type nor volume changes during the process (e.g. by feeding or medium exchange). Particularly at larger scales, heterogeneity in the culture environment due to gradients in dissolved gases and metabolites can become an issue. Another possible scale-up criterion is the mixing time [244]. Based on the turbulence theory, the mixing time was suggested to be independent of impeller type and was found to correlate with specific power input and geometrical parameters such as the impeller and vessel diameter and the filling height, which is valid for single impellers in draw and fill mode as well as multi-stage impellers [245]. In comparative studies with different stirred single-use bioreactors, mixing times of between approximately 44 s and 86 s for a constant tip speed of 0.5 m s^{-1} were found [246]. In single-use bioreactors there is still a lack of systematic comparisons of mixing that take geometric parameters into account. Furthermore, it should be emphasized that keeping mixing time constant during scaling-up results in significant increases in specific power input at larger scales [247,248].

For cells growing in suspension, the most frequently used scale-up criterion is the specific power input, which can be predicted from the impeller power number (Ne) according to Eq. 9, where ϱ_{fl}, n_R, d_R and V_L represent liquid density, the impeller speed, the impeller diameter and the liquid volume respectively.

$$P / V = \frac{Ne\, \rho_{fl}\, n_R^{\ 3}\, d_R^{\ 5}}{V_L} \tag{11}$$

The power number is specific for each impeller type and depends on the Reynolds number (Re), which was introduced in Section 2.2. In general, radial flow impellers exhibit greater power numbers than axial flow impellers, which therefore need to rotate more quickly in order to achieve the same power input (see Eq. 11). Reported power numbers for stirred single-use bioreactors range from 0.3 (Mobius® CellReady 3L) to 4.2 (Mobius® CellReady 50/200L), whereas the UniVessel SU 2L with two segment blade impellers has a power number of 1.1 [249].

For microcarrier-based processes, suspension criteria (N_{S1} and N_{S1U}) have been proposed for scaling-up, since they provide effective use of the available growth surface of the microcarriers. Both suspension parameters, which describe the impeller speed required to bring solid particles (the microcarriers) into suspension, were introduced more than 50 years ago [250-252], although their potential for microcarrier-based stem cell expansions and their scale-up was only reported quite recently [119]. The N_{S1} criterion displays the impeller speed at which all particles are just fully suspended, whereby a homogeneous dispersion of all microcarriers is not a necessary consequence. The N_{S1U} criterion is its lower limit and thus the impeller speed required to locate the particles at the bottom of the bioreactor with none of them at rest. Comparing different stirred SU bioreactors from small to pilot scale, the lowest impeller speeds, and, thereby power inputs required to lift up polystyrene-based microcarriers with solid fractions of up 0.2 % were found in the Mobius® CellReady bioreactor [182]. However, the suspension criteria also strongly depend on the microcarrier type used. Thus, they should be determined as part of process development for each individual process (see also Section 5.3).

Other agitation-related scale-up factors such as the Reynolds number or Kolmogorov's microscale of turbulence are more seldom used in scale-up studies. Because of the low oxygen demands of hMSCs (1.2-3.8$10^{-17}$ mol oxygen s^{-1} $cell^{-1}$) [253,254] and the relatively low cell density, the volumetric oxygen demand can often be covered by surface aeration. Therefore, scale-up factors related to oxygen mass transfer, such as the specific oxygen transfer coefficient (k_La), are also considered to be less important. However, they may increase in importance at increasing scales and higher cell densities.

5. A new scale-up approach for expanding hMSCs

5.1. Motivation

In this section, we introduce a new scale-up approach and show its suitability for rapid and efficient expansion of hADSCs (Figure 7 A) and hBM-MSCs (Figure 7 B). Both cell types are of therapeutic interest as the increasing number of clinical trials presented on www.ClinicalTrials.gov indicates. There is no doubt that alternatives to planar systems are required. Although, vessel versions with 10-or 40-layer units providing cell yields between $2 \cdot 10^8$ and $1 \cdot 10^9$ cells [31] have been established, a consistently high cell quality at high cell numbers is difficult to ensure. This explains the shift to suspension cultures, particularly using microcarriers and stirred instrumented bioreactors, when scalable hMSC production is the focus.

Our scale-up strategy (Figure 7), which was aimed at cell yields around $1\cdot10^{10}$ hMSCs and expansion factors exceeding 30, is based on two pillars: (i) screening studies performed in Corning spinners (Figure 7 C) at mL-scale and (ii) bioengineering investigations carried out for the spinner system and the pilot bioreactor system chosen (Figure 7 D and Figure 7 E). The bioengineering studies of both cultivation systems included CFD simulations and result verification by Particle Image Velocimetry (PIV). Their aim was to predict fluid flow pattern and fluid flow velocities to enable calculation of local shear stress distributions, turbulent dissipation rates and Kolmogorov`s micro-scales.

Together with the microcarrier distributions, these parameters were considered in order to predict optimum impeller speeds while avoiding high shear stresses, which finally result in cell differentiation or even cell death. As proposed by Kaiser et al. [255] the suspension criteria N_{S1} and N_{S1U} were employed to transfer the process from 100 mL to 35 L and 50 L working volumes, respectively.

It is worth mentioning that all the cultivations ran between 6 and 9 days and were executed in fed batch mode at low-serum conditions. Reducing the serum-content in the culture medium or using serum-free or chemically defined culture media contributes to process cost reductions, and simplifies downstream processing and product approval [65,256]. The two cell types used were provided by the Lonza Cologne GmbH, Germany. They were each cryopreserved (second passage, population doubling level of 10) and originated from a single consenting, informed donor. In all hMSC production processes, daily sampling and sample analyses in accordance with the descriptions of Schirmaier et al. [196] and Jossen et al. [197], were executed. In addition to the determination of cell density, viability, concentrations of glucose, lactate, glutamine and ammonia, flow cytometric and apoptotic investigations were performed.

5.2. Investigations at mL-scale

In total, seven different culture media from different suppliers with varying contents of FBS (0-10 %) and six different microcarrier types (different densities and diameters) were used in the comprehensive spinner screening. The results for both hMSC types (cell growth, substrate and metabolite courses and cell expansion factors) achieved in the Corning spinners and 2-layer Corning CellSTACKs, were compared with those of cultivations in standard DMEM (Dulbecco's Modified Eagle Medium) with 10 % FBS. While propagating the cells at microcarrier solid fractions of 1.43 % (polystyrene matrix-based microcarriers) and 0.3 % (gelatin matrix-based microcarriers) over 6 days, the culture broth was stirred intermittently or continuously at 60 rpm (37°C, 5 % CO_2 and 80 % humidity). This impeller speed had been previously determined in spinner experiments aimed at cell yield optimizations in standard DMEM medium containing 10 % FBS. Serum-free expansion (0 % FBS) of both cell types was possible, but the expansion factors achieved were below 10 and regarded as insufficient. Lonza

http://dx.doi.org/10.5772/59385

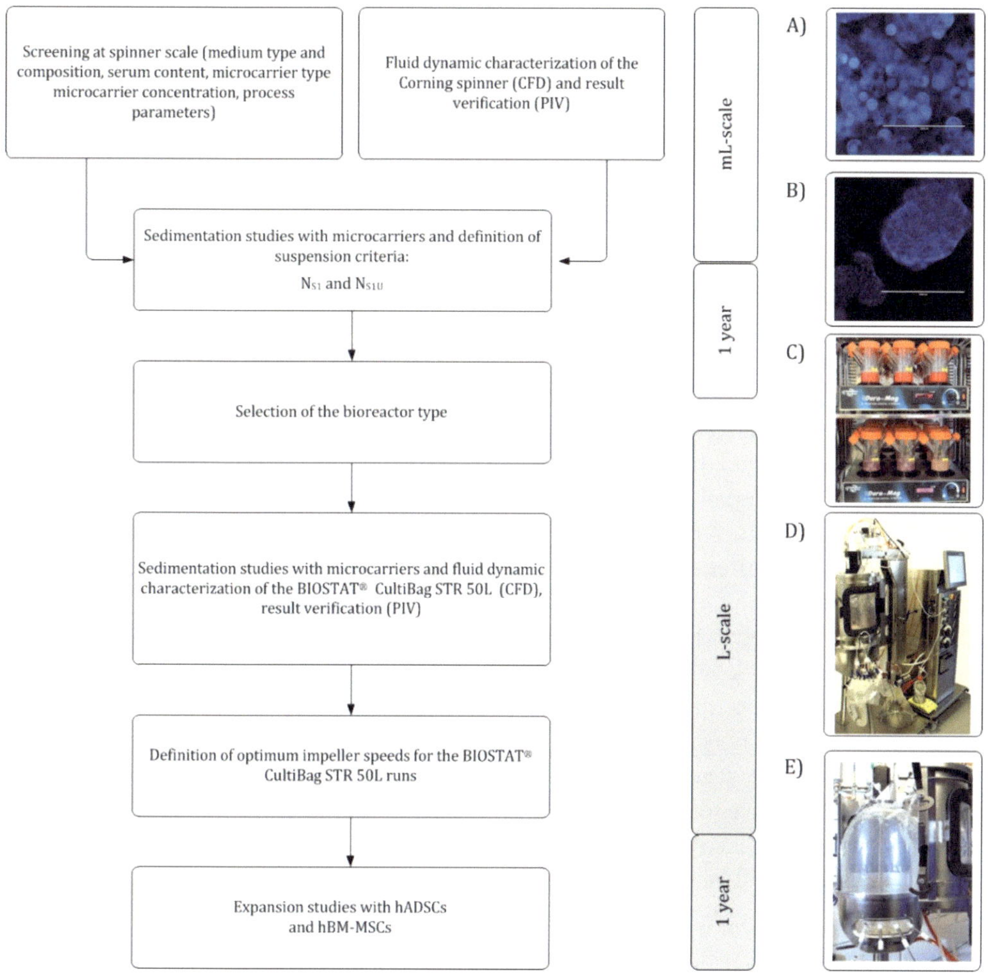

Figure 7. Our expansion approach with (A) DAPI staining of microcarriers realized in hADSC expansions with polystyrene-based carriers in the BIOSTAT® CultiBag STR 50L on cultivation day 7 (0.71 % solid fraction), (B) DAPI staining of microcarriers realized in hBM-MSC expansions with gelatin-based carriers in the BIOSTAT® CultiBag on cultivation day 7 (0.21 % solid fraction), (C) 125 mL spinner flasks equipped with top-mounted blade impeller with diameter of 41.5 mm and operated with 100 mL working volume, (D) BIOSTAT® CultiBag STR 50L during one experimental run and (E) its polyethylene multilayer bag in which two 3-segment blade impellers with a diameter of 143 mm, a microsparger, and single-use pH-and DO-probes were implemented. The bag have being fixed and shaped by a stainless steel support container. Scale bars in (A) and (B) indicate 1000 m.

stem cell medium provided the most promising results and allowed satisfactory growth at low serum-content (5 % FBS) and at continuous stirring, when hADSCs were propagated on the polystyrene carriers and hBM-MSCs expanded on the gelatin carriers with the lower density. A further improvement in the cell expansion factor (60 instead of 40) and the maximum cell yield ($1 \cdot 10^8$ cells instead of $4 \cdot 10^7$ cells) was achieved by: (1) modifying the Lonza stem cell medium and realizing one single feed (with a feeding solution) instead of performing a single 50 % medium exchange between day 3 and 4, (2) increasing the inoculum density and (3) increasing the microcarrier concentration.

The transfer of fluid flow at the optimum cultivation conditions found in the spinner presupposed CFD modeling. To date, several studies have demonstrated the usefulness of CFD in analyzing fluid dynamics at both micros-and macroscopic levels in stem cell bioreactors [132, 255,257]. We modeled the fluid flow inside the Corning spinner flask by using the finite volume solver Fluent from ANSYS (ANSYS Inc., version 14.0, USA) and calculated the three local velocity gradients as well as their magnitude. For more detailed information the interested reader is referred to Kaiser et al. [255]. As exemplarily shown in Figure 8 for 60 rpm and 100 mL working volume, the flow in the spinner flask was primary tangential with low axial and radial velocity components. Due to the wide impeller blade and the absence of baffles, the observation that the tangential velocity was the highest velocity component comes as no surprise. As also expected the velocities were highest at the edges of the impeller blade as well as at the tips of the impeller bars and correlated directly with u_{tip}. In contrast, the lowest velocities were determined below the impeller and near the vessel wall. It was also found that sedimentation and agglomeration of the microcarriers were most likely to occur below the impeller bar. This finding was also confirmed by the microcarrier distribution in the spinner flasks, which is depicted in Figure 9. Furthermore, at 60 rpm and 100 mL working volume the mean local shear stress level (τ_{nt}) was $4 \cdot 10^{-3}$ $N \cdot m^{-2}$ and the maximum value was estimated to be 0.2 $N \cdot m^{-2}$. The mean turbulent energy dissipation rate (ε) was $1.1 \cdot 10^{-3}$ $m^2 \cdot s^{-3}$ and the maximum turbulent energy dissipation rate was about 10 times higher. The mean and minimum Kolmogorov's microscales of turbulence were 60 μm and 191 μm, respectively. However, we would like to point out that Kolmogorov's theory is only valid for high turbulence ($Re \geq 10^4$) and because Re was 1722 under the process conditions used, the Kolmogorov's values estimated should be viewed critically.

As various mathematical models are available for the simulation of fluid flow in a stirred bioreactor system, the CFD-predicted fluid flow pattern and fluid velocities need to be verified. For this purpose different analytical methods can be used, PIV representing one of those most frequently applied. Its advantages are given by the contactless measuring principle, in which a 1 mm thick laser sheet is generated by a double-pulsed Nd:YAG laser (wavelength λ=532 nm). By adding rhodamine-coated fluorescent particles to the reactor volume, the fluid flow can be visualized and fluid velocities can be measured by recording a set of double images with a small time offset (cross correlation). More detailed representations of this method can be found in [255] and [132]. As exemplarily shown in Figure 8 C for the Corning spinner flask in two vertical positions (below and above the impeller), the CFD-predicted tangential fluid velocities are in good agreement with the PIV-measured data with relative deviations below 7 %. These results illustrate that the fluid velocities are well-captured by the CFD model and the model is therefore suitable.

In order to determine the microcarrier distribution, the single-phase model was extended by an additional Eulerian phase (Euler-Euler RANS approach), which also takes the microcarrier into account as a solid phase [255]. Figure 9 illustrates the microcarrier distribution of the

http://dx.doi.org/10.5772/59385

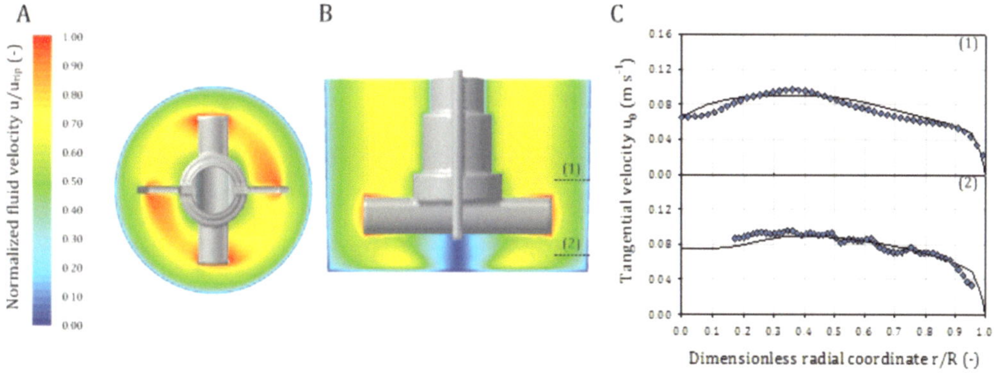

Figure 8. Fluid flow conditions in the Corning spinner flask at 60 rpm and 100 mL working volume. The fluid flow is presented as a combined vector-contour plot along a horizontal plane (A) and as normal contour plot along a vertical plane (B). The graph (C) illustrates the good agreement between the CFD-predicted (-) and PIV-measured (•) fluid velocities along the radial coordinates in two vertical positions below and above the impeller bar.

polystyrene-type carrier at stirrer speeds below and similar to the N_{S1U} criterion. As previously mentioned, the microcarrier sedimentation and agglomeration is most likely to be found below the impeller bar, especially at lower impeller speeds ($N < N_{S1U}$ and N_{S1}). At these conditions, the cells on the microcarriers, particularly those in the centre of the deposit are not well supplied with nutrients and oxygen. Thus, a mass transfer limitation and a gradient formation can occur, which may have a negative effect on cell quantity and quality. Both limitations can be reduced, or even prevented by increasing the impeller speed. However, this involves raising the specific power input in the bioreactor, thereby increasing the shear stress acting on cells. Consequently, impeller speeds are required that guarantee sufficient mixing at power inputs tolerated by cells without differentiation or dying off. Process conditions which fulfill these requirements are obtained at specific suspension criteria (N_{S1} and N_{S1U}). By increasing the impeller speed to the N_{S1U} criterion, the deposit directly below the impeller bar can be reduced to a state in which only a few microcarriers are situated directly below the impeller, with none of them at rest (local microcarrier movement along the bioreactor bottom). This can be seen in Figure 9 C, where a decrease in the local volume fraction of the microcarriers from around 0.4 % to 0.1 % below the impeller bar is shown. Compared to the N_{S1} criterion, where the microcarriers are not in contact with the reactor bottom for more than one second, the N_{S1U} provides more gentle conditions for the cells. However, both criteria strongly depend on the characteristics of the microcarrier type (density, diameter, porosity) and the bioreactor (dimension/configuration). Accordingly, the suspension criteria needed to be determined for each microcarrier and bioreactor used in the expansions.

The N_{S1U} and N_{S1} criteria determined for the polystyrene matrix-based microcarriers (solid fraction: 0.25 – 2 %) were in a range of 34 – 50 rpm and 44 – 61 rpm respectively, and for the gelatine matrix-based microcarriers (solid fraction: 0.25 – 1 %) 59 – 79 rpm and 80 – 93 rpm respectively. Indeed, the N_{S1u} criterion was always achieved at 20 ± 4 % lower impeller speeds, which indicates reproducible determinability of the N_{S1U} criterion.

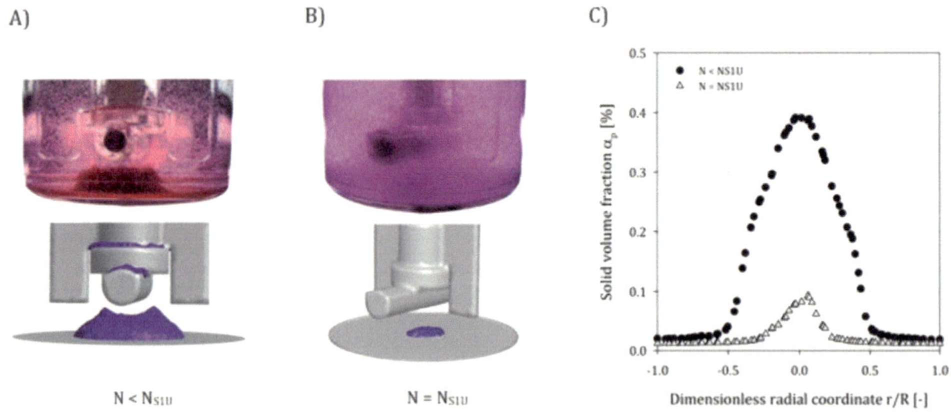

Figure 9. CFD-predicted, and experimentally observed microcarrier distributions for the Corning spinner flask at stirrer speeds below (A) and similar to (B) the N_{S1U} criterion. Good agreement is achieved between the CFD-predicted and experimentally observed microcarriers distributions. (C) The graph illustrates the reduction of the microcarrier volume fraction below the impeller from 0.4 % to 0.1 % by increasing the impeller speed to the suspension criterion (N_{S1U}).

5.3. Investigations at L-scale

Generation of hMSCs in clinically relevant numbers requires an increase in growth surface as well as in a bioreactor`s working volume. Because Sartorius`s BIOSTAT® CultiBag STR family is commercially available up to m^3 scale, and bioengineering data are available for the 50 L, 200 L, 1 m^3 and 2 m^3 versions [194,258], we decided to realize the process scale-up in the BIOSTAT® CultiBag STR 50L. For the preliminary investigations at mL-scale, which demonstrated the usefulness of the suspension criteria for the microcarrier-based stem cell cultivation [255], N_{S1U} and N_{S1} were experimentally determined for the BIOSTAT® CultiBag STR 50L in a first step. For this purpose a Plexiglas model with the dimensions of the bag holder (which allowed visual evaluation of the suspension criteria) was used instead of the flexible multilayer bag (Figure 7 E). More detailed information on the PIV measurement principle applied is given by Schirmaier et al. [196].

In order to achieve the target cell yields, the growth surface was increased accordingly, whereby microcarrier solid fractions of 0.20 – 1.25 % (polystyrene microcarriers) and 0.1 – 0.4 % (gelatin matrix-based microcarriers) were employed in the suspension studies. The impeller speeds required to ensure the suspension criteria (N_{S1U}, N_{S1}) were in a range of 40 – 66 rpm and 50 – 77 rpm for the polystyrene microcarriers. In the case of the gelatin matrix-based microcarriers, values ranging from 52 – 68 rpm and 63 – 79 rpm were required. Nevertheless, for both hMSC types a comparable correlation (20 ± 5 %) between the suspension criteria in the spinner flask and the BIOSTAT® CultiBag STR 50L was detected. Based on the estimated impeller speeds, single-and multi-phase CFD simulations were performed in order to investigate the fluid flow (flow pattern, fluid velocities) as well as to predict the specific power inputs and the shear gradients in the BIOSTAT® CultiBag STR 50L. As illustrated in Figure 10 A and B the fluid flow in the BIOSTAT® CultiBag STR 50L was primary axial with two flow

http://dx.doi.org/10.5772/59385

loops, whereby the axial fluid velocities at the bottom of the bioreactor enabled the microcarriers to swirl up. Due to the distinctive axial flow characteristics, which were caused by the two 3-segment blade impellers, as well as the relatively high axial velocities near the reactor bottom (0.04-0.1·u_{tip}), the microcarriers swirled up even at relatively low impeller speeds. Similarly, the prevalence of the axial fluid flow was obvious from PIV measurement data, where a good agreement between the CFD-predicted and the PIV-measured fluid velocities (relative deviation < 10 %) was achieved [258].

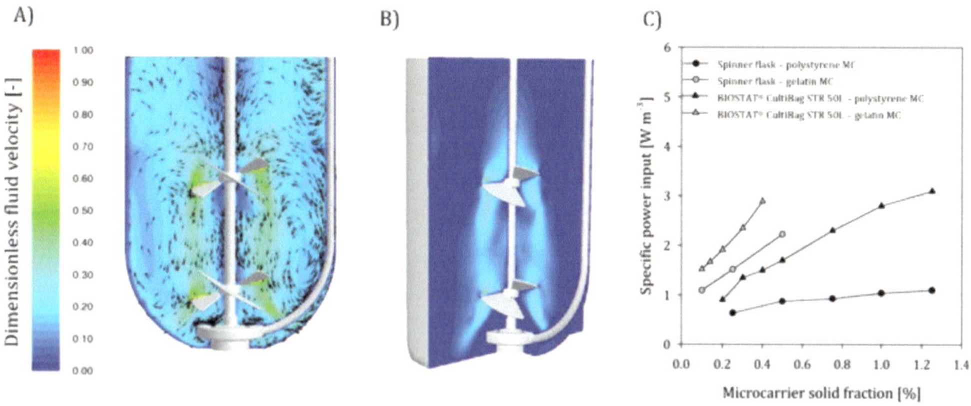

Figure 10. Fluid flow conditions and required specific power inputs for the N_{S1U} criterion in the BIOSTAT® CultiBag STR 50L. The contour and vector-plots are given along the mid-reactor plane. The contour plot (A) illustrates that the induced fluid flow in the BIOSTAT® CultiBag STR 50L is primary axial with two flow loops, whereby the axial fluid velocities (B) impinge on the reactor bottom and enable the microcarriers to swirl up. (C) The specific power inputs required to fulfill the N_{S1U} criterion are in a comparable range for both the spinner flask and the BIOSTAT® CultiBag STR 50L.

The suspension studies showed that the number of impellers has no significant effect on the suspension criteria. Thus, fed batch strategies can be realized which start with low working volumes and where only the lower impeller is immersed in culture medium. As is depicted in Figure 10 C, the CFD-predicted specific power inputs for the N_{S1} criterion at different microcarrier solid fractions in the BIOSTAT® CultiBag STR 50L were slightly higher (20 – 180 %) than in the spinner flask. However, the specific power inputs required were in the same magnitude and appropriate for the microcarrier solid fraction range desired. A closer look at the local maximum shear stress levels (1.06 N·m^{-2}) at 60 rpm (0.4 % and 0.2 % solid fraction of polystyrene and gelatine matrix-based microcarriers respectively) also revealed a slightly higher stress on the cells grown in the BIOSTAT® CultiBag STR 50L. However, these higher shear stresses occured only in a low percentage (3.15·10^{-5} %) of the volume and the cells were only exposed to these conditions for a short time. Beyond that, the local mean shear stress levels were in a range comparable to that in the spinner flask, which indicates a broader distribution of the local shear stresses in the BIOSTAT® CultiBag STR 50L [196,258]. The fluid flow in the BIOSTAT® CultiBag STR was more turbulent (Re=13591 – 26817) than in the spinner flask, which allowed a more reliable evaluation of Kolmogorov`s microscales under the conditions investigated. Compared to the diameters of the microcarriers, the predicted minimal (14 – 19 μm) and mean (197 – 268 μm) Kolmogorov`s microscales indicated that some turbulent eddies were in the same order of magnitude. However, the stem cells were only in contact with the critical turbulent eddies ($2/3 \cdot l_\lambda < d < l_\lambda$) for a restricted (short) time and cell death was not detected until day 8 of the cultivations.

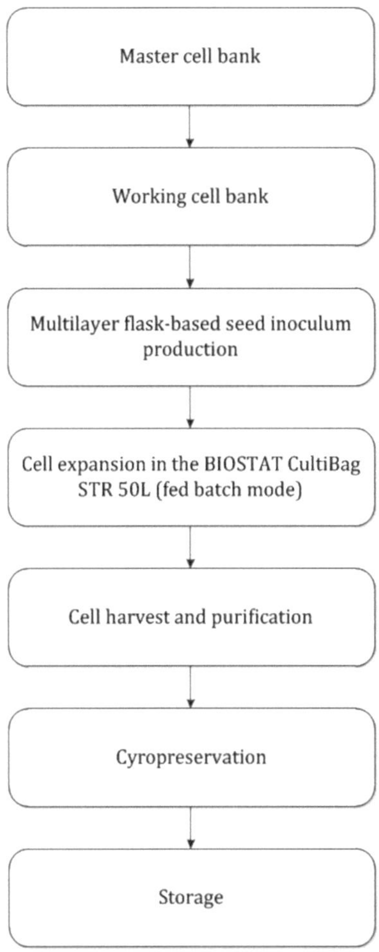

Figure 11. Flow chart showing the main steps of hBM-MSC expansions at pilot scale. hBM-MSCs from the working cell bank (second passage) were used to generate the seed inoculum (2-layer CellSTACKs®, 1272 cm^2 growth surface) within 4 days. After seed inoculum harvest and microcarrier preparation had been carried out according to the manufacturer`s recommendations, a 4-hour cell attachment phase was realized by transferring the inoculum to a BIOSTAT® CultiBag RM 20L basic (Sartorius Stedim Biotech) containing the equilibrated gelatin microcarriers. The bag was placed in an incubator and kept stationary to promote attachment of cells. For subsequent stem cell expansion, the microcarrier cell suspension was transferred to the BIOSTAT® CultiBag STR 50L. Pilot runs were started with 35 L working volume at 37 °C, a DO value above 20 %, and a maximum air flow rate of 0.03 vvm. pH ranged between 7.2 and 7.3, and impeller speeds were between 50 rpm and 66 rpm. After cell separation, harvest and purification, the cells were frozen for long-term storage (-196°C).

Based on the CFD-predicted engineering parameters and the optimal cultivation conditions derived for the two hMSC types, serum-reduced expansions at 35 L for the hADSCs and 50 L for the hBM-MSCs were carried out. Because of the higher clinical relevance of the hBM-MSCs, only one run (fed batch mode with 50 % medium exchange on day 4) was realized for the hADSCs [196], whereas an optimized feeding procedure was developed for the hBM-MSCs. The main process steps of the successful hBM-MSC-based feeding procedure, which is in agreement in the main lines for the 35 L hADSC production, are shown in Figure 11. A detailed description of the hADS expansion, where the multilayer flask-based seed inoculum production was omitted, can be found in [196].

For both hMSC types comparable cell growth was achieved in both the spinner flask and the BIOSTAT® CultiBag STR 50L runs (Figure 12 A and B). For the hADSCs, peak viable cell density in the spinner flask was $2.92 \cdot 10^5$ cells·mL^{-1} and $3.0 \cdot 10^5$ cells·mL^{-1} in the BIOSTAT® CultiBag STR 50L. The hADSCs grew with a growth rate of $33.6 \cdot 10^{-3}$ h^{-1} in the spinner flask and $30.4 \cdot 10^{-3}$ h^{-1} in the BIOSTAT® CultiBag STR 50L, which corresponds to doubling times of 20.6 h and 22.8 h respectively [196]. By realizing the optimized feeding approach, the peak viable cell densities of the hBM-MSCs were around 2 to 2.4 times higher ($6.8 \cdot 10^5$ cells·mL^{-1} in the spinner flask and $7.2 \cdot 10^5$ cells·mL^{-1} in the BIOSTAT® CultiBag STR 50L) compared to the hADSC expansions. The expansion factor was 39.6 ± 3.6 in the spinner flask and 51.5 ± 4.9 in the BIOSTAT® CultiBag STR 50 L. These are the highest expansion factors so far reported for serum-reduced hBM-MSC productions [64,191,192]. The hBM-MSCs grew with similar growth rates (μ) to the hADSCs ($34.8 \cdot 10^{-3}$ h^{-1} in the spinner flask and $30.4 \cdot 10^{-3}$ h^{-1} in the BIOSTAT® CultiBag STR 50L), whereby the peak viable cell densities were achieved one day earlier. Cell growth at the end of the cultivations was restricted by the limited microcarrier growth surface.

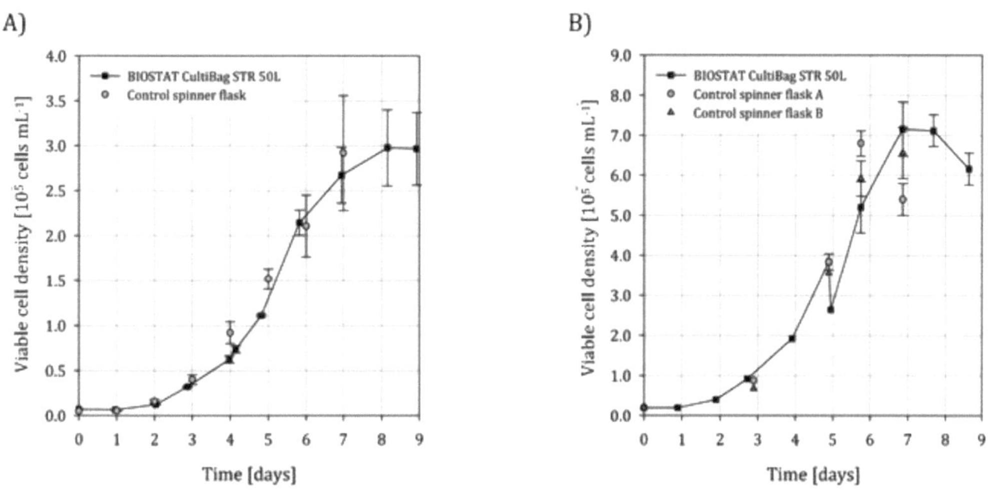

Figure 12. Courses of viable cell densities of hADSCs (A) and hBM-MSCs (B) in the BIOSTAT® CultiBag STR 50L and the control spinner flasks. (A) hADSC cultivation was performed at 35 L scale (50 % medium exchange on day 4). (B) Optimized hBM-MSC cultivation was performed at 50 L scale (feed on day 5).

During all the cultivations the expanded stem cells maintained their stem cell properties and qualities. Figure 13 shows the flow cytometric analysis of hBM-MSC samples. The specific surface markers ($CD34^-$, $CD45^-$, $CD73^+$, $CD95^+$, $CD105^+$) were determined according to the recommendations of the ISCT [25]. Comparable results were also found for the hADSC cultivations in the BIOSTAT® CultiBag STR 50L [196].

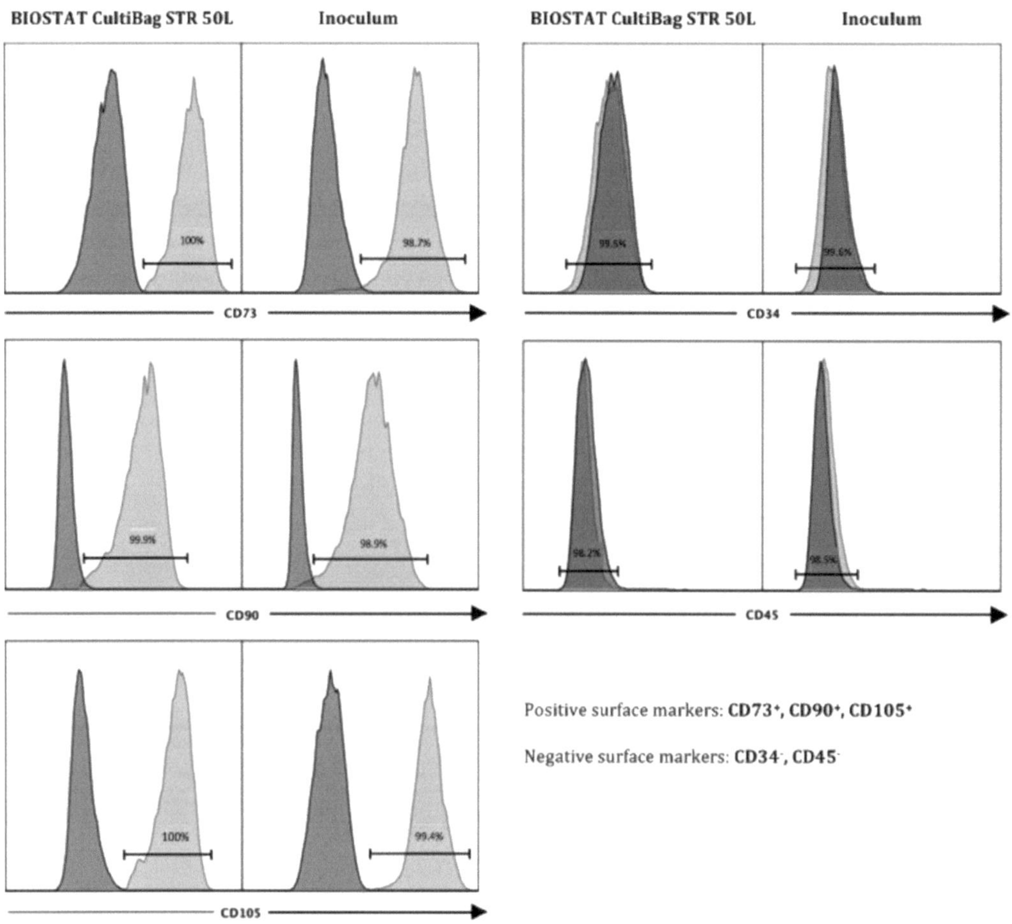

Figure 13. Flow cytometric analysis (FACS) of specific surface markers performed with hBM-MSC samples before (seed inoculum) and at the end (after harvest) of the expansion procedure. Positive surface markers: $CD73^+$, $CD90^+$, $CD105^+$. Negative surface markers: $CD34^-$, $CD45^-$.

6. Conclusion

The potential of MSCs is undoubted and explains the increasing clinical interest in high cell yields for therapy approaches. The successful expansion of MSCs to clinically relevant numbers and qualities strictly depends on the bioreactor type and its bioengineering parameters, in particular the fluid flow and shear stress. Non-instrumented multilayer vessels normally used for MSC expansion may be advantageously replaced by instrumented dynamic single-use bioreactors. Nowadays, users already have access to stirred, wave-mixed, parallel plate, hollow fiber and fixed bed versions, which confirm the applicability of these bioreactor types for different MSC expansion studies.

For autologous therapies, stirred singles bioreactors operating at benchtop scale and providing yields of $1 \cdot 10^9$ MSCs after harvest from microcarriers have proven themselves. Availability of bioreactor CFD models has supported the determination of bioengineering parameters as well as prediction of fluid flow and shear stress. Moreover, knowledge of fluid flow patterns and velocities, local shear stress distributions, turbulent dissipation rates and Kolmogorov`s microscales has contributed to an increase in expansion efficiency and supported avoidance of cell differentiation. Together with microcarrier distribution, these parameters can also be used to rapidly and successfully transfer hADSC-and hBM-MSC-based expansions from mL-up to pilot scale. They provide the basis for the production of MSC-derived allogeneic cellular therapeutic products, for which trillion of cells will be required annually.

Stem cell therapeutics are now in transition from development to clinical applications. A large number of stem cell therapeutics are in the clinical pipeline or already in clinical trials [14,31]. This requires clinical-grade bioprocessing and scale-up approaches for expansion of MSCs from human sources. However, the application of MSCs is hampered by a number of facts intrinsic to these cells. As summarized by Wuchter et al. [44], MSCs vary in composition, their cell source (e.g., bone marrow, adipose tissue and cord blood) and the protocols applied for cell isolation and expansion. The available markers are not regarded as sufficient for comparison of data and quality control. Therefore, experimental and clinical data from different laboratories involved in multicenter trials are difficult to compare. Criteria important for quality control include the immunophenotype of the cells, composition of the culture medium and the risk of malignant transformation, as well as the aging and the immunosuppressive potential of the manufactured MSCs [44].

Stem cells could be used in cell therapy either as mass-produced allogeneic cells or autologous cells [31]. In both cases scalable, controlled bioprocessing techniques are required. Establishing these processes is complicated by the fact that guide lines developed for the production of biopharmaceuticals (e.g. monoclonal antibodies) are generally unsuitable for cell therapeutics. Whereas in bioproduction the quality of the product is of major concern and the cells are typically discarded after production, for cell-based therapies the cell-specific properties are essential and require highly specific processes. An overview of the clinical and regulatory pathways for stem cell-based therapies and recent developments is given by [13,31,44].

Future availability of serum-free or even chemically defined culture media and microcarriers allowing simple cell harvest (e.g. after temperature-or pH-shift) and reproducible production will simplify the approval of cell therapeutics. Moreover, it will provide new findings in respect of operational demands and shear stress sensitivity of MSCs.

Abbreviations

1-D	One-dimensional
2-D	Two-dimensional
3-D	Three-dimensional
CD	Cluster of differentiation (e.g. CD105)
CFD	Computational Fluid Dynamics
CHO	Chinese hamster ovary
DAPI	4′,6-diamidino-2-phenylindole
DMEM	Dulbecco's Modified Eagle's Medium
DO	Dissolved oxygen
FBS	Fetal bovine serum
FCS	Fetal calf serum
FGF	Fibroblast growth factor
GMP	Good Manufacturing Practice
hADSC	Adipose tissue-derived human mesenchymal stem cell
hBM-MSC	Bone marrow-derived human mesenchymal stem cell
hMSC	Human mesenchymal stem cell
hMSC-TERT	Telomerase-immortalized human mesenchymal stem cells
ISCT	International Society for Cellular Therapy
MSC	Mesenchymal stem cell
PIV	Particle Image Velocimetry
RANS	Reynolds-averaged Navier-Stokes

Symbols

Latin symbols

Symbol	Unit	Description
b	m	Width of flow chamber
d_R	m	Impeller diameter
D_R	m	Vessel diameter
F	$m^3 s^{-1}$	Flow rate
h	m	Height of flow chamber

Symbol	Unit	Description
ISF	s^{-1}	Integrated shear factor
l_{Kol}	m	Turbulent Kolmogorov microscale
k_La	h^{-1}	volumetric mass transfer coefficient for oxygen or CO_2
n_R	s^{-1}	Rotational speed
Ne	-	Newton-Number
N_{S1} / N_{S1U}	rpm	Suspension criteria, describing the microcarrier distribution at just fully suspended conditions (N_{S1}) and below (N_{S1U})
P	W	Power input
P/V	W m^{-3}	Specific volumetric power input
Re	-	Reynolds-Number
u_{tip}	m s^{-1}	Maximum rotating velocity
U	m s^{-1}	Flow velocity
V_L	m^3	Liquid volume

Greek symbols

Symbol	Unit	Description
α_p	-	Solid volume fraction
γ	s^{-1}	shear rate
ε	m^2 s^{-3}	Turbulent energy dissipation rate
η_{fl}	Pa s	Dynamic viscosity
λ	nm	Wavelength
μ	h^{-1}	Specific growth rate
ν	m^2 s	Kinematic viscosity
ϱ_{fl}	kg m^{-3}	Liquid density
τ	Pa	Shear stress
τ_{nt}	Pa	Local shear stress
τ_w	Pa	Wall shear stress

Acknowledgements

We want to thank Carmen Schirmaier, Anita Ott and Lidija Lisica from the Group for Biochemical Engineering and Cell Cultivation Techniques Section of the Zurich University of Applied Sciences, and the LIFT team at Lonza Cologne GmbH (Ann Siehoff, Frank Jüngerkes,

Silke Brill, Agnieszka Safavi-Nab, Sonja Spicker, Barbara Reinisch, Claudia Rosenbaum, Christian van den Bos) for the excellent cooperation and, in particular, the technical support during development of the scale-up approach. Furthermore, we would like to mention that the results presented in Section 5 were generated in a two year research project financially supported by the Lonza and the Commission for Technology and Innovation in Switzerland.

Author details

Valentin Jossen[1], Ralf Pörtner[2*], Stephan C. Kaiser[1], Matthias Kraume[3], Dieter Eibl[1] and Regine Eibl[1]

*Address all correspondence to: poertner@tuhh.de

1 Zurich University of Applied Sciences, School of Life Sciences and Facility Management, Institute of Biotechnology, Switzerland

2 Hamburg University of Technology (TUHH), Institute of Bioprocess and Biosystems Engineering, Germany

3 Technische Universität Berlin (TU Berlin), Institute of Process Engineering, Germany

References

[1] Oppermann T, Leber J, Elseberg C, Salzig D, Czermak P. hMSC production in disposable bioreactors in compliance with cGMP guidelines and PAT. Am Pharm Rev. 2014;

[2] Nirmalanandhan VS, Sittampalam GS. Stem cells in drug discovery, tissue engineering, and regenerative medicine: emerging opportunities and challenges. J Biomol Screen. 2009;14(7):755–68.

[3] Kim SU, Lee HJ, Kim YB. Neural stem cell-based treatment for neurodegenerative diseases. Neuropathology. 2013;33(5):491–504.

[4] Mochizuki H, Choong CJ, Yasuda T. The promises of stem cells: stem cell therapy for movement disorders. Park Relat Disord. 2014;20(Suppl 1):128–31.

[5] Calaflore R, Montanucci P, Basta C. Stem cells for pancreatic β-cell replacement in diabetes mellitus: actual perspectives. Curr Opin Organ Transpl. 2014;19(2):162–8.

[6] Viswanathan C, Sarang S. Status of stem cell based clinical trials in the treatment for diabetes. Curr Diabetes Rev. 2013;9(6):429–36.

http://dx.doi.org/10.5772/59385

[7] Diekman BO, Gullak F. Stem cell-based therapies for osteoarthritis: challenges and opportunities. Curr Diabetes Rev. 2013;25(1):119–26.

[8] Wei L, Young JZ, Yan J. Potential of Stem Cells for Skin Regeneration Following Burns. Expert Rev Dermatol. 2009;4(2):97–9.

[9] Li SC, Wang LJ, Hong AJ, Chang AC, Loudon WG. Stem cell engineering for treatment of heart diseases: potentials and challenges. Cell Biol Int. 2009;33(3):255–67.

[10] Hall PA, Watt FM. Stem cells: The generation and maintenance of cellular diversity. Development. 1989;106(4):619–33.

[11] Lin H. The tao of stem cells in germline. Annu Rev Genet. 1997;31:455–91.

[12] Giordano A, Galderisi U, Marino IR. From the laboratory bench to the patient's bedside: an update on clinical trials with mesenchymal stem cells. J Cell Physiol. 2007;211:27–35.

[13] Trounson A, Thakar RG, Lomax G, Gibbons D. Clinical trials for stem cell therapies. BMC Med. 2011;9:52.

[14] Ratcliffe E, Glen K, Naing MW, William DJ. Current status and perspectives on stem cell-based therapies undergoing clinical trials for regenerative medicine: case studies. Br Med Bull. 2013;108:73–94.

[15] Inamadar AA, Inamadar AC. Culture conditions for growth of clinical grade human tissue derived mesenchymal stem cells: comparative study between commercial serum-free media and human product suspplemented media. J Regen Med Tissue Eng. 2013;2:10.

[16] Friedenstein AJ, Petrakova K V., Kursolesova AI, Frolova GP. Heterotropic of bone marrow. Analysis of precursor cells for osteogenic and hematopoietic tissues. Transplant. 1968;6:230–47.

[17] Friedenstein AJ, Chailakhjan RK, Lalykina KS. The development of fibroblast colonies in monolayer cultures of guinea-pig bone marrow and speel cells. Cell Tissue Kinet. 1970;3:393–403.

[18] Friedenstein AJ, Gorskaja JF, Kulagina NN. Fibroplast precursors in normal and irradiated mouse hematopoietic organs. Exp Hematol. 1976;4:267–74.

[19] Owen ME, Cave J, Joyner CJ. Clonal analysis in vitro of ostegenic differentiation of marrow CFU-f. J Cell Sci. 1987;Pt5:731–8.

[20] Owen ME, Friedenstein AJ. Stromal stem cells: Marrow derived osteogenic precursors. Giba Found Symp. 1988;136:42.

[21] Caplan AI. Mesenchymal stem cells. J Orthop Res. 1991;9:641–50.

[22] Pittinger MF, Mackay AM, Beck SC, Jaiswal RK, Douglas R, Mosca JD, et al. Multilineage potential of adult human mesenchymal stem cells. Science (80-). 1999;284:143–7.

[23] Goepfert C, Slobodianski A, Schilling AF, Adamietz P, Pörtner R. Cartilage Engineering from Mesenchymal Stem Cells. Adv Biochem Eng Biotechnol. 2010;123:163–200.

[24] Keating A. Mesenchymal stromal cells: New directions. Cell Stem Cell. 2012;10:709–16.

[25] Dominici M, Le Blanc K, Mueller I, Slaper-Cortenbach I, Marini F, Krause D, et al. Minimal criteria for defining multipotent mesenchymal stromal cells. The International Society for Cellular Therapy position statement. Cytotherapy. 2006;8(4):315–7.

[26] Zuk PA, Zhu M, Ashija P, Ugarte DA, Huabg JI, Mizuno H, et al. Human adipose tissue is a source of multipotent stem cells. Mol Biol Cell. 2002;13:4279–95.

[27] In't Anker PS, Scherjon SA, van der Keur K, de Groot-Swings GM, Claas FH, Fibbe WE, et al. Isolation of mesenchymal stem cells of fetal or maternal origin from human placenta. Stem Cells. 2004;22(7):1338-45

[28] Frices A, Conget P, Miguell JJ. Mesenchymal progenitor cells in human umbilical cord blood. Br J Haematol. 2000;109:235–42.

[29] Zvaifler NJ, Marinova-Mutafchieva L, Adams G, Edwards CJ, Moss J, Burger JA, et al. Mesenchymal precursor cells in blood of normal individuals. Arthritis Res. 2000;2(6):477–88.

[30] Schallmoser K, Rohde E, Reinisch A, Batmann C, Thaler D, Strunk D. Rapid large scale expansion of functional mesenchymal stem cells from unmanipulated bone marrow without animal serum. Tissue Eng Part C Methods. 2008;14(3):185–96.

[31] Van den Bos C, Keefe R, Schirmaier C, McCaman M. Therapeutic human cells: Manufacture for cell therapy/ regenerative medicine. Adv Biochem Eng Biotechnol. 2014;138:61–97.

[32] Ren G, Chen X, Dong F. Mesenchymal stem cells and translational medicine: Emerging Issues. Stem Cells Transl Med. 2012;1:51–8.

[33] Sharma S, Raju R, Shiu S. Stem cell culture engineering-process scale-up and beyond. Biotechnol. 2011;6(11):1317–29.

[34] Horwitz EM, Prather WR. Cytokines as the major mechanssm of mesenchymal stem cell clinical activity: Expanding the spectrum of cell therapy. Isr Med Assoc J. 2009;11(4):209–11.

[35] Stolzing A, Jones E, McGonagle D. Age-related changes in human bone marrow-derived mesenchymal stem cells: Consequence for cell therapies. Mech Ageing Dev. 2008;129(3):163–73.

[36] Uccelli A, Moretta L, Pistola V. Mesenchymal stem cells in health and disease. Nat Rev Immunol. 2008;8:726–36.

[37] Chen PM, Yen ML, Liu KJ, Sytwu HK, Yen BL. Immunomodulatory properties of human adult and fetal multipotent mesenchymal stem cells. J Biomed Sci. 2011;18:49.

[38] Caplan AI, Dennis JE. Mesenchymal stem cells as tropic mediators. J Cell Biochem. 2006;98(5):1076–84.

[39] Wang S, Qu X, Zhao RC. Clinical applications of mesenchymal stem cells. J Hematol Oncol. 2012;5:1–9.

[40] Nazarov C, Lo Surdo J, Bauer SR, Wei CH. Assessment of immunosuppressive activity of human mesenchymal stem cells using murine antigen specific CD4 and CD8 T cells in vitro. Stem Cell Res Ther. 2013;4:1–15.

[41] Mannello F, Tonti GA. Concise review: no breakthroughs for human mesenchymal and embryonic stem cell culture: conditioned medium, feeder layer, or feeder-free; medium with fetal calf serum, human serum, or enriched plasma; serum-free, serum replacement nonconditioned medium, or ad hoc formula? All that glitters is not gold! Stem Cells. 2007;25(7):1603-9.

[42] Ankrum J, Karp JM. Mesenchymal stem cell therapy: Two steps forward, one step back. Trends Mol Med. 2010;16:203–9.

[43] Hodges H, Pollock K, Stroemer P, Patel S, Stevanato L, Reuter I, et al. Making stem cell lines suitable for transplantation. Trends Mol Med. 2007;16(2):101–15.

[44] Wuchter P, Bieback K, Schrezenmeier H, Bornhäuser M, Müller LP, Bönig H, et al. Standardization of Good Manufacturing Practice-compliant production of bone marrow-derived human mesenchymal stromal cells for immunotherapeutic applications. Cytotherapy. 2014;

[45] Foley L, Whitaker M. Concise review: cell therapies: the route to widespread adoption. Stem Cells Transl Med. 2012;1(5):438–47.

[46] Shih D, Tzu B, Burnouf T. Preparation, quality criteria, and properties of human blood platelet lysate supplements for ex vivo stem cell expansion. N Biotechnol. 2014;

[47] Sotiropoulou PA, Perez SA, Salagianni M, Baxevanis CN, Papamichail M. Characterization of the optimal culture conditions for clinical scale production of human mesenchymal stem cells. Stem Cells. 2006;24(2):462–71.

[48] De Bruyn C, Najar M, Raicevic G, Meuleman N, Pieters K, Stamatopoulos B, et al. A Rapid, Simple, and Reproducible Method for the Isolation of Mesenchymal Stromal Cells from Wharton's Jelly Without Enzymatic Treatment. Stem Cells Dev. 2010;

[49] Rhodes NP, Srivastava JK, Smith RF, Longinotti C. Heterogeneity in proliferative potential of ovine mesenchymal stem cell colonies. J Mater Sci Mater Med. 2004;15(4): 397–402.

[50] Kretlow JD, Jin YQ, Wei Z, Wen J, Hong TH, Zhou G, et al. Donor age and cell passage affects differentiation potential of murine bone marrow-derived stem cells. BMC Cell Biol. 2008;9(60).

[51] Colter DC, Class R, Digirolamo CM, Prockop DJ. Rapid expansion of recycling stem cells in cultures of plastic-adherent cells from human bone marrow. Proc Natl Acad Sci U S A. 2000;97:3213–8.

[52] Ikebe C, Suzuki K. Mesenchymal stem cells for regenerative therapy: optimization of cell preparation protocols. Biomed Res Int. 2014; 951512.

[53] Stenderup K, Justesen J, Clausen C, Kassern M. Aging is associated with decreased maximal life span and accelerated senescence of bone marrow stromal cells. Bone. 2003;33(6):919–26.

[54] Pörtner R, Goepfert C, Wiegandt K, Janssen R, Ilinich E, Paetzhold H, et al. Technical Strategies to Improve Tissue Engineering of Cartilage Carrier Constructs-A Case Study. Adv Biochem Eng Biotechnol. 2009;112:145–82.

[55] Fu RH, Wang YC, Liu SP, Huang CM, Kang YH, Tsai CM, et al. Differentiation of stem cells: strategies for modifying surface biomaterials. Cell Transpl. 2011;20(1):37–47.

[56] Hudson JE, Mills RJ, Frith JE, Brooke G, Jaramillo-Ferrada P, Wolvetang EJ, et al. A defined medium and substrate for expansion of human mesenchymal stromal cell progenitors that enriches for osteo-and chondrogenic precursors. Stem Cells Dev. 2011;20(1):77–87.

[57] Hwang NS, Varghese S, Li H, Elisseeff J. Regulation of osteogenic and chondrogenic differentiation of mesenchymal stem cells in PEG-ECM hydrogels. Cell Tissue Res. 2012;344(3):499–509.

[58] Wang G, Zheng L, Zhao H, Miao J, Sun C, Ren N, et al. In vitro assessment of the differentiation potential of bone marrow-derived mesenchymal stem cells on genipin-chitosan conjugation scaffold with surface hydroxyapatite nanostructure for bone tissue engineering. Tissue Eng Part A. 2011;17(9-10):1341–9.

[59] He F, Pei M. Extracellular matrix enhances differentiation of adipose stem cells from infrapatellar fat pad toward chondrogenesis. J Tissue Eng Regen Med. 2013;7(1):73–84.

[60] Schmitz S. Der Experimentator: Zellkultur. 3rd ed. Spektrum Akad. Verl., Heidelberg; 2011.

[61] Weber C, Freimark D, Pörtner R, Pino-Grace P, Pohl S, Wallrapp C, et al. Expansion of human mesenchymal stem cells in a fixed-bed bioreactorcsystem based on non-po-

rous glass carrier-Part B: Modeling and scale-up of the system. Int J Artif Organs. 2010;33(11):782–95.

[62] Weber C, Freimark D, Pörtner R, Pino-Grace P, Pohl S, Wallrapp C, et al. Expansion of human mesenchymal stem cells in a fixed-bed bioreactor system based on non-porous glass carrier-Part A: inoculation, cultivation, and cell harvest procedures. Int J Artif Organs. 2010;33(8):512–25.

[63] Szczypka M, Splan D, Woolls H, Brandwein H. Single-Use Bioreactors and Microcarriers. BioProcess Tech. 2014;12(3):54–64.

[64] Rafiq QA, Coopman K, Hewitt CJ. Scale-up of human mesenchymal stem cell culture: current technologies and future challenges. Curr Opin Chem Eng. Elsevier Ltd; 2013 Feb;2(1):8–16.

[65] Rafiq Q a, Brosnan KM, Coopman K, Nienow AW, Hewitt CJ. Culture of human mesenchymal stem cells on microcarriers in a 5 l stirred-tank bioreactor. Biotechnol Lett. 2013 Aug;35(8):1233–45.

[66] Sart S, Errachid A, Schneider Y, Agathos SN. Controlled expansion and differentiation of mesenchymal stem cells in microcarrier-based stirred bioreactors. BMC Proc. 2011;5(Suppl 8):55

[67] Chen AKL, Reuveny S, Oh SKW. Application of human mesenchymal and pluripotent stem cell microcarrier cultures in cellular therapy: Achievments and future direction. Biotechnol Adv. 2013;31(7):1032–46.

[68] Elseberg CL, Leber J, Salzig D, Wallrapp C, Kassem M, Moustapha K, et al. Microcarrier-based expansion process for hMSCs with high vitality and undifferentiated characteristics. Int J Artif Organs. 2012;35(2):93–107.

[69] Ferrari C, Balandras F, Guedon E, Olmos E, Chevalot I, Marc A. Limiting cell aggregation during mesenchymal stem cell expansion on microcarriers. Biotechnol Prog. 2012;28:780–7.

[70] Freimark D, Pino-Grace P, Pohl S, Weber C, Wallrapp P, Geigle R, et al. Use of Encapsulated Stem Cells to Overcome the Bottleneck of Cell Avilability for Cell Therapy Approaches. Transfus Med Hemother. 2010;37:66–73.

[71] O'Heireamhoi S, Buckley CT, Jones E, McGonagle D, Mulhall KJ, Kelly DJ. Recapitulating aspects of the oxygen and substrate enviroment of the damaged joint milieu for stem cell-based cartilage tissue engineering. Tissue Eng Part C Methods. 2013;19(2):117–27.

[72] Thanos CG, Emerich DF. On the use of hydrogels in cell encapsulation and tissue engineering system. Recent Pat Drug Deliv Formul. 2008;2(1):19–24.

[73] Cheung HK, Han TTY, Marecak DM, Watkins JF, Amsden BG, Flynn LE. Composite hydrogel scaffolds incorporating decellularized adipose tissue for soft tissue engineering with adipose-derived stem cells. Biomaterials. 2014;35(6):1914–23.

[74] Kumar D, Gerges I, Taplenizza M, Lenardi C, Forsyth NR, Liu Y. Three-dimensional hypoxic culture of human mesenchymal stem cells encapsulated in a photocurable, biodegradable polymer hydrogel: a potential injectable cellular product for nucleus pulposus regeneration. Acta Biomater. 2014;10(8):3463–74.

[75] Ma J, Yang F, Both SK, Kersten-Niessen M, Bongio M, Pan J, et al. Comparison of cell-loading methods in hydrogel systems. J Biomed mater Res A. 2014;102(4):935–46.

[76] Mohan N, Gupta V, Sridharan B, Sutherland A, Detamore MS. The potential of encapsulating "raw materials" in 3D osteochondral gradient scaffolds. Biotechnol Bioeng. 2014;111(4):829–41.

[77] Naderi-Meshkin H, Andreas K, Matin MM, Sittinger M, Bidkhori HR, Ahmadiankia N, et al. Chitosan-based injectable hydrogel as a promising in situ forming scaffold for cartilage tissue engineering. Cell Biol. 2014;38(1):72–84.

[78] Raghothaman D, Leong MF, Lim TC, Toh JKC, Wan ACA, Yang Z, et al. Engineering cell matrix interactions in assembled polyelectrolyte fiber hydrogels for mesenchymal stem cell chondrogenesis. Biomaterials. 2014;35(9):2607–16.

[79] Tzouanas SN, Ekenseair AK, Kasper FK, Mikos AG. Mesenchymal stem cell and gelatin microparticle encapsulation in thermally and chemically gelling injectable hydrogels for tissue engineering. J Biomed mater Res A. 2014;102(5):122–1230.

[80] Serra M, Brito C, Leite SB, Gorjup E, Von BH, Carondo MJ, et al. Stirred bioreactors for the expansion of adult pancreatic stem cells. Ann Anat. 2009;191:104–15.

[81] Frith JE, Thomson B, Genever PG. Dynamic three-dimensional culture methods enhance mesenchymal stem cell properties and increase therapeutic potential. Tissue Eng Part C Methods. 2010;16:735–49.

[82] Cao C, Zhang H, Gong L, He Y, Zhang N. High glucose conditions suppress the function of bone marrow-derived endothelial progenitor cells via inhibition of the eNOS-caveolin-1 complex. Mol Med Rep. 2012;5(2):341–6.

[83] Dhanasekaran M, Indumathi S, Rajkumar JS, Sudarsanam D. Effect of high glucose on extensive culturing of mesenchymal stem cells derived from subcutaneous fat, omentum fat and bone marrow. Cell Biochem Funct. 2013;31(1):20–9.

[84] Ferrari C, Olmos E, Balandras F, Tran N, Chevalot I, Guedon E, et al. Investigation of growth conditions for the expansion of procine mesenchymal stem cells on microcarriers in stirred cultures. Appl Biochem Biotechnol. 2014;172(2):1004–17.

[85] Kim SJ, Choi YS, Ko ES, Lim SM, Lee CW, Kim DI. Glucose-stimulated insulin secretion of various mesenchymal stem cells adter insulin-producing cell differentiation. J Biosci Bioeng. 2012;113(6):34–40.

[86] Saki N, Jalalifar MA, Soleimani M, Hajizamani S, Rahim F. Adverse effect of high glucose concentration on stem cell therapy. Int J Hematol Oncol Stem Cell Res. 2013;7(3):34–40.

[87] Sawangmake C, Pavasant P, Chansiripornchai P, Osathanon T. High glucose condition suppresses neurosphere formation by human periodontal ligament-derived mesenchymal stem cells. J Cell Biochem. 2014;115(5):928–39.

[88] Weil BR, Abarbanell AM, Hermann JL, Wang Y, Meldrum DR. High glucose concentration in cell culture medium does not actuely affect human mesenchymal stem cell growth factor production or proliferation. Am J Physiol Regul Integr Comp Physiol. 2009;296(6):1735–43.

[89] Deorosan B, Nauman EA. The role of glucose, serum, and three-dimensional cell culture on the metabolism of bone marrow-derived mesenchymal stem cells. Stem cells Int. 2011;429187.

[90] Li YM, Schilling T, Benisch P, Zeck S, Meissner-Weigl J, Schnider D, et al. Effects of high glucose on mesenchymal stem cell proliferation and differentiation. Biochem Biophys Res Commun. 2007;363(1):209–15.

[91] Stolzing A, Bauer E, Scutt A. Suspension culture of bone-marrow derived mesenchymal stem cells: effects of donor age and glucose level. Stem Cells Dev. 2012;21(14): 2718–23.

[92] Jung MR, Min YL, Pil Y, Han J. High glucose regulates cyclin D1/E of human mesenchymal stem cells through TGF-beta1 expression via Ca2 /PKC/MAPKs and PI3K/Akt/mTOR signal pathways. J Cell Physiol. 2010;224(1):59–70.

[93] Schop D, Janssen FW, van Rijn LD, Fernandes H, Bloem RM, Al. E. Growth, metabolism, and growth inhibitors of mesenchymal stem cells. Tissue Eng Part A. 2009;15(8): 1877–1866.

[94] Lo T, Ho JH, Yang MH, Lee OK. Glucose reduction prevents replicative senescence and increases mitochondrial respiration in human mesenchymal stem cells. Cell Transpl. 2011;20(6):813–25.

[95] Coutu DL, Galipeau J. Roles of FGF signaling in stem cell self-renewal, senescence and aging. Aging (Albany NY). 2011;3(10):920–33.

[96] Clark DA, Coker R. Transforming growth factor-beta (TGF-beta). Int J Biochem Cell Biol. 1998;30(3):293–8.

[97] Yamachika E, Tsujingiwa H, Matsubara M, Hirata Y, Kita K, Takabatake K, et al. Basic fibroblast growth factor supports expansion of mouse compact bone-derived mes-

enchymal stem cells (MSCs) and regeneration of bine from MSC in vivo. J Mol Histol. 2012;43(2):223–33.

[98] Vallier L, Morgan A, Pedersen R. Activin/Nodal and FGF pathways cooperate to maintain pluripotency of human embryonic stem cells. J Cell Sci. 2005;118(19):4495–509.

[99] Jianguo W, Tianhang L, Hong Z, Zhengmao L, Jianwei B, Xuchao X, et al. Optimization of culture conditions for endothelial progenitor cells from porcine bone marrow in vitro. Cell Prolif. 2010;43(4):418–26.

[100] Harmouch C, El-Omar R, Labrude P, Decot V, Menu P, Kerdjoudj H. Influence of serum percentage on the behavior of Wharton's jelly mesenchymal stem cells in culture. Biomed Mater Eng. 2013;23(4):273–80.

[101] Chase LG, Yang S, Zachar V, Yang Z, Lakshimpathy U, Bradford J, et al. Development and characterization of a clinicalls compliant xeno-free culture medium in good manufacturing practice for human multipotent mesenchymal stem cells. Stem Cells Transl Med. 2012;1(10):750–8.

[102] Sotiropoulou PA, Perez SA, Salagianni M, Baxevanis CN, Papamichail M. Cell culture medium composition and translational adult bone marrow-derived stem cell research. Stem Cells. 2006;24(5):1409–10.

[103] Gottipamula S, Sanjay M, Manujunatha S, Chaansa S, Al. E. Isolation, expansion and characterization of bone marrow-derived mesenchymal stromal cells in serum-free conditions. J Tissue Eng Regen Med. 2013;

[104] Jung S, Sen A, Rosenberg L, Behie LA. Human mesenchymal stem cell culture: rapid and efficient isolation and expansion in a defined serum-free medium. J Tissue Eng Regen Med. 2012;6(5):391–403.

[105] Liu Q, Luo Z, He SP, Xinlei X, Al. E. Conditioned serum-free medium from umbilical cord mesenchymal stem cells has anti-photoaging properties. Biotechnol Lett. 2013;35(10):1707–14.

[106] Mark P, Kleinsorge M, Gaebel R, Lux C, Al. E. Human Mesenchymal Stem Cells Display Reduced Expression of CD105 after Culture in Serum-Free Medium. Stem Cells Int. 2013;

[107] Miwa H, Hashimoto Y, Tensho K, Al. E. Xeno-free proliferation of human bone marrow mesenchymal stem cells. Cytotechnology. 2012;64(3):301–8.

[108] Rajaraman G, Whit J, Tan KS, Ulrich D, Al. E. Jerome; Gargett, Caroline E. (2013): Optimization and scale-up culture of human endometrial multipotent mesenchymal stromal cells: potential for clinical application. Tissue Eng Part C Methods. 2013;19(1):80–92.

[109] Roy S, Arora S, Kumari P, Ta M, Al. E. A simple and serum-free protocol for cryopreservation of human umbilical cord as source of Wharton's jelly mesenchymal stem cells. Cryobiology. 2014;68(3):467–72.

[110] Simöes IN, Boua JS, dos Santos F, Al. E. Human mesenchymal stem cells from the umbilical cord matrix: successful isolation and ex vivo expansion using serum-/xeno-free culture media. J Biotechnol. 2013;8(4):448–58.

[111] Swamynathan P, Venugopal P, Kannan S, Al. E. Are serum-free and xeno-free culture conditions ideal for large scale clinical grade expansion of Wharton's jelly derived mesenchymal stem cells? A comparative study. Stem Cell Res Ther. 2014;5(4):88.

[112] Wang X, Zheng F, Liu O, Zheng S, Al. E. Epidermal growth factor can optimize a serum-free culture system for bone marrow stem cell proliferation in a miniature pig model. Vitr Cell Dev Biol Anim. 2013;49(10):815–25.

[113] Wu M, Han ZB, Liu JF, Wang Y, Al. E. Serum-free media and the immunoregulatory properties of mesenchymal stem cells in vivo and in vitro. Cell Physiol Biochem. 2014;33(3):569–80.

[114] Dos Santos F, Andrade PZ, Boura JS, Al. E. Ex vivo expansion of human mesenchymal stem cells: a more effective cell proliferation kinetics and metabolism under hypoxia. J Cell Physiol. 2010;223(1):27–35.

[115] Dos Santos F, Campbell A, Fernandes-Platzgummer A, Al. E. Xenogeneic-Free bioreactor system for the clinical-scale expansion of human mesenchymal stem/stromal cells. BiotechnolBioeng. 2014;111(6):1116–27.

[116] Carrancio S, Lopez-Holgado N, Sanchez-Guijo FM, Al. E. Optimization of mesenchymal stem cell expansion procedures by cell separation and culture conditions modification. Exp Hematol. 2008;36(8):1014–21.

[117] Both SK, Van der Muijsenber AJC, van Blitterswijk CA, Boer J, Al. E. A rapid and efficient method for expansion of human mesenchymal stem cells. Tissue Eng. 2007;13(1):3–9.

[118] Bartmann C, Rohde E, Schallmoser K, Al. E. Two steps to functional mesenchymal stromal cells for clinical applications. Transfusion. 2007;47(8):1426–35.

[119] Hewitt CJ, Lee K, Nienow AW, Thomas RJ, Smith M, Thomas CR. Expansion of human mesenchymal stem cells on microcarriers. Biotechnol Lett. 2011 Nov;33(11):2325–35.

[120] Neuhuber B, Swanger SA, Howard L, Mackay A, Fischer I. Effects of plating density and culture time on bone marrow stromal cell characteristics. Exp Hematol. 2008;36(9):1176–85.

[121] Stolberg S, McCloskey KE. Can shear stress direct stem cell fate? Biotechnol Prog. 2009;25(1):10–9.

[122] Eibl R, Eibl D, Pörtner R, Catapano G, Czermak P. Cell and Tissue Reaction Engineering. Springer-Verlag Berlin Heidelberg; 2009.

[123] Cherry RS, Papoutsakis TE. Hydrodynamic effects on cells in agitated tissue culture reactors. Bioprocess Eng. 1986;1:29–41.

[124] Chisti Y. Hydrodynamic damage to animal cells. Crit Rev Biotechnol. 2001;21(2):67–110.

[125] Croughan MS, Hamel JF, Wang DIC. Hydrodynamic effects on animal cells grown in microcarrier cultures. Biotechnol Bioeng. 1987;29:130–41.

[126] Platas OB, Jandt U, Phan LM, Villanueva ME, Schaletzky M, Rath A, et al. Evaluation of criteria for bioreactor comparison and operation standardisation for mammalian cell culture. Eng Life Sci. 2012;12(5):518–28.

[127] Löffelholz C, Husemann U, Greller G, Meusel W, Kauling J, Ay P, et al. Bioengineering Parameters for Single-Use Bioreactors: Overview and Evaluation of Suitable Methods. Chemie Ing Tech. 2013;85(1-2):40–56.

[128] Zhang H, Wang W, Quan C, Al. E. Engineering considerations for process development in mammalian cell cultivation. Curr Pharm Biotechnol. 2010;11(1):103–12.

[129] Oncul AA, Kalmbach A, Genzel Y, Al. E. Characterization of flow conditions in 2 L and 20 L wave bioreactors using computational fluid dynamics. Biotechnol Prog. 2010;26(1):101–10.

[130] Singh H, Hutmacher DW. Bioreactor studies and computational fluid dynamics. Adv Biochem Eng Biotechnol. 2009;112:231–49.

[131] Patrachari AR, Podichetty JT, Madihally S V. Application of computational fluid dynamics in tissue engineering. J Biosci Bioeng. 2012;114(2):123–32.

[132] Liovic P, Šutalo ID, Stewart R, Glattauer V, Meagher L. Fluid flow and stresses on microcarriers in spinner flask bioreactors. Ninth Int Conf CFD Miner Process Ind. 2012;1–6.

[133] Song K, Wang H, Zhang B, Al. E. Numerical simulation of fluid field and in vitro three-dimensional fabrication of tissue-engineered bones in a rotating bioreactor and in vivo implantation for repairing segmental bone defects. Cell Stress Chaperones. 2013;18(2):193–201.

[134] Burdick JA, Vunjak-Novakovic G. Engineered microenvironments for controlled stem cell differentiation. Tissue Eng Part A. 2009;15(2):205–19.

[135] Anderson EJ, Kaliyamoorthy S, Iwan J, Alexander D, Knothe T, Melissa L. Nano-microscale models of periosteocytic flow show differences in stresses imparted to cell body and processes. Ann Biomed Eng. 2005;33(1):52–62.

[136] Zhang H, Dau S, Bi J, Liu KK. Biomimetic three-dimensional microenvironment for controlling stem cell fate. Interface Focus. 2011;1(5):792–803.

[137] Higuera GA, van Boxtel A, van Blitterswijk CA, Moroni L. The physics of tissue formation with mesenchymal stem cells. Trends in Biotechnol. 2012;30(11):583–90.

[138] Yeatts AB, Chouquette DT, Fisher JP. Bioreactors to influence stem cell fate: augmentation of mesenchymal stem cell signaling pathways via dynamic culture systems. BiochimBiophysActa. 2013;1830(2):2470–80.

[139] Adamol L, Garcia-Cardena G. Directed stem cell differentiation by fluid mechanical forces. Antioxid Redox Signal. 2011;15:1463–73.

[140] Chang H, Wang Y. Cell Responses to Surface and Architecture of Tissue Engineering Scaffolds. Regenerative Medicine and Tissue Engineering-Cells and Biomaterials. InTech; 2011.

[141] Prockop DJ, Brenner M, Fibbe WE, Horwitz E, Le BK, Phinney DG, et al. Defining the risks of mesenchymal stromal cell therapy. Cytotherapy. 2010;12:576–8.

[142] Prockop DJ, Oh JY. Medical therapies with adult stem/progenitor cells (MSCs): a backward journey from dramatic results in vivo to the cellular and molecular explanations. J Cell Biochem. 2012;113:1460–9.

[143] Luo W, Xiong W, Zhou J, Fang Z, Chen W, Fan Y, et al. Laminar shear stress delivers cell cycle arrest and anti-apoptosis to mesenchymal stem cells. Acta Biochim Biophys Sin. 2011;43:210–6.

[144] Maul TM, Chew DW, Nieponice A, Vorp DA. Mechanical stimuli differentially control stem cell behavior: morphology, proliferation, and differentiation. Biochem Model Mechanobiol. 2011;10:939–53.

[145] Yi W, Sun Y, Wei X, Gu C, Dong X, Kang X, et al. Proteomic profiling of human bone marrow mesenchymal stem cells under shear stress. Mol Cell Biochem. 2010;341:9–16.

[146] Zhang H, Kay A, Forsyth NR, Liu KK, El Haj AJ. Gene expression of single human mesenchymal stem cell in response to fluid shear. J Tissue Eng. 2012;3(1).

[147] Zheng W, Xie Y, Zhang W, Wang D, Ma W, Wang Z, et al. Fluid flow stress induced contraction and re-spread of mesenchymal stem cells: a microfluidic study. Integr Biol. 2012;4(9):1102–11.

[148] Barron MJ, Tsai CJ, Donahue S, Al. E. Mechanical stimulation mediates gene expression in MC3T3 osteoblastic cells differently in 2D and 3D environments. J Biomech Eng. 2010;132(4).

[149] Kim J, Ma T. Bioreactor strategy in bone tissue engineering: pre-culture and osteogenic differentiation under two flow configurations. Tissue Eng Part A. 2012;18(21-22):2354–64.

[150] Li D, Tang T, Lu J, Al. E. Effects of flow shear stress and mass transport on the construction of a large-scale tissue-engineered bone in a perfusion bioreactor. Tissue Eng Part A. 2009;15(10):2773–83.

[151] Liu YB, Chen J, Tang Z, Al. E. Different effects of intermittent and continuous fluid shear stresses on osteogenic differentiation of human mesenchymal stem cells. Biomech Model Mechanobiol. 2012;11(3-4):391–401.

[152] Yourek G, McCormick SM, Mao JJ, Reilly GC. Shear stress induces osteogenic differentiation of human mesenchymal stem cells. Regen Med. 2010;5(5):713–24.

[153] Yu X, Botchwey EA, Levine EM, Al. E. Bioreactor-based bone tissue engineering: the influence of dynamic flow on osteoblast phenotypic expression and matrix mineralization. Proc Natl Acad Sci. 2004;101(31):11203–8.

[154] Grellier M, Bareille R, Bourget C, Al. E. Responsiveness of human bone marrow stromal cells to shear stress. J Tissue Eng Regen Med. 2009;3(4):302–9.

[155] Da Silva AML, Marins A, Costa-Pinto AR, Al. E. Chondrogenic differentiation of human bone marrow mesenchymal stem cells in chitosan-based scaffolds using a flow-perfusion bioreactor. J Tissue Eng Regen Med. 2011;5(9):722–32.

[156] Wescoe KE, Schugar RC, Chu CR, Al. E. The role of the biochemical and biophysical environment in chondrogenic stem cell differentiation assays and cartilage tissue engineering. Cell Biochem Biophys. 2008;52(2):85–102.

[157] Pörtner R, Hsu HH, Goepfert C. Bioreaktoren für Knochen Tissue Engineering. Osteologie. 2013;22(3):188–95.

[158] Weinbaum S, Cowin SC, Zeng Y. A model for the excitation of osteocytes by mechanical loading induced bone fluid shear stresses. J Biomech. 1994;27:339–60.

[159] Holtorf HL, Sheffield TL, Ambrose CG, Al. E. Flow perfusion culture of marrow stromal cells seeded on porous biphasic calcium phosphate ceramics. Ann Biomed Eng. 2005;33(9):1238–48.

[160] Bjerre L, Bünger CE, Kassern M, Al. E. Flow perfusion culture of human mesenchymal stem cells on silicate-substituted tricalcium phosphate scaffolds. Biomaterials. 2008;29:2616–27.

[161] Datta N, Pham QP, Sharma U, Al. E. In vitro generated extracellular matrix and fluid shear stress synergistically enhance 3D osteoblastic differentiation. PNAS. 2006;103(8):2488–93.

[162] Zhao F, Chella R, Ma T. Effects of shear stress on 3-D human mesenchymal stem cell construct development in a perfusion bioreactor system: Experiments and hydrodynamic modeling. Biotechnol Bioeng. 2007;96:584–95.

[163] Wang LH, Hu YY, Wang Z, Al. E. Flow perfusion of human fetal bone cells in large beta-tricalcium phosphate scaffold with controlled architecture. J Biomed Mat Res Part A. 2008;

[164] Maes F, van Ransbeeck P, van Oosterwyck H, Al. E. Modeling fluid flow through irregular scaffolds for perfusion bioreactors. BiotechnolBioeng. 2009;103(3):621–30.

[165] Chen Y, Zhou S, Cadman J, Al. E. Design of cellular porous biomaterials for wall shear stress criterion. Biotechnol Bioeng. 2010;107(4):737–46.

[166] Xu S, Pigan D, Youthuan X, Al. E. Cell distribution in a scaffold with random architectures under the influence of fluid dynamics. J Biomater Appl. 2008;23(3):229–45.

[167] Bjerre L, Bünger C, Baatrup A, Al. E. Flow perfusion culture of human mesenchymal stem cells on coralline hydroxyapatite scaffolds with various pore sizes. J Biomed mater Res A. 2011;97(3):251–63.

[168] Nienow AW, Qasim A, Rafiq K, Al. E. A Potentially Scalable Method for the Harvesting of hMSCs from Microcarriers. Eng J. 2014;

[169] Cormier JT, Zur Nieden NI, Rancourt DE, Kallos MS. Expansion of undifferentiated murine embryonic stem cells as aggregates in suspension culture bioreactors. Tissue Eng. 2006;12(11):3233–45.

[170] Keller KC, Rodrigues B, Zur Nieden NI. Suspension culture of pluripotent stem cells: effect of shear on stem cell fate. Crit Rev Eukaryot Gene Expr. 2014;24(1):1–13.

[171] Kresnowati MTAP, Forde G, Chen XD. Model-based analysis and optimization of bioreactor for hematopoietic stem cell cultivation. Bioprocess Biosyst Eng. 2011;34(1): 81–93.

[172] Sen A, Kallos MS, Behie LA. Expansion of mammalian neural stem cells in bioreactors: effect of power input and medium viscosity. Brain Res Dev Brain Res [Internet]. 2002 Mar 31;134(1-2):103–13. Available from: http://www.ncbi.nlm.nih.gov/pubmed/11947941

[173] Blüml G. Microcarrier cell culture technology. In: Pörtner R, editor. Animal cell biotechnology. 2007. p. 149–78.

[174] Rowley J, Abraham E, Campbell A, Brandwein H, Oh S. Meeting lot-size challenges of manufacturing adherent cells for therapy. BioProcess Int. 2012;10:16–22.

[175] Aunins JG, Bader B, Caola A. Fluid mechanics, cell distribution and environment in CellCube Bioreactors. Biotechnol Prog. 2003;19:2–8.

[176] Goltry KL, Hampson BS, Venturi NA, Batel RL. Adult stem cell therapies for tissue regeneration: Ex vivo expansion in an automated system. In: Yanhong S, Clegg DO, editors. Stem Cell research and Therapeutics. Springer; 2008. p. 251–74.

[177] Reichardt A, Polchow B, Shakibaei M, Henrich W, Hetzer R, Lueders C. Large scale expansion of human umbilical cord cells in a rotating bed system for cardiovascular tissue engineering applications. Open Biomed Eng J. 2014;7:50–61.

[178] Chen HC, Hu YC. Bioreactors for tissue engineering. Biotechnol Lett. 2006;28:1415–23.

[179] Eibl R, Kaiser S, Lombriser R, Eibl D. Disposable bioreactors: the current state-of-the art and recommended applications in biotechnology. Appl Microbiol Biotechnol. 2010;86:41–9.

[180] Yan ST, Liu X. Cell culture processes for biologics manufacturing: recent developments and trends. Pharm Bioprocess. 2013;1:133–6.

[181] Shukla AA, Gottschalk U. Single-use disposable technologies for biopharmaceutical manufacturing. Trends in Biotechnol. 2013;31:147–54.

[182] Kaiser SC, Kraume M, Eibl D, Eibl R. Single-use bioreactors for animal and human cells. In: Rubeai M, editor. Cell engineering vol 9. Springer; 2014. p. (in press).

[183] Eibl R, Löffelholz C, Eibl D. Single-use bioreactors: an Overview. In: Eibl R, Eibl D, editors. Single Use Technology in Biopharmaceutical Manufacture. Wiley VCH; 2011. p. 33–51.

[184] Horvath B, Tsang VL, Lin W, Dai XP, Kunas K, Frank GA. A generic growth test method for improving quality control of disposables in industrial cell culture. BioPharm Int. 2013;23:34–41.

[185] Wood J, Mahaian E, Shiratori M. Strategy for selecting disposable bags for cell culture media applications based on root-cause investigations. Biotechnol Prog. 2013;

[186] Hammond N, Nunn H, Rogers G, Lee H, Marghitou AL, Nashed-Samuel Y, et al. Identification of a leachable compound detrimental to cell growth in single-use bioprocess containers. PDA J Pharm Sci Technol. 2014;67:123–34.

[187] Steiger N, Eibl R. Interlaboratory test for detection of cytotoxic leachables arising from single-use bags. Chemie Ing Tech. 2013;85:26–8.

[188] Hammond M, Marghitoiu L, Lee H, Perez L, Rogers G, Nashed-Samuel Y. A cytotoxic leachable compound from single-use bioprocess equipment that causes poor cell growth performance. Biotechnol Prog. 2014;30(2):332–7.

[189] Ott A. Kultivierung und Zellrückgewinnung von humanen mesenchymalen Stammzellen in einem modifizierten Bioreaktor. Zurich University of Applied Science, Switzerland, Master's Thesis; 2014.

[190] Kaiser SC, Eibl R, Eibl D. Engineering characteristics of a single-use stirred bioreactor at bench-scale: The Mobius CellReady 3L bioreactor as a case study. Eng Life Sci. 2011 Aug 28;11(4):359–68.

[191] Cierpka K, Elseberg CL, Niss K, Kassem M, Salzig D, Czermak P. hMSC Production in Disposable Bioreactors with Regards to GMP and PAT. Chemie Ing Tech. 2013 Feb 19;85(1-2):67–75.

[192] Jing D, Sunil N, Punreddy S, Aysola M. Growth Kinetics of Human Mesenchymal Stem Cells in a 3-L Single-Use, Stirred-Tank Bioreactor. BioPharm Int. 2013;26:28–38.

[193] Stadler M. Stammzellkultivierung auf Microcarriern und Zellernte. Zurich University of Applied Sciences, Switzerland, Bachelor thesis; 2013.

[194] Kaiser S, Löffelholz C, Werner S, Eibl D. CFD for characterizing standard and single-use stirred cell culture bioreactors. Computational Fluid Dynamics. InTech; 2011. DOI:10.5572/23496

[195] Löffelholz C. CFD als Instrument zur bioverfahrenstechnischen Charakterisierung von single-use Bioreaktoren und zum Scale-up für Prozesse zur Etablierung und Produktion von Biotherapeutika. Brandenburg University of Technology, Germany, PhD thesis; 2013.

[196] Schirmaier C, Jossen V, Kaiser SC, Jüngerkes F, Brill S, Safavi-Nab A, et al. Scale-up of adipose tissue-derived mesenchymal stem cell production in stirred single-use bioreactors under low-serum conditions. Eng Life Sci. 2014;14(3):292–303.

[197] Jossen V, Kaiser SC, Schirmaier C, Herrmann J, Tappe A, Eibl D, et al. Modification and qualification of a stirred single-use bioreactor for the improved expansion of human mesenchymal stem cells at benchtop scale. Pharm Bioprocess. 2014;2(4):311-322.

[198] Eibl R, Werner S, Eibl D. Bag bioreactor based on wave-induced motion: characteristics and applications. Disposable bioreactors. Springer-Verlag Berlin Heidelberg; 2010.

[199] Werner S, Eibl R, Lettenbauer C, Röll M, Eibl D, Jesus M. Innovative, non-stirred bioreactors in scales from milliliters up to 1000 liters for suspension cultures of cells using disposable bags and containers. A Swiss contribution. Chimia (Aarau). 2010;64:819–23.

[200] Hundt B, Best C, Schlawin N, Kassner H, Genzel Y, Reichel U. Establishment of a mink eneritis vaccine production process in stirred-tank reactor and Wave Bioreactor microcarrier culture in 1-10 L scale. Vaccine. 2007;25:3987–95.

[201] Genzel Y, Behrendt I, König S, Sann H, Reichl U. Metabolism of MDCK cells during cell growth and influenza virus production in large-scale microcarrier culture. Vaccine. 2004;22:2202–8.

[202] Genzel Y, Dietzsch C, Rapp E, Schwarzer J, Reichl U. MDCK and Vero cells for influenza virus vaccine production: a one-to-one comparison up to lab-scale bioreactor cultivation. Appl Microbiol Biotechnol. 2010;88:461–75.

[203] Slivac I, Srcek VG, Radoservic K, Kmetic I, Kniewald Z. Aujeszky's disease virus production in disposable bioreactor. J Biosci. 2006;31(3):363–8.

[204] Hami LS, Green C, Leshinsky N, Markham E, Miller K, Craig S. GMP production and testing of xcellerated t cells for the treatment of patients with cell. Cytotherapy. 2004;6:554–62.

[205] Akerström. Expansion of adherent cells for cell therapy. Uppsala University, Sweden; 2009.

[206] Timmins NE, Kiel M, Günther M, Heazlewood C, Doran MR, Brooke G, et al. Closed system isolation and scalable expansion of human placental mesenchymal stem cells. Biotechnol Bioeng. 2012;109(7):1817–26.

[207] Shyu Y, Stratheam KE, Eglen RM. Large-scale expansion of stem cells for therapy and screening. Drug Discov World Winter. 35–9.

[208] Peng CA, Palsson BO. Stem cell replication and differentiation in tissue engineering bioreactors strongly influenced by bioreactor geometry. Biotechnol Bioeng. 1996;50:479–92.

[209] Nielsen LK. Bioreactors for Hematopoietic Cell Culture. Annu Rev Biomed Eng. 1999;1:129–52.

[210] Whitford WG, Hardy JC, Cadwell JJS. Single-Use, Continous Processing of Primary Stem Cells. Cell Ther Process. 2014;12(3):27–33.

[211] Knazek RA, Wu Y-W, Aebersold PM, Rosenberg SA. Culture of himan tumor infiltrating lymphocytes in hollow fibre bioreactors. Immunol Methods. 1990;127(1):29–37.

[212] Malone CC, Scholtz PM, Mackintosh AD, Beutel LD, Heinemann FS, Dillman RO. Characterization of human tumor-infiltrating lymphocytes expanded in hollow-fiber bioreactors for immunotherapy of cancer. Radiopharm. 2001;16:381–90.

[213] De Bartolo L, Piscioneri A, Cotroneo G, Salerno S, Tasselli F, Campana C, et al. Human lymphocyte peek-wc hollow fiber membrane bioreactor. J Biotechnol. 2007;132(1):65–74.

[214] Pan D, Whitley CB. Closed hollow-fiber bioreactor: a new approach to retroviral vector production. Gene Med. 1999;1:433–40.

[215] Gardner TA, Ko SC, Yang L, Cadwell J j. Serum-free recombinant production of adenovirus using a hollow fiber capillary system. Biotechniques. 2001;30:422–7.

[216] Gramer MJ, Maas J. Optimal NS0 cell growth in a hollow fiber bioreactor requires increased serum concentration or a cholesterol supplement on the cell side of the fiber. Biotechnol Prog. 2003;19:1762–6.

[217] Kalbfuss B, Grenzel Y, Wolff M, Zimmermann A, Morenweiser R, Reichl U. Harvesting and concentration of human influenza A virus produced in serum-free mammalian cell culture for the production of vaccines. Biotechnol Bioeng. 2007;97(1):73–85.

[218] Kuhn J, Mölle K, Brinkmann T, Götting C, Kleesiek K. High-density tissue-like cultivation of JAR chorocarcinoma cells for the in vitro production of human xylosyltransferase. J Biotechnol. 2003;103:191–6.

[219] Hopkinson J. Hollow fibre cell culture systems for economical cell-product manufacturing. Biotechnology. 1985;3:225–30.

[220] Gorter A, van de Gried R, van Eendenburg JD, Haasnot WH, Fleuren G. Production of bispecific monoclonal antibodies in a hollow-fibre bioreactor. J Immunol Methods. 1993;161:145–50.

[221] Marx U. Membrane-based cell culture technologies: A scientifically, economically satisfactory alternative malignant ascites production for monoclonal antibodies. Res Immunol. 1998;6:557–9.

[222] Abu-Abs SF, Seth G, Narayana RA, Groehler K, Lai P, Anderson ML, et al. Characterization of a hollow fiber bioartificial liver device. Atif Organs. 2005;29(5):419–22.

[223] Curcio E, De Bartolo L, Barbieri G, Rende M, Giorno L, Morelli S. Diffusive and convective transport through hollow fiber membranes for liver cell culture. J Biotechnol. 2005;117:309–21.

[224] Men Q, Zhang G, Wu D. Hepatocyte culture in bioartifical livers with different membrane characteristics. Biotechnol Lett. 2004;26:1407–12.

[225] Nordon RE, Schindhelm K. Design of hollow fiber modules for uniform shear elution affinity cell separation. Artif Organs. 1997;21(2):107–15.

[226] Godara P, McFarland CD, Nordon RE. Design of bioreactors for mesenchymal stem cell tissue engineering. J Chem Technol. 2008;83(4):408-20

[227] Hambor J. Bioreactor design and bioprocess controls for industrialized cell processing. BioProcess Int. 2012;10(6):22–33.

[228] Rojewski MT, Fekete N, Baila S, Nguyen K, Fürst D, Antwiler D, et al. GMP-compliant isolation and expansion of bone marrow-derived MSCs in the closed, automated device Quantum Cell Expansion System. Cell Transplant. 2013;22(11):1981–2000.

[229] Jones M, Varella-Garcia M, Skokan M, Bryce S, Schowinsky J, Peters R, et al. Genetic stability of bone marrow-derived human mesenchymal stromal cells in the Quantum System. Cytotherapy. 2013;15(11):1323–39.

[230] Hanley PJ, Mci Z, Durette AG, Cabreira-Harrison MD, Klis M, Li W, et al. Efficient manufacturing of therapeutic mesenchymal stromal cells with the use of the Quantum Cell Expansion System. Cytotherapy. 2014;16(8):1048–58.

[231] Wang TY, Brennan JK, Wu JH. Multilineal hematopoiesis in a three-dimensional murine long-term bone marrow culture. Exp Hematol. 1995;23(1):26–32.

[232] Warnrock JN, Bratch K, Al-Rubeai M. Packed bed bioreactors. In: Chaudhuri J, Al-Rubeai M, editors. Bioreactors for tissue engineering. Springer; 2005. p. 87–113.

[233] Mantalaris A, Keng P, Bourne P, Cheng A, Wu J. Engineering a human bone marrow model: a case study on ex vivo erythropoiesis. Biotechnol Prog. 1998;14(1):126–33.

[234] Meissner P, Schröder B, Herfurth C, Biselli M. Development of a fixed bed bioreactor for the expansion of human hematopoietic progenitor cells. Cytotechnology. 1999 Jul; 30(1-3):227–34.

[235] Jelinek N, Schmidt S, Hilbert U, Thoma S, Biselli M, Wandrey C. Novel bioreactors for the ex vivo cultivation of hematopoietic cells. Eng Life Sci. 2002;7(1):15–8.

[236] Meuwly F, Loviat F, Ruffieux PA, Bernard AR, Kadouri A, von Stocker U. Oxygen supply for CHO cells immobilized on a packed-bed of Fibra-Cell disks. Biotechnol Bioeng. 2006;93:791–800.

[237] Meuwly F, Ruffieux PA, Kadouri A, von Stocker U. Packed-bed bioreactors for mammalian cell culture: bioprocess and biomedical applications. Biotechnol Adv. 2007;25:45–56.

[238] Lennaertz A, Knowles S, Drugmand JC, Castillo J. Viral vector production in the Integrity iCELL is single-use fixed-bed bioreactor, from bench-scale to industrial scale. BMC Proc. 2013;7:59.

[239] Weber C, Pohl S, Pörtner R, Wallrapp C, Kassern M, Geigle P, et al. Cultivation and differentiation of encapsulated hMSC-TERT in a disposable small-scale syringe-like fixed bed reactor. Open Biomed Eng J. 2007;1:64–70.

[240] Kunas KT, Papoutsakis TE. Damage mechanisms of suspended animal cells in agitated bioreactors with and without bubble entrainment. Biotechnol Bioeng. 1990;36(5): 476–83.

[241] Shiragami N. Effect of shear rate on hybridoma cell metabilism. Bioprocess Eng. 1997;16(6):345–7.

[242] Nienow AW. Reactor engineering in large scale animal cell culture. Cytotechnology. 2006;50(1):9–33.

[243] Huang Y, Xiaoling L, Bai K, Gong X, Fan Y. Effect of Fluid Shear Stress on Cardiomyogenic Differentiation of Rat Bone Marrow Mesenchymal Stem Cells. Arch Med Res. 2010;41:497–505.

[244] Thomassen YE, van der Well JE, van Eikenhorst G, van der Pol LA, Bakker WAM. Transfer of an adherent Vero cell culture method between two different rocking motion type bioreactors with respect to cell growth and metabolic rates. Process Biochem. 2012;47(2):288–96.

[245] Nienow AW. On impeller circulaton and mixing evectiveness in the turbulent flow regime. Chem Eng Sci. 1997;52(15):2557–65.

[246] Chaubard JF, Dessoy S, Ghislain Y, Gerkens P, Barhier B, Battisti R, et al. Disposable bioreactors for viral vaccine production: challenges and opportunities. BioPharm Int Suppl. 2010; http://www.biopharminternational.com/biopharm/Disposables/Disposable-Bioreactor-for-Viral-Vaccine-Production/ArticleStandard/Article/detail/695153.

[247] Junker BH. Scale-up methologies for Escherichia coli and yeast fermentation processes. J Biosci Bioeng. 2004;97(6):347–64.

[248] Xing Z, Kenty BM, Li ZJ, Lee SS. Scale-up analysis for a CHO cell culture process in large-scale bioreactors. Biotechnol Bioeng. 2009;103(4):733–46.

[249] Löffelholz C, Kaiser SC, Kraume M, Eibl R, Eibl D. Dynamic Single-Use Bioreactors Used in Modern Liter-and m3-Scale Biotechnological Processes: Engineering Characteristics and Scaling UP. AdvBiochemEngBiotechnol. 2014;138:1-41

[250] Zwietering T. Suspending of solid particles in liquid by agitators. Chem Eng Sci. 1958;8(3-4):244–53.

[251] Liepe F, Sperling R, Solomon J. Rührwerke: Theoretische Grundlagen, Auslegung und Bewertung. Eigenverlag Fachhochschule Köthen; 1998.

[252] Ibrahim S, Nienow AW. Suspension of Microcarriers for Cell Culture with Axial Flow Impellers. Chem Eng Res Des. 2004;82(9):1082–8.

[253] Van Heimburg D, Hemmrich K, Zachariah S, Staiger H, Pallua N. Oxygen consumption in undifferentiated versus differentiated adipogenic mesenchymal precursor cells. Respir Physiol Neurobiol. 2005;146(3):107–16.

[254] Zhao F, Pathi P, Garyson W, Xing Q, Locke BR, Ma T. Effects of oxygen transport on 3-d human mesenchymal stem cell metabolic activity in perfusion and static cultures: experiments and mathematical model. Biotechnol Prog. 2005;21(4):1269–80.

[255] Kaiser SC, Jossen V, Schirmaier C, Eibl D, Brill S, van den Bos C, et al. Fluid Flow and Cell Proliferation of Mesenchymal Adipose-Derived Stem Cells in Small-Scale, Stirred, Single-Use Bioreactors. Chemie Ing Tech. 2013;85(1-2):95–102.

[256] Jung S, Panchalingam KM, Rosenberg L, Behie LA. Ex vivo expansion of human mesenchymal stem cells in defined serum-free media. Stem Cells Int. 2012;2012:21 pages.

[257] Ponnuru K, Wu J, Ashok P, Tzanakakis E, Furlani EP. Analysis of Stem Cell Culture Performance in a Microcarrier Bioreactor System. Proc Int NSTI Nanotech Conf. 2014;2:132–5.

[258] Löffelholz C, Kaiser SC, Kraume M, Eibl R, Eibl D, Löffelholz C, et al. Dynamic Single-Use Bioreactors Used in Modern Liter-and m 3-Scale Biotechnological Processes: Engineering Characteristics and Scaling Up. Biochem Eng Biotechnol. 2014;138:1–41.

Chapter 6

Non-Invasive Imaging Modalities for Clinical Investigation in Regenerative Medicine

Deana G. Haralampieva, Simon M. Ametamey,
Tullio Sulser and Daniel Eberli

Additional information is available at the end of the chapter

http://dx.doi.org/10.5772/59356

1. Introduction

Recent developments in the field of regenerative medicine highlighted the emerging need for cellular therapies. This novel approach is at the doorstep to its clinical application. Nevertheless, there is still a lack of consistent and unbiased data on the survival, distribution and safety of implanted cells. The tight collaboration between pharmacology, chemistry, biology and physics expanded the collection of methods for *in vivo* imaging and allowed for real time assessment of functional read-outs, localization and viability of the implants. The successful implementation of stem cell visualization will provide for improvements in many healthcare fields. Of key importance is the identification of the most valuable method for non-invasive monitoring of cells and affected tissue. In this chapter we will assess the advantages and drawbacks of different imaging techniques, focusing specifically on cell therapy and methods utilized for tissue engineering and regenerative medicine.

Organ transplantation still remains the gold standard for the treatment of terminally damaged organs. This method brings along a collection of drawbacks like the shortage of available donor organs and the high morbidity of immunosuppressive therapy. Regenerative medicine offers an alternative for the replacement of organs and tissues overcoming pitfalls and proposing a therapy with patient's own cells. The discovery of the self-renewal and differentiation capacities of stem cells accelerated the development of regenerative medicine and inspired researchers and clinicians to search for novel treatment options for the improvement of the quality of life of patients [1]. The tremendous success in medical sciences and biotechnology allowed the development of many ideas, some of which are now being approved by clinical trials. Nearly 26'000 studies utilizing cell therapy are currently ongoing [2] trying to establish the safety and effectiveness of this new method for applications in neurology, cardiology or oncology. Autologous stem cell therapy is a novel promising approach with increasing impact on regenerative medicine. Preliminary data of many preclinical studies on different human disorders showed a beneficial effect of this treatment [3]. The usage of the patient's own cells to rebuild tissues and regain functionality after trauma is currently investigated, as it may have a profound influence on many human diseases once its application is approved. Several hurdles are yet to be overcome before the safety of this method is guaranteed. One possibility to avoid the post mortem read-out of a study or the invasiveness of a biopsy is the use of *in vivo* imaging. Succeeding in this would also significantly reduce the number of animal experiments to be performed. The emerging need for "real time" visualization of cells and tissue gives rise to many questions in the field of regenerative medicine. Many different cellular therapies are on the door step into clinics and a method for non-invasive tracking of transplanted cells and defining their fate, functionality and differentiation is of great importance. Of high interest is also the visualization of their effect on the damaged tissue or organ.

In the past two decades many researchers concentrated on developing new modalities for better visualization on the cellular and molecular level. As none of the available imaging systems fulfils all needs, many efforts have been put into expanding the potential of the available methods by testing the application of a variety of vectors, genes, proteins and cells for precise read-outs. The visualization techniques relevant to clinical cell tracking include positron emission tomography (PET), single-photon emission computed tomography (SPECT) and magnetic resonance imaging (MRI) [4]. Hybridizations of computed tomography (CT) and MRI with PET and SPECT allow simultaneous acquisition of anatomical information and reveal new state-of-the-art modalities for successful molecular imaging (SPECT/CT, PET/CT, PET/MRI) [5]. Generally, the utilization of radioactive tracers provides high sensitivity and requires short image acquisition time. Nevertheless, the production of the radioactive isotopes is very expensive and its application comes along with radiation exposure for the patient. Therefore, MRI seems to be an attractive alternative devoid of harmful radiation. This modality has spectacular image resolution and anatomical read-out capabilities [6]. However, it is more time-consuming and lacks the high sensitivity of PET and SPECT [7] (see Table 1). There are also a few optical imaging modalities using bioluminescence (BLI) or fluorescence (FI), which have been widely utilized in small animal models [8]. These are cost-saving methods (using fluorescent proteins or luciferase activity) which allow for very rapid result obtainment. Nevertheless, their spatial resolution is low and decreases with depth, which is the main reason why they are not applicable for life cell imaging in humans.

Many cells are potentially accessible by imaging and developments in this field already support the diagnosis and therapy of several human diseases. However, a feasible technique for non-invasive monitoring of the tissue engineering process is still missing. A variety of cells was shown to be applicable for tissue and organ restoration, but yet little is known about the safety, viability, functionality and migration of these cells *in vivo* and their direct or indirect involvement in the healing process. Despite the abundance of new techniques for non-invasive visualization of cells, their translation into the clinic is challenging. Hence, here we will discuss the advantages and drawbacks of different imaging modalities applicable for humans and their implementation in the diagnostics and monitoring of the treatment of various diseases located in the hot spot of current clinical trials. Moreover, we will consider potential strategies for expansion of the available techniques, in particular for successful tissue engineering and regenerative medicine.

Imaging Modality	Spatial Resolution	Temporal Resolution	Sensitivity	Safety	Clinically Applied	Cost
CT	50-200 μm (preclinical) 0.5-1 mm (clinical)	minutes	not defined	Ionizing radiation	Yes	$$
MRI	25-100 μm (preclinical) ~1 mm (clinical)	minutes-hours	$10^{-3} – 10^{-5}$M poor	No ionizing radiation	Yes	$$$
PET	1-2 mm (preclinical) 5-7mm (clinical)	seconds-minutes	$10^{-11} – 10^{-12}$M excellent	Ionizing radiation	Yes	$$$
SPECT	1-2 mm (preclinical) 8-10 mm (clinical)	minutes	$10^{-10} – 10^{-11}$M excellent	Ionizing radiation	Yes	$$

Table 1. Characterizations of clinical imaging modalities (adapted from [9]).

2. MR imaging

2.1. MRI basics

The phenomenon of Magnetic Resonance Imaging (MRI) was discovered observing the reaction of certain nuclei, mostly ^1H in the body, when placed in a strong magnetic field (B0) [10]. Within a magnetic field, the sum of spins is a net magnetization aligned with the applied field. The macroscopic magnetization is an effect of a slight excess of spins in "low energy" state, in parallel to the B0 direction. The precession frequency of each atom with magnetic moment is directly proportional to the strength of the magnetic field. After applying a radiofrequency (RF) pulse which matches the precession frequency, some of the "excess" atoms absorb energy and flip, thereby changing the direction of the net magnetization vector. The flip angle depends on the intensity, waveform and duration of the RF pulse [11]. These atoms then re-emit the energy during transition to their original orientation (relaxation) (Figure 1).

As already shown, MRI is based on a unique combination of radio waves and strong magnetic fields, which allows the generation of cross-sectional images of the body in any plane. Depending on the relaxation time after the RF pulse, one can distinguish between T1-and T2-weighted images.

MRI is an imaging modality, which uses non-ionizing electromagnetic radiation and is thus not harmful for patients. This is one of the strongest advantages of this technology over the other methods described in this chapter.

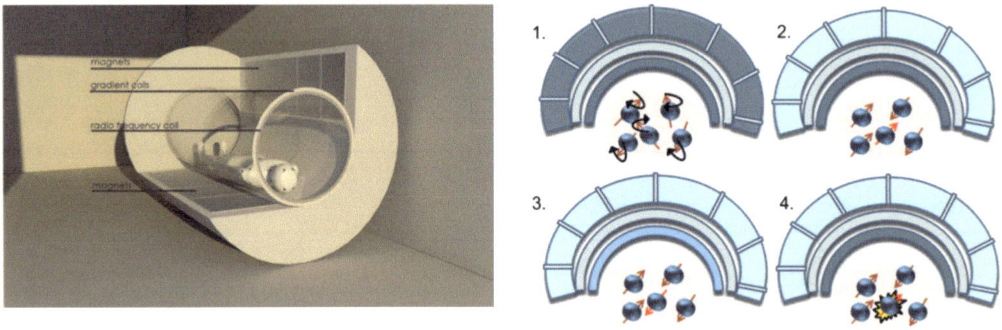

Figure 1. MRI scanner uses radiowaves and strong magnetic fields to obtain anatomical and functional body images. 1) Atoms spin in random directions. 2) Applying a magnetic field makes atoms line up either in the direction of B0, or in the opposite direction. The distribution is not exactly equal, as more atoms prefer to turn in "low energy" state, thereby creating a net magnetization parallel to B0. 3) Introducing a RF pulse makes half of the "excess" atoms change the direction of their magnetization moment-to the "high energy" state. 4) Turning the RF off makes these atoms turn back to their "low energy" position (relaxation), thereby emitting energy.

2.2. MRI applications

One of the biggest advantages of MRI is its high soft tissue contrast. This allows for very precise images and simultaneous detection of functional, molecular and anatomical data. The tight collaboration of science and medicine led to the clinical implementation of the MRI modality for various diseases. Nevertheless, most of its potential remains at the pre-clinical state, as the application safety of many methods is still under investigation. Cerebral perfusion imaging, MR angiography flow imaging, cardiac MRI, functional MRI and MR spectroscopy are just a few of the MRI sub-groups. However, up-to-date, MR is clinically mainly used for the imaging of soft tissues.

A clinical trial study with patients suffering from acute myocardial infarction showed successful implementation of MRI. Intracoronary infusions of bone marrow-derived stem cells at different time points helped with cardio regeneration, which could be visualized by this technique [12]. Another study is on its doorstep to clinics, gathering breast cancer patients for breast restoration. The aim is to use MRI of autologous adipose-derived stem cells to visualize the engraftment into the host tissue. The outcome of this study may have a great impact on the utilization of MRI for cell tracking as a novel tool for breast tissue regeneration [13]. A further example for the implementation of MRI for cell therapy studies was the evaluation in a multiple sclerosis clinical trial. In this case, adult bone marrow derived stromal cells were intravenously administered to the patients and MRI scans were performed to follow the occurring changes. The main objective was to evaluate the effect of mesenchymal stem cell transplantation on the number of Gadolinium (Gd)-positive lesions [14]. Another way of addressing the safety and efficacy of using autologous mesenchymal stem cells as a possible therapy for multiple sclerosis is being proposed by a group, using MR imaging for the assessment of the therapeutic effect of the cell treatment [15]. All in all, MRI seems to be a promising tool for monitoring therapy and diagnostic outcomes. Nevertheless, most of the clinical studies using stem cell therapy in combination with MRI are focused on the visuali-

http://dx.doi.org/10.5772/59356

zation of the regenerating soft tissue, rather than on the transplanted stem cells, their localization, viability and safety.

Further efforts have been put into the development of strategies for bone and cartilage regeneration. Recent studies are concentrating on developing a MRI-based method for visualization of knee osteoarthritis by assessing the number and location of lesions, cartilage thickness, and subchondral bone alteration at different time points [16]. Others went further by using autologous adipose tissue derived mesenchymal stem cells (AD MSCs) as the most hopeful candidate progenitor cell source for cartilage tissue engineering in patients with degenerative arthritis. Promising results showed improved knee functions and reduced pain without any adverse events. The cartilage regeneration process was captured by MRI at different time points, showing decreased size of cartilage defect, while the volume of cartilage increased over time in the high-dose patients (Figure 2) [17]. This depicts an excellent example of the gradual morphologic regeneration process, visualized by means of MRI.

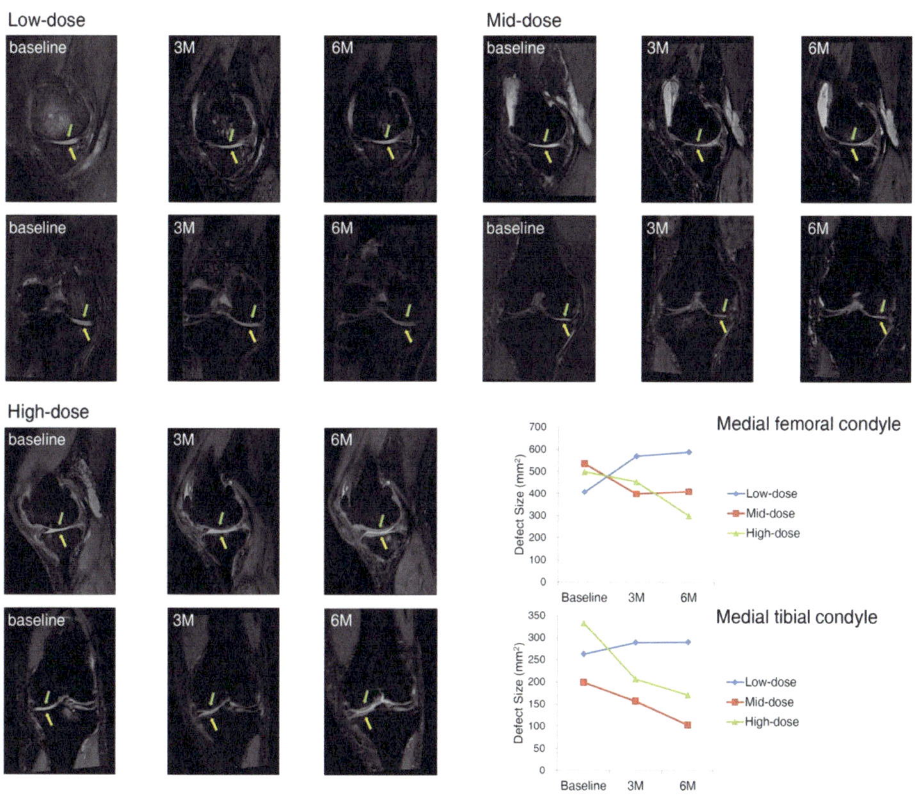

Figure 2. MRI evaluation of articular cartilage regeneration in the medial and femoral condyles after intra-articular injection of autologous AD MSCs. Sagittal and coronal MRIs before, 3, and 6 months after AD MSCs injection depicted the beneficial effect of the high-dose stem cell therapy upon cartilage regeneration by visualizing increase in thickness, and significantly decreased defect size. (Reprinted from [17], Copyright (2014), with permission from "John Wiley and sons" and Kang Sup Yoon, M.D., Ph.D.)

Besides its tremendous spatial resolution for visualization of soft tissues, MRI has proven successful in the imaging of the spinal cord and column. There are several clinical studies concentrating on the regeneration of this region using mesenchymal stem cell (MSC) transplants after injury. One group is trying to evaluate the change of MSC-treated spinal cord injury using MRI. Scans will be performed at 3 and 6 months after transplantation in order to track the effects of the possible cell treatment and the different ways of application (intravenous, intrathecal, into the spinal cord) [18]. The outcome of this trial will allow suggesting the most beneficial cell delivery method for further studies. Another project was concentrating on the visualization of spinal cord lesions inducing limb paralysis. The proposed treatment strategy included injections of bone marrow derived stem cells. Again, the preferred non-invasive imaging method for localization of the injuries was MRI [19]. It seems that MRI is a suitable imaging tool to assess spinal cord lesions and their repair by stem cells. One group addressed the safety and efficacy of autologous MSC injections in patients with chronic spinal cord injury. The outcome of this study revealed changes in MRI such as decreased cavity size and the appearance of fiber-like low signal intensity streaks [20]. An increasing amount of research is concentrating on the transplantation of stem cells as a therapeutic tool, but still, most of the data derives from pre-clinical models. Nevertheless, promising studies in humans reveal new possible solutions for many disorders.

The implementation of contrast agents for enhancement of MRI offers higher contrast and resolution. Gadolinium (Gd) and superparamagnetic iron-oxide (SPIO) are most commonly used for direct stem cell labeling in clinical and preclinical trials, with SPIO particles offering the highest sensitivity amongst all MR contrast agents. Despite initial reports that SPIO labeling of stem cells is safe and does not affect cell biology, recent studies indicate certain metabolic and functional changes. Therefore, some essential issues need to be considered before accepting such agents for clinical trials. A recent study [21] used Gd-or SPIO-labeled mesenchymal stem cells, transduced with a luciferase vector to monitor their viability via BLI. The outcome showed a distinction between viable and non-viable Gd cells, correlating with the luciferase signal. However, dead SPIO-cells indicated a persistent signal void *in vivo*, with simultaneous decrease of luciferase activity. This demonstrates that the absence or presence of a contrast agent signal alone does not prove for cell viability. Often, dead cells are being phagocytized by macrophages and the labeling persists in the new "host cells". Although Gd-labeling of stem cells seems feasible [22], its main disadvantages are the large amounts of Gd chelates needed to produce a signal, the toxic effect of Gd if released from the complex (affects kidneys) and the lower sensitivity compared to SPIO. These issues should be taken into consideration, as they might lead to false quantification of the imaging data. Recently, a new Gd-based contrast agent (Dotarem®) was approved by the FDA for Phase III clinical studies for MRI in brain, spine and associated tissues in adult and pediatric patients to detect areas with disruption of the blood-brain barrier and/or abnormal vascularity [23]. These clinical studies showed superiority of the enhanced over the unenhanced images of the CNS lesions. All in all, contrast agents improve the quality of the images, but their side effects have to be taken into consideration.

Safety and efficacy of autologous stem cell therapies remains a big issue and a non-invasive visualization tool like MRI is facilitating the refinement of the procedures. Since MRI provides more biological and functional data than CT-without radiation-the system provides a huge advantage for pediatric patients as well as patients needing multiple scans. This makes it a superior imaging modality in many clinical studies and encourages more patients to sign up for the trials, as the risk for unwanted complications is decreased. However, MRI does not come with excellent sensitivity and abundant molecular probes compared to other techniques, following in this chapter.

3. PET imaging

3.1. PET basics

Radionuclide-based positron emission tomography (PET) is one of the most sensitive molecular imaging techniques. Contrary to MRI, this technique uses radiation. The agents used for the visualization are labeled with a positron emitting radionuclide (^{11}C, ^{13}N, ^{15}O, ^{18}F etc.) which decays by emitting a positively charged particle (β^{+}). A positron is a particle with the same mass as an electron, but with an opposite charge. Every positron-emitting radionuclide has its own positron range (traveling distance), depending on the energy of its β^{+}-particle. The higher the energy, the longer the distance and, therefore, the larger the loss of spatial resolution. Once most of its energy is lost, the positron eventually annihilates with an electron from the surrounding tissue, whereby their mass is converted into electromagnetic energy in the form of high-energy photons [24]. The PET Imaging technique is based on the coincidence detection of two 511 keV photons emitted simultaneously in opposite directions (180° apart). They are collected by a ring system of photon-sensitive detectors (scintillation crystals), surrounding the object of interest (Figure 3). Although the exact location of a single annihilation is unknown, the acquisition of a large number of coincidence events over many angles can provide enough information to reconstruct an image of the spatial distribution of radioactivity as a function of time [5].

The idea for the development of the PET technique was developed in the 1950s and was introduced only a few decades later [25] as a clinical device for neurology and cardiology. Later, the need of precise anatomic localization and metabolic readout led to the invention of new imaging strategies. Combining two complementary modalities can add anatomical and/or physiological information to molecular imaging studies using software fusion of data. This technology is most successful for studies of organs and tissues that do not move with time, e.g. the brain [26]. Another approach allowing for simultaneous and/or sequential obtainment of data by a single device is the fusion of PET with CT or MRI. The immense value of combining the benefits of two imaging modalities in one instrument in order to assess molecular as well as morphologic information is of major interest nowadays. These devices allow for more exact attribution of the radioactivity to a specific anatomic site within the patient. Another great impact on the expansion of possible PET applications is the increasing availability of PET radiopharmaceuticals. Many researchers are working towards the devel-

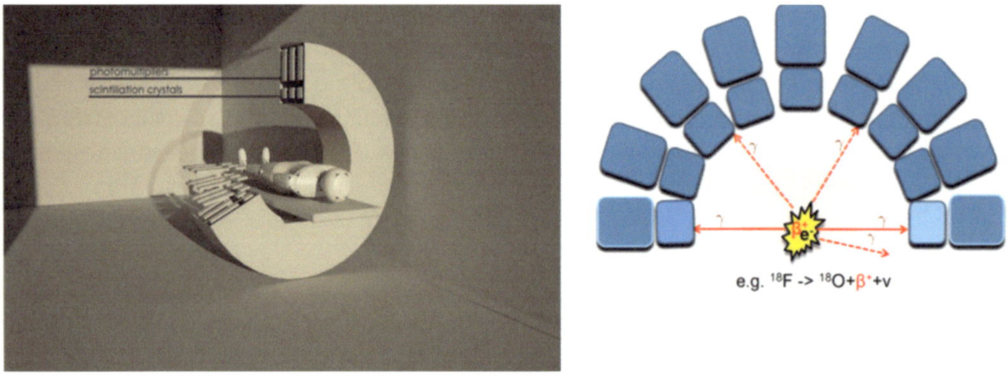

Figure 3. PET scanner detects coincidence of two 511 keV photons emitted in opposite directions after annihilation of a positron (β^+) with an electron (e^-) in the surrounding media. The signals are captured by scintillation crystals and amplified by photomultipliers. These events are then collected to construct images depicting the activity distribution.

opment of novel radiotracers for diagnosis and/or therapy of various diseases. Combinations of new technologies and radioligands have already been applied in diverse studies and thus provided many insights for the further development of personalized medicine.

3.2. PET applications

Great improvements have been made towards the diagnosis and the therapy monitoring of cancer, brain and heart diseases, as well as for a variety of malignancies and organ/tissue restorations. [^{18}F]fluorodeoxyglucose ([^{18}F]FDG) is the most widely used PET radiopharmaceutical. It is a metabolic tracer and the mechanism of uptake is based on phosphorylation to [^{18}F]FDG-6-phosphate, which is not further metabolized. [^{18}F]FDG has a wide range of applications in oncology, neurology, cardiology and inflammation imaging.

In the past few years many researchers concentrated on cell-based therapies for the restoration of damaged myocardium showing the involvement of different cell types in this process such as: skeletal myoblasts, embryonic stem cells, bone-marrow derived stem cells, cardiac resident cells, mesenchymal stem cells, and circulating progenitor cells. [27]. The introduction of cells into the damaged myocardium seems to be a promising technique and therefore several tools for assessing the viability, localization and metabolism of the injected cells have been evaluated [28]. The implanted cells should be suitably labeled for detection via imaging, in order to be distinguishable from the surrounding cells and structures. This can be achieved by PET imaging using [^{18}F]FDG. It was already shown to be feasible for dynamic tracking of [^{18}F]FDG-labeled progenitor cells during intracoronary injection after acute myocardial infarction [29]. One of the main glucose uptake sites in the body are skeletal muscles. Using PET imaging of this glucose derivate allows for longitudinal assessment of skeletal muscle regeneration and degeneration [30]. This allows for easy visualization by [^{18}F]FDG uptake in response to physical exercise, thus making muscle cells a good diagnostic tool for therapy of muscle disorders [31]. Whether this technology is suitable for regenerative medicine purposes has to be further evaluated.

http://dx.doi.org/10.5772/59356

Another application of this tracer was shown in a clinical study using bone marrow cells for autologous stem cell therapy after myocardial infarction. They included the PET modality in their experimental design and investigated the myocardial flow ([^{13}N]NH_3) and perfusion/metabolism mismatch ([^{18}F]FDG) [32]. The usage of PET for tracking *in vivo* cell retention of adipose tissue-derived stem cells for myocardial regeneration also showed promising results [33]. It seems that these cells support the cardiomyocyte regeneration and angiogenesis in the implanted area. The PET/CT modality appears to be a promising technique for effective cell-tracking *in vivo*, which is vital for a more in-depth investigation into future clinical applications.

Another possibility for PET application is the utilization of reporter-gene systems (e.g. Herpes Simplex Virus type 1 thymidine kinase (HSV1-tk)) [34]. The main advantage of this method is that the obtained signal is specific only to viable cells, as they are able to express the reporters. This allows not only for visualization of acute cell retention, but also for viability read-outs [7]. HSV1-tk is currently the most investigated enzyme reporter gene for long term *in vivo* visualization of cells applied in the treatment of various cardiovascular diseases [35-37]. Recent studies also investigated the use of reporter gene tracking via PET imaging to track viable stem cells in large animal models [38-40]. These studies confirm the potential of this novel method to allow for a better understanding of the cell faith after implantation. Using cell-based therapies for tissue engineering with stem and/or progenitor cells for boosting regeneration seems to offer great possibilities in regenerative medicine. Therefore, the development of novel methods for non-invasive imaging of implants *in vivo* is of high priority.

PET/CT has also become an essential tool for assessing prognosis and establishing treatment decisions in oncology. This system found broad application, mainly in defining different stages/locations of metastatic cancer. In this case PET is used for visualizing the increased metabolic activity of the tumor and the addition of CT provides anatomical information on its exact location in the body [41]. This helped immensely in the management of a broad spectrum of malignancies, their diagnosis, staging and therapy-response-assessment. The most commonly used radiopharmaceutical in clinics [^{18}F]FDG was shown to be essential to the treatment of a great range of pathologies. This can significantly contribute to precise therapy response assessment possibly influencing the therapeutic management and treatment planning [42]. Another study revealed a promising application of fluorine-18 labeled amino acid-based radiotracers for small cell lung cancer detection via PET imaging [43]. Furthermore, a recently developed categorization method for differentiation of osteosarcoma phenotypes by comparing SUV values of PET radioligands for glucose metabolism ([^{18}F]FDG), hypoxia ([^{18}F]FMISO) and bone remodelling ([^{18}F]fluoride) showed promising results. These findings can improve the future evaluation of the treatment strategy depending on the specificity of the malignancy [44]. These tracers can be applicable also for regenerative purposes, studying the healing process of e.g. bones. A current clinical study for bone reconstruction of the skull is examining the regeneration after introducing an implant using PET/CT via [^{18}F]fluoride. Succeeding in this would help out many patients with cranial defects and significantly improve their quality of life [45].

Cardiology also gained immense benefits from the development of the hybrid imaging systems (e.g. correction of attenuation and evaluation of coronary classifications), CT for coronary angiography, and acquisition of 3-D hybrid images for the definition of heart viability and diagnosis of cardiac inflammations and infections [46]. Recent advances in stem cell therapy show promising results for myocardium restoration, neovascularization and enhancement of perfusion [47-50].

Research on the central nervous system (CNS) mainly emphasized on mapping different aspects of neurotransmitter activity, e.g. dopaminergic, cholinergic, serotonergic and glutamatergic systems as well as beta-amyloid visualization for Alzheimer's disease (AD) [5, 51, 52]. PET Imaging has been applied widely to detect and quantify subtle abnormalities in CNS diseases. Therefore, this technology is also immensely popular as a modeling tool in CNS drug discovery and development. It is used for evaluation of drug deposition, as well as for studying disease biomarkers to monitor drug effects on brain pathologies [24]. Another CNS utilization of this modality was shown in a promising study regarding spinal cord injury [53]. The combined efforts of stem cell and neuroscience researchers made it possible to visualize the spinal cord regeneration process on the cellular and molecular level. In order to explore the effect of *in vivo* PET on tracking dopamine receptor positive stem cells transplanted into the spinal cord, [^{11}C]Raclopride was injected intravenously, followed by a scan. The accumulation of signal in the site of injury showed successful tracking and localization of the injected cells. These results suggest that PET with a radiotracer is a useful technique for functional studies in developing cell-based therapies [54]. Similar experimental setups allowed for more rapid progression of many pre-clinical and clinical studies of various CNS pathologies (e.g. [^{11}C]Methionine PET/CT in patients with cancer; [^{18}F]FLT in pediatric patients with CNS tumors [55]).

After the successful introduction of the PET/CT system to the clinics, the integration of PET with MRI was the next step. This new hybrid technique allows for simultaneous functional PET imaging combined with soft tissue morphological MR images. The novel approach brings better contrast among soft tissues as well as functional imaging capabilities, when compared to PET/CT. Nevertheless, this technology is still not clinically mature at this point. Technically, the implementation of PET/MR faces three major issues: 1. The photomultiplier-based PET scanners do not work in the proximity of the magnetic field of the MR scanner; 2. Metallic objects (e.g. surface coils used to get the best MR quality images) interfere with gamma rays from PET, resulting in unwanted attenuation; 3. MR data cannot be used for attenuation correction, as CT data, which is a limiting factor for the use of PET in therapy response monitoring [56].

As PET signal processing might disturb the high-frequency signals of MRI, and conventional PET electronics fail in an even weak magnetic field, it soon became clear that new solutions had to be found to circumvent the joint interferences. One possibility would be to leave just the PET crystals in the MR field, whereas the PMTs are positioned outside and connected to the crystals by optical fibres. Another approach to avoid the disturbance would be to place the PET and MR scanners remote to each other, but still linked by a common bed, so that the patient does not move between the two scans. Last, but not least, the PMTs could be replaced by solid state electronics which are not disturbed by the MR field [57]. Hence, various approaches were developed during the last fifteen years for application in small animal studies, where the usage of more PET tracers is allowed and where use is not as legally restricted, as in human studies. For a review, see Herzog, 2012 [57]. Improvements in pre-clinical study design enable smooth transfer of knowledge and molecular measurements between species thereby facilitating clinical translation. Recently, PET/MRI for human imaging became available as well, towing to support form the industry. Siemens Healthcare first designed prototypes of a BrainPET/MRI device and the first reports of patients have been communicated [58, 59]. This opened new possibilities in the field of molecular imaging and prepared the development of an integrated whole-body PET/MRI scanner. A recent study compared the performance of whole-body PET/ MRI to PET/CT of breast cancer patients and claimed that PET/MRI is a feasible technique for usage in clinics, forgoing gamma radiation for high quality imaging with short examination time [60]. Another example for implementation of this new technique was shown in a sarcoma patient, who underwent chemotherapy and autologous stem cell injection. MRI showed lesions in the upper and lower leg, whereas PET revealed high [^{18}F]FDG uptake only in one of the lesions (Figure 4). This case illustrates the benefit of combining different imaging modalities in order to get more precise outcomes, thereby improving the application fields of personalized medicine. This methodology might also be useful for the investigation of regenerative processes in damaged organs or tissues, for example after autologous stem cell therapy for boosting the healing progression.

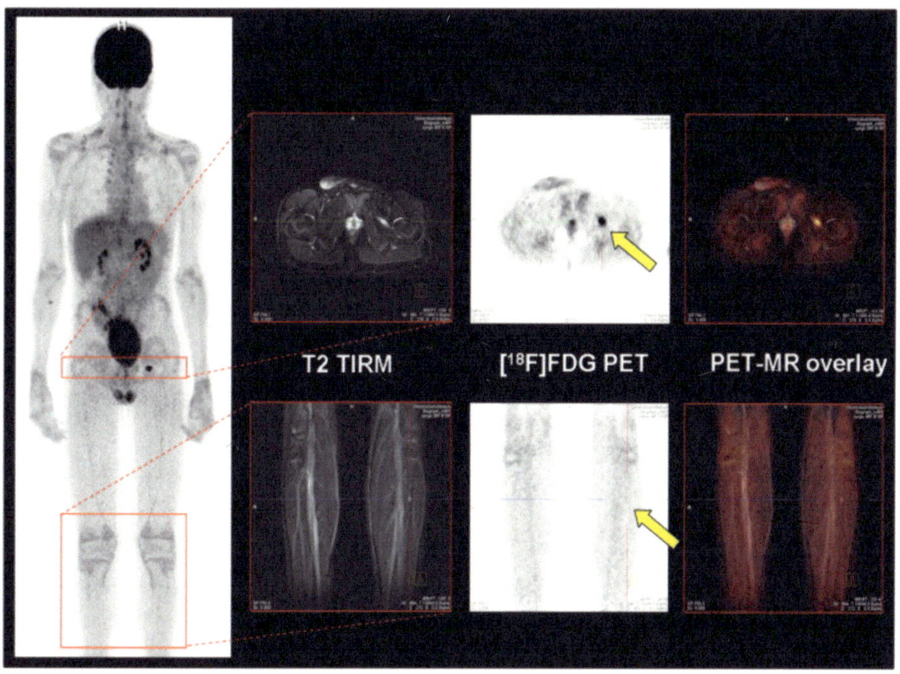

Figure 4. Simultaneous whole-body PET/MRI acquired with a molecular MR (mMR) scanner in a 13-year old boy with a Ewing sarcoma known for six years. The patient underwent standard chemotherapy together with autologous stem cell transplantation. After *i.v.* injection of [^{18}F]FDG whole body PET-MRI was carried out. MR resulted in two suspicious lesions: One in the left upper leg muscle region, and one in the left lower leg. While the lesion in the left upper leg muscles showed intense [^{18}F]FDG uptake, the other one was PET-negative (arrows). Follow-up confirmed a Ewing sarcoma relapse in the left upper leg muscle region. (Reprinted from [60], Copyright (2014), with permission from "Elsevier" and Prof. Dr. med. O. Sabri)

The clinical application of the PET/MRI technique has to be further established. Nevertheless, if MRI can replace CT for anatomical screening, it adds value as well as decreases radiation exposure, which would be of great benefit for the patients. However, the ongoing question of choosing the best suitable visualization method remains unanswered, as there is still insufficient knowledge about which PET/MRI applications are superior to the well-known PET/CT for a variety of clinical cases.

4. SPECT imaging

4.1. SPECT basics

Unlike PET, single photon emission computer tomography (SPECT) uses radiopharmaceuticals labeled with gamma emitters instead of positron emitters. The concept of transmission and emission tomography, later developed into SPECT, was first introduced in the late 1950s. This period is associated with phenomenal growth of Nuclear Medicine, when some of the most prominent gamma-radiation emitting radionuclides (^{131}I and ^{99m}Tc) were produced [61]. While the primary use of ^{131}I was dedicated to thyroid cancer treatment, its use was later expanded for imaging of the thyroid gland itself, its function and the therapy of hyperthyroidism. The development of a generator system for ^{99m}Tc (1960s) was an important breakthrough and today it is the most utilized element in the field of Nuclear medicine. Other typical gamma emitting radionuclides utilized in SPECT are ^{123}I, ^{67}Ga and ^{111}In. The most common SPECT systems consist of a gamma camera with rotating NaI(Tl) detector modules located between the photomultipliers and collimators (Pb) (Figure 5), an on-line computer for acquisition and processing of data and a display system [62]. Generally, a target-specific SPECT radiopharmaceutical consists of two parts: a gamma-emitting radionuclide and a targeting biomolecule (e.g. peptide, antibody fragment) [63]. Similar to PET, SPECT also enables 3-D imaging and hybrid systems (SPECT/CT) are available as well.

4.2. SPECT applications

Various stem cells hold promise for the treatment of many human diseases. However, little is known about the exact function, location and survival of the transplanted cells. The usage of the SPECT imaging modality to answer these questions revealed promising results. Clinical studies with SPECT indicated evidence for efficacious imaging of transplanted radiolabeled cells in cardiac disease. Encouraging results were obtained from studies using [^{99m}Tc]hexamethylpropylenamineoxine (HMPAO)-labeled stem cells. Comparison of intracoronary (IC) and transendocardial (TE) delivery of the cells to the heart showed more intense retention signal 18h after cell injection in the TE patients [64]. Another study could illustrate homing of bone marrow mononuclear cells TE injection after acute myocardial infarction. Cell retention was successfully visualized in the damaged heart tissue [65]. Similar results were obtained by a group showing homing of bone marrow progenitor cells after intracoronary transfer using the same radiotracer [66]. Controversially, further experiments investigating the homing of the same cells after myocardial infarction indicated very heterogeneous uptake of the ^{99m}Tc-labeled

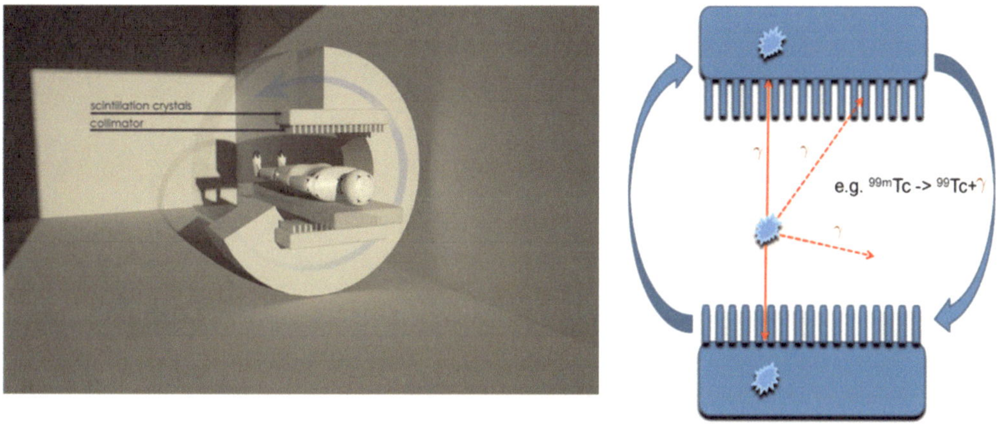

Figure 5. Underlying principle of SPECT: (1) A radioisotope decays by emitting gamma rays. (2) A rotating gamma-sensitive detector with a collimator in front indicates the scintillation site where gamma rays are parallel to the collimator pinholes. (3) The collected data is then transferred to photomultipliers and finally to a computer for analysis.

cells, but for all patients there was lack of or decreased uptake of cells in walls with perfusion defects. The perfusion imaging was performed with thallium (^{201}Tl) [67]. Although the safety of autologous stem cell transplantation to the heart after infarction has been shown, it is still not known how many of these cells actually remain in the damaged site, or in near proximity to it. To address their homing and apoptotic rate, autologous bone marrow cells were labeled with [^{111}In]oxine and injected *via* intracoronary route. SPECT imaging was performed 24h after the transplantation and showed homogenous tracer accumulation in liver and spleen, while uptake in the heart was focal. Up to 10% of the injected cells were retained in the myocardium [68]. Similar results regarding cell retention were obtained by [^{18}F]FDG labeling of the cells and detection by PET [69]. Further insights in this field were obtained from a trial using ^{111}In-oxine-labelled pro-angiogenic cells after intracoronary injection in patients at different time points after myocardial infarction. SPECT imaging was utilized to monitor the cell retention, showing highest values in patients with recent onset of the disease, progressively decreasing in patients treated in an intermediate phase or at chronic stage [70]. The substantial amount of radioactivity in the myocardium suggests correct homing of these cells. Nevertheless, most of the injected cells still remain locked in other locations. One possibility for cell homing improvement might be by significantly reducing their entrapment in filter organs.

Further investigations were made to detect angiogenesis after myocardial infarction utilizing SPECT. This was assessed using a novel radiolabelled peptide ([^{99m}Tc]NC100692, Maraciclatide), which has high affinity to an angiogenesis-related integrin ($\alpha_v\beta_3$) [71] (Figure 6). The successful implementation of this tracer led to a follow-up study, addressing the effect of bone marrow stem cells transplantation on angiogenesis in chronic ischaemic heart failure patients [72]. The patients underwent SPECT imaging using the same [^{99m}Tc]RGD imaging peptide. This new imaging tracer was shown to be feasible in patients with heart failure. However, the changes with stem cell therapy did not reach statistical significance.

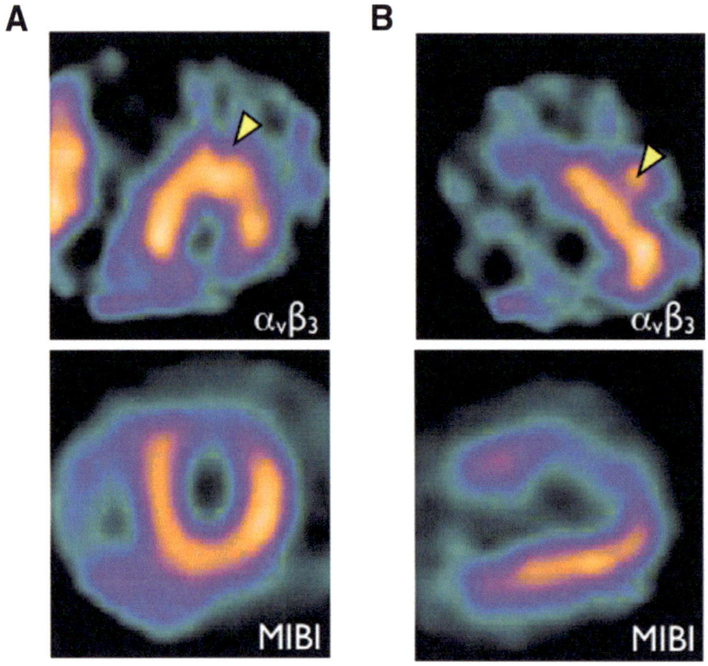

Figure 6. Illustration of localized angiogenesis-tracer uptake (Maraciclatide:top) in a region of perfusion deficit (MIBI: bottom). A: short axis slice; B: long axis slice. Figure was provided by courtesy of Prof. Brian Hutton.

Despite the immense progress in the field of stem cell imaging research, the pre-labeling approach for cell visualization remains challenging for now. Most of the stem cell clinical trials tend to utilize techniques for following the regenerative process (e.g. tissue growth, vascularization and functional recovery), rather than direct tracking of the injected cells [73]. The use of autologous stem cell therapy for heart regeneration is of high importance for cardiomyopathy patients and is showing promising results. However, the therapeutic effect is by now mostly measured by an enhancement of the heart regional and global contractility as a cell treatment outcome [73]. One other factor, often taken into consideration when testing a new approach for cardio-regeneration, is the heart perfusion efficacy. This could be measured by *i.v.* injection of [^{99m}Tc]sestamibi at rest, followed by an induction of pharmacologic stress by *i.v.* adenosine application. Successful SPECT imaging at rest and post stress suggested a trend towards improvement in cell-treated patients [74]. These non-invasive approaches are able to effectively show the outcome of the therapy. Still, they do not answer the questions regarding proper homing, or viability of the injected cells.

Another possibility to assess the regeneration process is by using direct labeling of certain proteins, known to be connected to basic organ functions. This method was performed to estimate the liver self-regeneration after hepatectomy via ^{99m}Tc-labeled galactosyl-human-serum-albumin. The liver uptake value could be calculated from SPECT/CT images of patients at different time points after surgery [75]. An additional visualization possibility for the hepatic function was shown by [^{99m}Tc]phytate colloid uptake by the liver. As colloid uptake by

http://dx.doi.org/10.5772/59356

perfused Kupffer cells is proportional to perfused hepatocyte mass, this could be used for direct quantification of perfused liver mass by quantitative SPECT in cirrhotic patients [76]. Although the liver is well-known for its self-regeneration capacity, stem cell therapy seems to boost this capacity. Several studies addressed the therapeutic effect of autologous stem cells for liver regeneration [77-79]. Nevertheless, none of them assessed the fate of these cells *in vivo*. Recently, the proper homing of injected autologous mesenchymal stem cells in the liver of patients with advanced cirrhosis could be illustrated, as these cells have previously been shown to have beneficial effect. The cells were labeled with [^{111}In]oxine before intravenous infusion and could be detected by SPECT/CT up to 10 days after application. Shortly after injection, accumulation of signal was observed in the lungs, and gradually shifted to the liver and spleen during the following hours to days [80]. Using SPECT for evaluation of stem cell homing seems a feasible technique. However, the safety, exact mechanism of action and timing of cell delivery have to be further estimated. There is need for more essential information about the long-term benefit of stem cell therapies. Critical for the future success of SPECT is the design of new and specific tracers for the detection, localization, and staging of a disease and for monitoring of the regeneration process. In the past few decades, there is an increase in the awareness of the importance of this imaging modality, its significant impact on diagnosis and management of various diseases and successful visualization of potential treatments.

5. Conclusions and future directions

There is no perfect imaging modality for non-invasive cell tracking in clinics. The research findings of the last few decades on the arena of imaging have prompted an intense interest of many investors, who facilitated the translation of these ideas into clinics. Nevertheless, to date no state-of-the-art imaging tool is capable of delivering an all-in-one solution. Many factors have to be considered before choosing the proper visualization modality, depending on the question to be addressed. There is emerging need for versatile imaging methods capable of monitoring the autologous stem cells *in vivo*. Although developments in the field of non-invasive visualization add significant costs and have to overcome many regulatory roadblocks, it is the only way of translating the research into clinical treatments and to support future medicine by providing new tailored treatments to patients.

Despite the immense progress in the field of personalized medicine, there are still many black boxes, on which non-invasive imaging is beginning to shed light. One can observe an increased awareness of the importance of these imaging modalities and their significant impact on diagnosis and management of human disorders are being recognized. Brilliant ideas and developments in this field are catalyzing the design of new strategies every day, giving rise to possible solutions. A summary of the characteristics of the "perfect clinical imaging modality" for stem cell tracking is given in Table 2. Nevertheless, such an all-in-one technique is not available yet, but the development of hybrid imaging devices for multimodal imaging was a big step towards the ideal system.

1. Safe, non-toxic and biocompatible
2. Single-cell detection at any anatomic location
3. Limitless depth of penetration
4. High spatial and temporal resolution, excellent molecular sensitivity
5. No genetic modifications of the stem cells
6. Insignificant/no dilution with cell division
7. Possible quantification of cell number and viability
8. Serial and long term monitoring in patients up to several months/years
9. Insignificant/no uptake of tracer by non-target cells
10.Multimodality-based approach
11.Low costs

Table 2. Requirements for the design of the "perfect clinical imaging modality" for stem cell tracking

The techniques of SPECT/CT, PET/CT and PET/MRI are the state-of-the-art modalities for successful biomedical molecular imaging. Various diseases related to metabolism, apoptosis, tumors, genetics, and stem cells have been identified by these systems [62]. The same approaches are already being successfully transferred in numerous clinical studies using autologous stem cells for organ and tissue restoration. Nevertheless, the selection of a given visualization technique depends on its strengths and weaknesses with respect to the intended use. All currently developed molecular imaging techniques for stem cell tracking have inherent limitations. For imaging of delivery and short-term homing of the injected stem cells in different organs, a direct labelling approach may be the answer (iron oxide or [^{18}F]FDG), even though any potential toxicity must be taken into account. MRI offers the highest spatial resolution and near real-time image guidance for cell delivery, although with significantly lower molecular sensitivity than other modalities, such as PET or SPECT. For long-term cell monitoring, reporter gene imaging, using PET or SPECT, appears to be a better choice. Ultimately, finding the most suitable system or combination of systems for proper diagnosis and/or treatment remains the essence of health care and personalized medicine.

Acknowledgements

The authors would like to thank Arch. Teofana Haralampieva for the great support in the visualization of the different imaging modalities and Damina Balmer for her editorial assistance in the preparation of this manuscript. A special thank you goes to the team members of the Laboratory for Urologic Tissue Engineering and Stem Cell Therapy at USZ.

Author details

Deana G. Haralampieva[1,2,3*], Simon M. Ametamey[2,3], Tullio Sulser[1] and Daniel Eberli[1,3]

*Address all correspondence to: deana.haralampieva@usz.ch

1 Laboratory for Tissue Engineering and Stem Cell Therapy, Division of Urology, University Hospital, Zurich, Switzerland

2 Institute of Pharmaceutical Sciences, ETHZ, Zurich, Switzerland

3 Zurich Center for Integrative Human Physiology (ZIHP), Switzerland

References

[1] Piscaglia AC. Stem cells, a two-edged sword: risks and potentials of regenerative medicine. World J Gastroenterol. 2008 Jul 21;14(27):4273-9.

[2] Health USNIo. www.clinicaltrials.gov [accessed 5 Mai 2014].

[3] Kraitchman DL, Bulte JW. Imaging of stem cells using MRI. Basic research in cardiology. 2008 Mar;103(2):105-13.

[4] Srinivas M, Aarntzen EH, Bulte JW, Oyen WJ, Heerschap A, de Vries IJ, et al. Imaging of cellular therapies. Adv Drug Deliv Rev. 2010 Aug 30;62(11):1080-93.

[5] Ametamey SM, Honer M, Schubiger PA. Molecular imaging with PET. Chem Rev. 2008 May;108(5):1501-16.

[6] Dominietto M, Rudin M. Could magnetic resonance provide histology? Front Genet. 2014 Jan 13;4:298.

[7] Collins SA, Hiraoka K, Inagaki A, Kasahara N, Tangney M. PET imaging for gene & cell therapy. Curr Gene Ther. 2012 Feb 1;12(1):20-32.

[8] Sutton EJ, Henning TD, Pichler BJ, Bremer C, Daldrup-Link HE. Cell tracking with optical imaging. Eur Radiol. 2008 Oct;18(10):2021-32.

[9] James ML, Gambhir SS. A molecular imaging primer: modalities, imaging agents, and applications. Physiological reviews. 2012 Apr;92(2):897-965.

[10] Rinck PA. Magnetic Resonance Imaging – How It All Began http://www.rinckside.org/Rinckside%20Columns/1990%20MRI%20How%20it%20all%20began.htm2003 [accessed 10 June 2014].

[11] Nuclear Magnetic Resonance [21 July 2014]. Available from: http://www.imaios.com/de/e-Courses/e-MRI/NMR.

[12] SWiss Multicenter Intracoronary Stem Cells Study in Acute Myocardial Infarction (SWISS-AMI) [Internet]. [cited 10 June 2014]. Available from: https://clinicaltrials.gov/ct2/show/NCT00355186?term=MRI+stem+cells&rank=2.

[13] 19F Hot Spot MRI of Human Adipose-derived Stem Cells for Breast Reconstruction (CS-1000) [Internet]. [cited 10 June 2014]. Available from: https://clinicaltrials.gov/ct2/show/NCT02035085?term=MRI+stem+cells&rank=5.

[14] Evaluation of Autologous Mesenchymal Stem Cell Transplantation (Effects and Side Effects) in Multiple Sclerosis [Internet]. [cited 15 July 2014]. Available from: http://clinicaltrials.gov/ct2/show/NCT01377870?term=stem±cells±MRI&rank=47.

[15] Safety and Efficacy Study of Autologus Bone Marrow Mesenchymal Stem Cells in Multiple Sclerosis [Internet]. [cited 15 July 2014]. Available from: http://clinicaltrials.gov/ct2/show/NCT01895439?term=stem±cells±MRI&cond=%22Multiple±Sclerosis%22&rank=8.

[16] Treatment of Knee Osteoarthritis by Intra-articular Injection of Bone Marrow Mesenchymal Stem Cells [Internet]. [cited 10 July 2014]. Available from: http://clinicaltrials.gov/ct2/show/NCT02123368?term=stem±cells±MRI&rank=13.

[17] Jo CH, Lee YG, Shin WH, Kim H, Chai JW, Jeong EC, et al. Intra-articular injection of mesenchymal stem cells for the treatment of osteoarthritis of the knee: a proof-of-concept clinical trial. Stem Cells. 2014 May;32(5):1254-66.

[18] Safety and Effect of Adipose Tissue Derived Mesenchymal Stem Cell Implantation in Patients With Spinal Cord Injury [Internet]. [cited 18 July 2014]. Available from: http://clinicaltrials.gov/ct2/show/NCT01769872?term=stem±cells±MRI&no_unk=Y&rank=118.

[19] Safety and Efficacy of Autologous Bone Marrow Stem Cells in Treating Spinal Cord Injury (ABMST-SCI) [Internet]. [cited 17 July 2014]. Available from: http://clinicaltrials.gov/ct2/show/NCT01186679?term=stem±cells±MRI&no_unk=Y&rank=95.

[20] Park JH, Kim DY, Sung IY, Choi GH, Jeon MH, Kim KK, et al. Long-term results of spinal cord injury therapy using mesenchymal stem cells derived from bone marrow in humans. Neurosurgery. 2012 May;70(5):1238-47; discussion 47.

[21] Guenoun J, Ruggiero A, Doeswijk G, Janssens RC, Koning GA, Kotek G, et al. In vivo quantitative assessment of cell viability of gadolinium or iron-labeled cells using MRI and bioluminescence imaging. Contrast media & molecular imaging. 2013 Mar-Apr;8(2):165-74.

[22] Modo M, Mellodew K, Cash D, Fraser SE, Meade TJ, Price J, et al. Mapping transplanted stem cell migration after a stroke: a serial, in vivo magnetic resonance imaging study. NeuroImage. 2004 Jan;21(1):311-7.

[23] GUERBET Press releases: FDA approves Dotarem [29 July 2014]. Available from: http://www.imaios.com/de/Technical-Exhibition/Vendors/GUERBET/Press-releases/FDA-Approves-Dotarem-R.

[24] Varnas K, Varrone A, Farde L. Modeling of PET data in CNS drug discovery and development. Journal of pharmacokinetics and pharmacodynamics. 2013 Jun;40(3):267-79.

[25] Phelps ME, Hoffman EJ, Mullani NA, Ter-Pogossian MM. Application of annihilation coincidence detection to transaxial reconstruction tomography. J Nucl Med. 1975 Mar;16(3):210-24.

[26] Hill DL, Batchelor PG, Holden M, Hawkes DJ. Medical image registration. Phys Med Biol. 2001 Mar;46(3):R1-45.

[27] Segers VF, Lee RT. Stem-cell therapy for cardiac disease. Nature. 2008 Feb 21;451(7181):937-42.

[28] Zhang Y, Ruel M, Beanlands RS, deKemp RA, Suuronen EJ, DaSilva JN. Tracking stem cell therapy in the myocardium: applications of positron emission tomography. Curr Pharm Des. 2008;14(36):3835-53.

[29] Doyle B, Kemp BJ, Chareonthaitawee P, Reed C, Schmeckpeper J, Sorajja P, et al. Dynamic tracking during intracoronary injection of 18F-FDG-labeled progenitor cell therapy for acute myocardial infarction. J Nucl Med. 2007 Oct;48(10):1708-14.

[30] Ahmad N, Welch I, Grange R, Hadway J, Dhanvantari S, Hill D, et al. Use of imaging biomarkers to assess perfusion and glucose metabolism in the skeletal muscle of dystrophic mice. BMC Musculoskelet Disord. 2011;12:127.

[31] Goncalves MD, Alavi A, Torigian DA. FDG-PET/CT assessment of differential chemotherapy effects upon skeletal muscle metabolism in patients with melanoma. Annals of nuclear medicine. 2014 May;28(4):386-92.

[32] Colombo A, Castellani M, Piccaluga E, Pusineri E, Palatresi S, Longari V, et al. Myocardial blood flow and infarct size after CD133+cell injection in large myocardial infarction with good recanalization and poor reperfusion: results from a randomized controlled trial. Journal of cardiovascular medicine. 2011 Apr;12(4):239-48.

[33] Yang JJ, Liu ZQ, Zhang JM, Wang HB, Hu SY, Liu JF, et al. Real-time tracking of adipose tissue-derived stem cells with injectable scaffolds in the infarcted heart. Heart Vessels. 2013 May;28(3):385-96.

[34] Chen IY, Wu JC. Molecular imaging: the key to advancing cardiac stem cell therapy. Trends Cardiovasc Med. 2013 Aug;23(6):201-10.

[35] Schwaiger M, Bengel FM. From thallium scan to molecular imaging. Mol Imaging Biol. 2002 Nov;4(6):387-98.

[36] Pei Z, Lan X, Cheng Z, Qin C, Wang P, He Y, et al. A multimodality reporter gene for monitoring transplanted stem cells. Nuclear medicine and biology. 2012 Aug;39(6): 813-20.

[37] Roelants V, Labar D, de Meester C, Havaux X, Tabilio A, Gambhir SS, et al. Comparison between adenoviral and retroviral vectors for the transduction of the thymidine kinase PET reporter gene in rat mesenchymal stem cells. J Nucl Med. 2008 Nov; 49(11):1836-44.

[38] Gyongyosi M, Blanco J, Marian T, Tron L, Petnehazy O, Petrasi Z, et al. Serial noninvasive in vivo positron emission tomographic tracking of percutaneously intramyocardially injected autologous porcine mesenchymal stem cells modified for transgene reporter gene expression. Circ Cardiovasc Imaging. 2008 Sep;1(2):94-103.

[39] Willmann JK, Paulmurugan R, Rodriguez-Porcel M, Stein W, Brinton TJ, Connolly AJ, et al. Imaging gene expression in human mesenchymal stem cells: from small to large animals. Radiology. 2009 Jul;252(1):117-27.

[40] Zhang WY, Ebert AD, Narula J, Wu JC. Imaging cardiac stem cell therapy: translations to human clinical studies. J Cardiovasc Transl Res. 2011 Aug;4(4):514-22.

[41] Schoder H, Erdi YE, Larson SM, Yeung HW. PET/CT: a new imaging technology in nuclear medicine. Eur J Nucl Med Mol Imaging. 2003 Oct;30(10):1419-37.

[42] Krause BJ, Schwarzenbock S, Souvatzoglou M. FDG PET and PET/CT. Recent Results Cancer Res. 2013;187:351-69.

[43] Chiotellis A, Mu L, Muller A, Selivanova SV, Keller C, Schibli R, et al. Synthesis and biological evaluation of (1)(8)F-labeled fluoropropyl tryptophan analogs as potential PET probes for tumor imaging. European journal of medicinal chemistry. 2013 Dec; 70:768-80.

[44] Campanile C, Arlt MJ, Kramer SD, Honer M, Gvozdenovic A, Brennecke P, et al. Characterization of different osteosarcoma phenotypes by PET imaging in preclinical animal models. J Nucl Med. 2013 Aug;54(8):1362-8.

[45] Bone Reconstruction of the Skull Using a Metal Ceramic Implant After Previously Failed Reconstruction [Internet]. [cited 16 June 2014]. Available from: http://clinicaltrials.gov/ct2/show/NCT01899807?term=PET±regeneration&rank=2.

[46] Garcia-Orta R, Mahia-Casado P, Gomez de Diego JJ, Barba-Cosials J, Rodriguez-Palomares JF, Aguade-Bruix S, et al. Update on Cardiac Imaging Techniques 2013. Rev Esp Cardiol. 2014 Jan 3.

[47] Mignone JL, Kreutziger KL, Paige SL, Murry CE. Cardiogenesis from human embryonic stem cells. Circulation journal : official journal of the Japanese Circulation Society. 2010 Nov;74(12):2517-26.

[48] Schuster MD, Kocher AA, Seki T, Martens TP, Xiang G, Homma S, et al. Myocardial neovascularization by bone marrow angioblasts results in cardiomyocyte regeneration. Am J Physiol Heart Circ Physiol. 2004 Aug;287(2):H525-32.

[49] Li Z, Lee A, Huang M, Chun H, Chung J, Chu P, et al. Imaging survival and function of transplanted cardiac resident stem cells. J Am Coll Cardiol. 2009 Apr 7;53(14): 1229-40.

[50] Gong Z, Niklason LE. Use of human mesenchymal stem cells as alternative source of smooth muscle cells in vessel engineering. Methods Mol Biol. 2011;698:279-94.

[51] Reiman EM, Jagust WJ. Brain imaging in the study of Alzheimer's disease. NeuroImage. 2012 Jun;61(2):505-16.

[52] Varrone A, Halldin C. New developments of dopaminergic imaging in Parkinson's disease. Q J Nucl Med Mol Imaging. 2012 Feb;56(1):68-82.

[53] Song F, Tian M, Zhang H. Molecular imaging in stem cell therapy for spinal cord injury. BioMed research international. 2014;2014:759514.

[54] Bai JZ LZ, Ding WM, Xu GH, Shen L, Wang F, Tian JH. Tracking neural progenitor cells transplanted into rabbit spinal cord by detection of dopamine receptor 2 with positron emission computed tomography. Zhonghua Yi Xue Za Zhi. 2006;86(29): 2060-4.

[55] clinical trials. [cited 12 June 2014]. Available from: https://clinicaltrials.gov/ct2/results?term=PET+CNS+.

[56] von Schulthess GK, Kuhn FP, Kaufmann P, Veit-Haibach P. Clinical positron emission tomography/magnetic resonance imaging applications. Semin Nucl Med. 2013 Jan;43(1):3-10.

[57] Herzog H. PET/MRI: challenges, solutions and perspectives. Zeitschrift fur medizinische Physik. 2012 Dec;22(4):281-98.

[58] Schlemmer HP, Pichler BJ, Schmand M, Burbar Z, Michel C, Ladebeck R, et al. Simultaneous MR/PET imaging of the human brain: feasibility study. Radiology. 2008 Sep; 248(3):1028-35.

[59] Herzog H, Langen KJ, Weirich C, Rota Kops E, Kaffanke J, Tellmann L, et al. High resolution BrainPET combined with simultaneous MRI. Nuklearmedizin. 2011;50(2): 74-82.

[60] Pace L, Nicolai E, Luongo A, Aiello M, Catalano OA, Soricelli A, et al. Comparison of whole-body PET/CT and PET/MRI in breast cancer patients: lesion detection and quantitation of 18F-deoxyglucose uptake in lesions and in normal organ tissues. Eur J Radiol. 2014 Feb;83(2):289-96.

[61] History of Nuclear Medicine [Internet]. [cited 20 July 2014]. Available from: http://www.news-medical.net/health/History-of-Nuclear-Medicine.aspx.

[62] Saha GB. Fundamentals of Nuclear Pharmacy. 6th ed: Springer; 2010.

[63] Muller C, Schibli R. Single photon emission computed tomography tracer. Recent Results Cancer Res. 2013;187:65-105.

[64] Vrtovec B, Poglajen G, Lezaic L, Sever M, Socan A, Domanovic D, et al. Comparison of transendocardial and intracoronary CD34+cell transplantation in patients with nonischemic dilated cardiomyopathy. Circulation. 2013 Sep 10;128(11 Suppl 1):S42-9.

[65] Silva SA, Sousa AL, Haddad AF, Azevedo JC, Soares VE, Peixoto CM, et al. Autologous bone-marrow mononuclear cell transplantation after acute myocardial infarction: comparison of two delivery techniques. Cell Transplant. 2009;18(3):343-52.

[66] Goussetis E, Manginas A, Koutelou M, Peristeri I, Theodosaki M, Kollaros N, et al. Intracoronary infusion of CD133+and CD133-CD34+selected autologous bone marrow progenitor cells in patients with chronic ischemic cardiomyopathy: cell isolation, adherence to the infarcted area, and body distribution. Stem Cells. 2006 Oct;24(10): 2279-83.

[67] Barbosa da Fonseca LM, Xavier SS, Rosado de Castro PH, Lima RS, Gutfilen B, Goldenberg RC, et al. Biodistribution of bone marrow mononuclear cells in chronic chagasic cardiomyopathy after intracoronary injection. International journal of cardiology. 2011 Jun 16;149(3):310-4.

[68] Kurpisz M, Czepczynski R, Grygielska B, Majewski M, Fiszer D, Jerzykowska O, et al. Bone marrow stem cell imaging after intracoronary administration. International journal of cardiology. 2007 Oct 1;121(2):194-5.

[69] Hofmann M, Wollert KC, Meyer GP, Menke A, Arseniev L, Hertenstein B, et al. Monitoring of bone marrow cell homing into the infarcted human myocardium. Circulation. 2005 May 3;111(17):2198-202.

[70] Schachinger V, Aicher A, Dobert N, Rover R, Diener J, Fichtlscherer S, et al. Pilot trial on determinants of progenitor cell recruitment to the infarcted human myocardium. Circulation. 2008 Sep 30;118(14):1425-32.

[71] Verjans J, Wolters S, Laufer W, Schellings M, Lax M, Lovhaug D, et al. Early molecular imaging of interstitial changes in patients after myocardial infarction: comparison with delayed contrast-enhanced magnetic resonance imaging. Journal of nuclear cardiology : official publication of the American Society of Nuclear Cardiology. 2010 Dec;17(6):1065-72.

[72] Mozid AM, Holstensson M, Choudhury T, Ben-Haim S, Allie R, Martin J, et al. Clinical feasibility study to detect angiogenesis following bone marrow stem cell transplantation in chronic ischaemic heart failure. Nucl Med Commun. 2014 Aug;35(8): 839-48.

[73] Dimmeler S, Zeiher AM. Cell therapy of acute myocardial infarction: open questions. Cardiology. 2009;113(3):155-60.

[74] Perin EC, Silva GV, Henry TD, Cabreira-Hansen MG, Moore WH, Coulter SA, et al. A randomized study of transendocardial injection of autologous bone marrow mononuclear cells and cell function analysis in ischemic heart failure (FOCUS-HF). American heart journal. 2011 Jun;161(6):1078-87 e3.

[75] Yoshida M, Shiraishi S, Sakamoto F, Beppu T, Utsunomiya D, Okabe H, et al. Assessment of hepatic functional regeneration after hepatectomy using Tc-GSA SPECT/CT fused imaging. Annals of nuclear medicine. 2014 Jul 8.

[76] Zuckerman E, Slobodin G, Sabo E, Yeshurun D, Naschitz JE, Groshar D. Quantitative liver-spleen scan using single photon emission computerized tomography (SPECT) for assessment of hepatic function in cirrhotic patients. Journal of hepatology. 2003 Sep;39(3):326-32.

[77] Mohamadnejad M, Namiri M, Bagheri M, Hashemi SM, Ghanaati H, Zare Mehrjardi N, et al. Phase 1 human trial of autologous bone marrow-hematopoietic stem cell transplantation in patients with decompensated cirrhosis. World J Gastroenterol. 2007 Jun 28;13(24):3359-63.

[78] Levicar N, Pai M, Habib NA, Tait P, Jiao LR, Marley SB, et al. Long-term clinical results of autologous infusion of mobilized adult bone marrow derived CD34+cells in patients with chronic liver disease. Cell Prolif. 2008 Feb;41 Suppl 1:115-25.

[79] Pan XN, Shen JK, Zhuang YP, Chen XL, Li YX, Chen LJ, et al. [Autologous bone marrow stem cell transplantation for treatment terminal liver diseases]. Nan fang yi ke da xue xue bao=Journal of Southern Medical University. 2008 Jul;28(7):1207-9.

[80] Gholamrezanezhad A, Mirpour S, Bagheri M, Mohamadnejad M, Alimoghaddam K, Abdolahzadeh L, et al. In vivo tracking of 111In-oxine labeled mesenchymal stem cells following infusion in patients with advanced cirrhosis. Nuclear medicine and biology. 2011 Oct;38(7):961-7.

Section 2

Biomaterials

Chapter 7

Materials for Central Nervous System Tissue Engineering

Manuel Pérez-Garnes, Juan A Barcia, Ulises Gómez-Pinedo,
Manuel Monleón Pradas and Ana Vallés-Lluch

Additional information is available at the end of the chapter

http://dx.doi.org/10.5772/59339

1. Introduction

Neurodegenerative disorders and traumatic injuries to the central nervous system (CNS) can cause severe and irreversible damage. Nowadays there are few clinical therapies to treat brain damage, and they achieve low functional recovery; most of them are based on drug administration for neurodegenerative diseases with well-defined targets as Parkinson's disease. For traumatic injuries such as traumatic brain injury (TBI), spinal cord injury (SCI), and stroke there is no effective treatment to restore lost functions and the clinical approaches are symptomatic and based on minimizing the damage progression or enhance local plasticity with rehabilitation. On the whole, the limitations of current therapies underline the need to search for novel strategies.

Brain damage has constituted up to now a theoretically unsolvable problem. This was because three of the most important tenets in neurobiology were against the restoration of the brain after damage: 1) there are no new neurons, 2) axons cannot grow within the CNS, and 3) neurons cannot produce new connections between them. Although these tenets have been questioned due to research done during the last decades, there is currently no clinical application useful to produce reconstruction of brain damage. This is probably due to several causes: endogenous regeneration in the adult human brain is, although present, insufficient to recover the lost neuronal population; also, axonal elongation is highly repressed by inhibitory factors, and the lack of the appropriate neurotropic factors and guidance cues. Furthermore, compared to other organs, the brain is a very complex system and it does not suffice to get a high number of new neural cells and connections between them, but these connections between new and old neurons must be exactly organized.

Thus, the problem of neural regeneration includes mainly three processes: the need to have new neurons to replace the lost ones, the need to promote axonal regeneration within the CNS to the appropriate targets, and the need to make functionally effective connections with the same connectivity of the lost ones.

Despite the existence of neurogenesis in the adult mammalian [1] and human [2] brain, cell replacement cannot be accomplished because of the massive death of new cells. After injury, there is an increased neurogenesis and migration to the damaged area. In an experimental model of stroke, new neurons appeared at the core ischemic area [3]. However, 80% of these neurons disappear from the core 5 weeks after stroke, probably because they die or migrate to other areas. A probable hypothesis for this phenomenon is that the core ischemic zone is not a proper environment for cell survival, since it lacks vascularization, stromal structure or trophic sustain for these cells.

In humans with amyotrophic lateral sclerosis and frontotemporal dementia, an increased proliferation of neural progenitors can be seen which is not able to restore the disease condition [4], probably because these cells die because of the action of a toxic environment produced by the disease.

Biomaterials could constitute a proper environment to facilitate vascularization, provide structural scaffold and improve the milieu in order to facilitate the survival of the newly formed cells [5].

2. Neurogenesis in the adult mammalian brain

It is currently well accepted that most of the neurons in the CNS are produced both in the prenatal as well as in the early postnatal stage. Moreover, during the past decades, neurogenesis in the adult brain of mammals has been reported, humans included. Adult neurogenesis has been shown to occur in two main areas of the adult mammalian brain: the olfactory bulb (OB) and the dentate gyrus (DG) of the hippocampus. While in the hippocampus stem cells and new neurons coexist together, the new neuronal cells in the OB are generated from neural progenitor cells in the anterior part of the subventricular zone (SVZ). The SVZ is a narrow region of tissue within the wall of the lateral ventricle in the forebrain. The neural progenitor cells of the SVZ migrate to the OB where they differentiate into interneurons such as granule and periglomerular cells [6]. It is known that a different migration exists from the SVZ to corpus callosum, where cells differentiate preferentially into oligodendrocytes wich could be a target for treatment of neurodegenerative diseases where myelin regeneration is required [7]. There is little evidence of the existence of alternative pathways for cell migration under normal conditions, from neurogenic areas to other regions of the brain. However, it is known that when damage occur such as in ictus, TBI or SCI it is possible to find migration from neurogenic niches to the injured region [8].

2.1. Structure and organization of subventricular zone and dentate gyrus in mammals

As mentioned above, stem cells responsible of adult neurogenesis in the OB, have their origins in the SVZ, and migrate along a restricted pathway, called the rostral migratory stream (RMS).

The RMS in mice has a length between 5 y 6 mm. Two principal cellular types were observed in the RMS: type A cells (migrating cells) and type B cells (astrocytes). In sagittal sections, type B cells and their processes flanked the chains of type A cells. In frontal sections, type B cells ensheathed the chains of migratory cells [6].

In the SVZ, have been identified at least four different cell types: type A (migrating cells), type B (astrocytes), type C (proliferative precursors or neuroblast) and type E (ependymal cells), these cells were defined by their immunological markers and morphology. Using proliferation markers such as 3HT or BrdU, the most actively dividing cell in the SVZ corresponded to Type C cells [9].

Briefly, the characteristics of the 4 types of the SVZ are: Type A cells (migrating cells), equivalent to those observed in the RMS. Their major characteristics were an elongated cell body with one or two processes, abundant lax chromatin with two to four small nucleoli, and a scant, dark cytoplasm containing many free ribosomes and many microtubules oriented along the long axis of the cells. The nuclei of Type A cells were occasionally invaginated. Furthermore, their membranes showed cell junctions intercalated with large intercellular spaces that allow cell movement. These cells were grouped forming chains and show immunoreactivity to PSA-NCAM, Dlx-2 and TuJ1.

Type B cells (astrocytes) formed a wide network adjacent to ependymal cells surrounding the chains of migrating cells (Type A cell). The cytoplasm of Type B cells was light and contained few free ribosomes. One of the most important characteristic of type B cells was the presence of extensive intermediate filaments in their cytoplasm, and they have also irregular contours that profusely filled the spaces between neighbouring cells. These cells had irregular nuclei that frequently contained invaginations.

Currently type B cells are divided into two subtypes: B1 and B2. Type B1 cells make contact with the ventricular cavity while B2 cells do not. B1 cells show one short cilium to the ventricular cavity, named primary cilium. The number of B1 cells in contact with the ventricle increases drastically when growth factors such as Epidermal Growth Factor (EGF) or ephrins are injected [10,11], possibly the necessary signals for activation are in the cerebrospinal fluid where, probably, the primary cilia could play an important role. Both B1 and B2 cells are immunoreactive against GFAP, vimentin and nestin and no molecular marker can differentiate them. Type B2 was similar to those described in the RMS.

Type C cells (precursor cells) were larger, more spherical (less elongated), and more electron-lucent than Type A cells but more electron-dense than Type B cells. Their nuclei contained deep invaginations and mostly lax chromatin, although sometimes the chromatin was clumped. Type C cells had a typical large reticulated nucleolus. Their cytoplasm contained fewer ribosomes than Type A cells, and no bundles of intermediate filaments of Type B cells. Groups of C cells (2-3 cells) are associated with the chains of A cells. C cells are immunoreactive against Dlx-2, a transcription factor, although this marker is not exclusive to this type of cell.

Type E (ependymal cells) formed an epithelial monolayer separating the SVZ from the ventricular cavity. Their main distinguishing characteristics were as follow: the lateral processes of adjacent ependymal cells were heavily interdigitated and contained apical junctional complexes. The surface exposed to the ventricular cavity contained microvilli and is ciliated. The cytoplasm was electron-lucent and contained many mitochondria and basal bodies located in the apical cytoplasm. Their nuclei were spherical, and the chromatin was non-clumped. These cells are immunoreactive against vimentin, S-100 and CD-24. In addition, microglia, a few pyknotic cells, mitoses, and large neurons were observed sporadically in the SVZ. Furthermore the presence of a blood vessel network plays an important role in the SVZ and there are evidences that the activation of the neurogenic niches is directly related with this network [12,13].

The DG is the other region within the brain where neurogenesis takes place in mammals [14,15]. The function of neurogenesis in the hippocampus has been correlated with memory and learning. The DG is composed of small neurons called granular cells that are typically packed in a V-shaped band. Stem cells are located in the inner of this band and these are responsible for generating new neurons, which will be placed at the cell body layer. Moreover, astrocytes have been identified in the DG as neuronal stem cells [16]. These cells are identified as radial cells, crossing the entire thickness of the cell layer. It is also known that before immature cells transform into mature neurons, they are partially isolated from the environment by astrocytes, and it is thought that probably astrocytes will allow the differentiation to mature and functional neurons. At ultrastructural level, these stem cells are similar to the astrocytes in the SVZ. Before the transformation to mature neurons, stem cells within the DG go through an intermediate stage of indifferentiation, named Type D cells. These cells are characterized by lax cromatin, a crescent number of ribosomes and mytochondrias in the cytoplasm.

Although OB and hippocampus are the most studied and widely believed neurogenic regions, it is thought they are not the only ones. Probably more cells exist and keep their proliferative capacity within the brain, maybe in the surroundings of the ventricular cavities. Recently neurospheres have been obtained from the channel spinal cord of rodents, confirming the existence of stem cells in the channel, although the nature of the cell responsible for such proliferation is unknown [17].

2.2. Ventricular zone in humans: Organization and structure

After demonstrating the neurogenesis exist in the human [18], the new step has been the identification and location of the stem cells responsible for adult neurogenesis in humans. They have been located in the walls of the lateral ventricles displaying markers of astrocytary type cells with GFAP. These cells proliferate *in vivo* and *in vitro* and give rise to neurospheres that, in certain conditions of culture, differentiate in neurons, astrocytes and oligodendrocytes proving the existence of stem cells [19].

The organization of the SVZ in the human brain is very different from the one in rodents. Basically, three layers have been identified which have been denominated: 1) ependymal cell layer, 2) Gap layer and 3) Ribbon layer. The ependymal cells layer is formed by a monolayer of cubic cells in contact with the ventricular light that show long radial expansions into the neurophil, containing intermediate filaments and some mitochondria and constituting the second or Gap layer. In the ventricular wall, facing the light, numerous cilia and microvilli can be seen as well as the unions between them [20].

The second layer (Gap) is formed by expansions of astrocytes from the layer ribbon and ependymal cells, and is known as hipocellular layer because of its low quantity of cell bodies. The astrocyte expansions contain abundant intermediate filaments and numerous desmosomes between them. Another interesting fact is the existence, in this layer, of ependymal cell aggregates. These cells make up little groups of 5 to 20 cells that have the typical organization of the ependymal cells.

Finally, the third layer, also known as Astrocyte Ribbon, is formed by large astrocytes, myelinated axons, oligodendrocytes and a progressive increase of synapses, as we move away from the ventricular cavity. Astrocytes of this layer have been considered stem cells. Also, there is an interesting fact about the astrocytes that occasionally send a long prolongation, which makes contact with the ventricular light [20]. This fact agrees with what has been seen in the mice SVZ, where presumably, there has been an activation of the neurogenesis by the cells in contact with the ventricular cavity and the appearance of neuronal soma limits the following layer [9]. The main difference with rodents is the lack of precursor or Type C cells and migrating cells.

2.3. The human brain with neurogenic features: Other regions?

The use of trophic factors such as basic fibroblast growth factor (bFGF), brain-derived neurotrophic factor (BDNF), platelet-derived growth factor (PDGF), epidermal growth factor (EGF) allows us to prove the presence of "sleeping" stem cells in other regions of the brain [21–25] like the third and fourth ventricles [26–28]. But, not only proliferation has been found in these walls, since there are authors that support the existence of neurogenic places distributed in the neurophil [29,30]. Therefore, in these last years there have been studies in which the formation of new neurons in the adult cerebral cortex of mice [29–31] and primates [32] has been proved. These findings have originated controversy between diverse investigation groups because the results cannot be totally duplicated, existing experimental variations to find the "new neurons" [33] and, sometimes, it is even possible that immature cells change to stem cells due to a certain stimulus [34].

Furthermore, the neurogenic term has been questioned since some authors think that, to consider a place as neurogenic, new neurons have to be produced or recruited in normal conditions and without the presence of any stimulus like an injury or an external trophic factor [33,35]. The neocortex, striatum, amygdala, substantia nigra (SN), channel spinal cord, third and fourth ventricle are some of the areas that are being proposed as neurogenic niches in mammalians including the human [26,28,33,36–38].

3. The injured central nervous system and the regenerative process

Regeneration in the adult nervous system is limited compared with other tissues. This limitation is due to the lack of proliferation and regeneration of neurons to accomplish correct connections. The regenerative response is different between the peripheral (PNS) and the CNS. Regeneration in the PNS is more successful due to the presence of Schwann cells (SCs), which provide nutrients, guide and myelinate the peripheral axons and synthesize growth factors. Moreover, a unique structure exists in the PNS: the bands of Büngner, consisting in clusters of oriented SCs and their processes, which remain after axon injury facilitating its regeneration. On the contrary, the predominant glial cells in the CNS are the oligodendrocytes, which provide axon support and myelinate axons, and the astrocytes, which principal function is to provide nutrients to the neurons. Oligodendrocytes express inhibitory proteins for axon growth, such as the myelin associated glycoprotein (MAG) and tenascin R, while SCs lack of these inhibitory proteins. In addition, astrocytes are activated to clustered layers of hypertrophic astrocytes giving place to the glial scar, which constitutes a physical barrier for axon outgrowth. Moreover, hypertrophic astrocytes secrete inhibitory molecules such as sulphated proteoglycans, MAGs, Nogo, etc., constituting a chemical barrier for axon outgrowth [39].

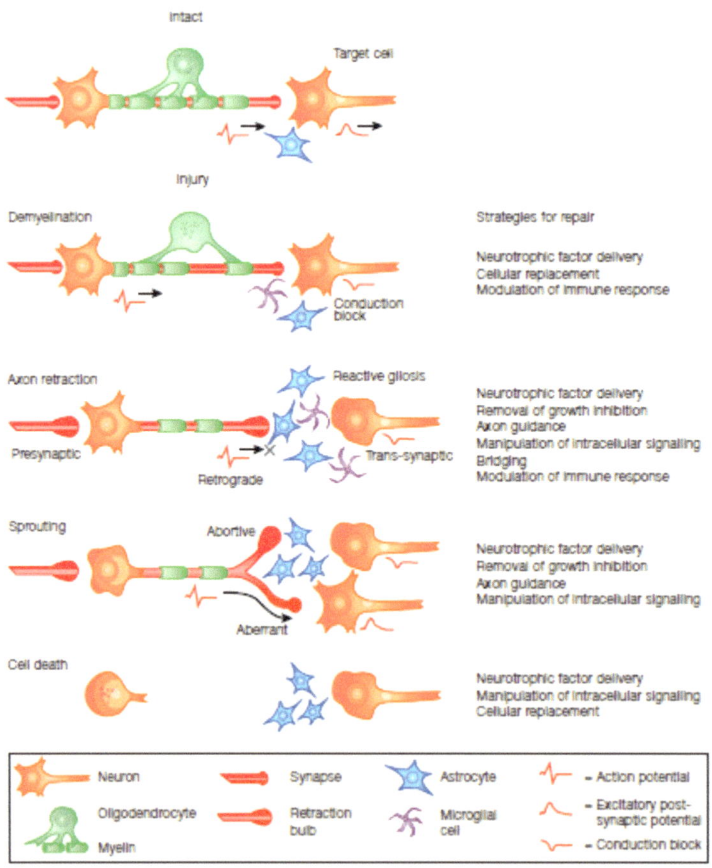

Figure 1. Consequences of neural injury and strategies for repair (from P. J. Horner and F. H. Gage [39], with authorization).

http://dx.doi.org/10.5772/59339

In the case of traumatic injuries as TBI, stroke and SCI the primary damage generates an abrupt loss of cells and may result in the disruption of the blood brain barrier (BBB) and blood spinal cord barrier (BSCB), respectively. The breach in the BBB and BSCB permits infiltration of foreign cells such as macrophages and fibroblasts, and inflammatory molecules activating astrocytes. In the case of neurodegenerative disorders as Parkinson's disease (PD), the loss of neurons is not abrupt but progressive, but the injured site is dominated by a toxic microenvironment as well. The neuronal injury and the inhibitory microenvironment result in degeneration at a cellular level: demyelination, axonal retraction and aberrant sprouting, and cell death. The tissue engineering strategies proposed aiming to repair injured tissues in the CNS must take into account these mechanisms of cell degeneration.

Thus, the main difficulty to regenerate neural tissues in the CNS after an injury is the toxic microenvironment generated and the lack of neuron replacement, although some neurogenesis has been observed after trauma. Indeed, despite the dramatic situation following a CNS injury, several studies have demonstrated that CNS axons could regenerate with an appropriate microenvironment. A favourable environment for CNS regeneration takes into account aspects as the normal function of glial cells, a correct system of nutrient distribution, a permissive physical pathway for axon growth and reconnection and a correct balance and distribution between inhibitory and promoting molecules for axon regeneration.

4. Cell therapy in the central nervous system

Cell therapy aims to supply the damaged tissue with glial cells, neurons, and even extracellular matrix components, adhesion molecules and/or neurotrophic factors. Cells could lead to a neuroprotective effect, help glial and neuronal remodelling, promote angiogenesis and synaptogenesis and modulate the inflammatory response. A wide range of cells has been employed to help regeneration in the injured CNS: glial cells, adult stem cells, embryonic stem cells and pluripotent stem cells [40]; the particular characteristics thereof are discussed next.

SCs are cells that promote regeneration in the PNS because of their ability to generate a permissive environment for axon regeneration and provide neuroprotection. SCs support growth and produce neurotrophic factors, including fibroblast growth factor (FGF), neurotrofin-3 (NT-3), brain derived neurotrophic factor (BDNF), nerve growth factor (NGF) and ciliary neurotrophic factor (CNTF), and secrete proteins such as laminin (LN) to promote axon growth. Moreover, SCs provide pathways for regenerating axons assisting the growth cone extension and guidance, and remove myelin debris following damage and remyelinate the regenerating axons. In some spinal cord lesions, SCs infiltrate into the CNS from the peripheral ganglions and assist the regeneration process. For this reason, some researchers have investigated the implantation of SCs in the damaged spinal cord, and they have demonstrated their ability to promote axonal regeneration and myelination [41]. However, their use to address CNS regeneration is limited because SCs activate astrocytes, which increase the expression of inhibitory chondroitin sulphated proteoglycans (CSPGs) for axon regrowth and overexpress glial fibrillary acidic protein (GFAP) contributing to the formation of the glial scar. Moreover, the neuroprotective effect of transplanted SCs is limited because they not prevent degeneration and secondary injuries. Other glia (astrocytes, oligodendrocytes, and microglia) has been also studied in connection with CNS regeneration. Implantation of macrophages in the injured spinal cord has shown some degree of regeneration due to their effect clearing myelin debris decreasing the inhibitory MAG levels for axons guidance [42]. However, the regeneration process assisted by macrophages is unclear, because they can inhibit axonal growth after SCI. On the other hand, the use of other glial cells, including astrocytes and oligodendrocytes, is usually addressed by employing precursors limited to differentiate to glial cells. These progenitor cells difficult the formation of gliotic scar than SCs and induce axon regeneration and remielynation [43].

Olfactory ensheathing cells (OECs) constitute an interesting type of cells in CNS regeneration strategies; they have been employed in animal models of SCI and stroke with several advantages over SCs [44,45]. These cells, in their natural environment, guide axons from the PNS to the CNS, and do not induce unfavourable interactions with the glial scar components. Furthermore, OECs express many of the same neurotrophic factors and proteins as SCs, including NGF, BDNF, GDNF, NT-3 and L1, facilitating axon regeneration and providing neuroprotection. After their transplantation in different SCI models OECs have demonstrated their ability to generate a more permissive microenvironment; they even lead to a limited regeneration, electrophysiological and functional recovery [46–48]. Infusions of OECs have been employed to address stroke, resulting in benefits to repair the damaged cerebral tissue, providing neuroprotection, facilitating neurite outgrowth, activating stem cells, reducing the lesion cavity and decreasing cell apoptosis, even improving the neurological function [44,49]. However, the implantation of OECs alone in the nigrostriatal tract of PD models is not sufficient to promote regeneration and functional recovery, because OECs do not release L-dopa to stimulate dopaminergic neurons [50]. Nonetheless, their role could be to permit the reentry of dopaminergic axons into the striatum once they have grown through a guiding structure, such a peripheral nerve graft [51]. Their transplantation together with other cells, including dopaminergic neurons derived from neural stem cells (NSCs), has demonstrated though ability to provide a slight functional recovery in PD models [52]. Moreover, some studies have employed genetically modified OECs overexpressing some neurotrophic factors in SCI models, resulting in a modest improvement of tissue sparing and functional recovery [53]. Some researchers have transplanted OECs addressing spinal cord repair and observed remyelination of axons, but this could be due to a contamination of the OECs suspension by other cells such as SCs since myelination by OECs has been debated [54]. OECs have been also employed in some controversial clinical studies of stroke and SCI, showing some degree of functional recovery, but these results must be corroborated in further studies [55,56].

Adult stem cells from different origins, including mesenchymal stem cells (MSCs), NSCs and neural progenitor cells (NPCs) have been transplanted in several traumatic injuries and neurodegenerative diseases in preclinical studies, and in clinical trials of stroke and SCI [57]. MSCs and NSCs are multipotent cells; the particular feature exploited in MSCs is its ability to differentiate into cells secreting neurotrophic factors, while NSCs can differentiate in both glial cells and neurons [58,59]. MSCs have been implanted in the spinal cord, showing the formation of bundles that bridge the lesions, even increase the locomotor function, suggesting that MSCs provide trophic factors and support for host cells and themselves [60]. Meanwhile, NSCs implanted in the injured spinal cord have eventually differentiated into astrocytes [61], thus these cells should be stimulated to favour their differentiation into neurons. The implantation of NSCs in the injured spinal cord may also be combined with growth factors promoting the neural lineage or limiting the effect of the inhibitory molecules in the environment of the damaged spinal cord. This technique has been implemented combining NSCs with agonists of astrocyte differentiation, accomplishing differentiation of NSCs into neurons and functional recovery after SCI [62]. Furthermore, MSCs and NSCs have been transplanted in ischemic brain and stroke rat models showing promising results, including stimulation of endogenous cells and neurogenesis from the SVZ, promotion of angiogenesis, reduction of glial scar, even functional benefits [63–65]. In addition, MSCs have been tested in a clinical trial of stroke to prove their safety in humans, showing promising results [66]. MSCs and NSCs have also been studied to address neurodegenerative diseases, including Huntington's disease (HD) and PD. MSCs have demonstrated migration to the injured striatum, neuroprotection through the release of neurotrophic factors, even functional recovery over months [67–70]. Besides, NSCs implantation into the striatum has resulted in neuron differentiation and some functional improvements [71,72], suggesting that infusions of these cells could constitute a possible therapy for cell replacement in neurodegenerative diseases. In this sense, NSCs have been employed in clinical trials of both PD and HD, resulting in functional improvements over 36 months and 10 months, respectively [72,73]. However, these cells presented problems to engraft the lesion site and showed a poor survival. Other types of adult cells such as the adrenal chromaffin cells and human retinal pigment epithelium cells (hRPEs) have been employed in clinical trials to restore functions in PD, since they secrete dopamine or its precursor L-dopa, and can be employed as suppliers of L-dopa *in situ*. Both adrenal chromaffin cells and hRPEs have demonstrated modest functional improvements [74,75]. However, these cells have a poor survival and only a small fraction is able to secrete L-dopa or dopamine, or is secreted in small amounts.

Embryonic stem cells (ESCs) are derived from the blastocyst and possess interesting properties, including pluripotency and self-renewal, allowing their differentiation into cells from the three germ layers. For this reason, ESCs may constitute a source of different cells, including neuronal ones, with many possibilities in CNS regeneration. ESCs from mouse and human origin have been employed to address experimental studies in different CNS damages. For instance, ESCs have been induced to differentiate, preferably, into oligodendrocytes and implanted in the injured spinal cord of different animal models, including mouse, rats and chick embryos; they have shown to stimulate remyelination and promote locomotor improvements [76–78]. In another work, large amounts of motor neurons have been obtained *in vitro* from human ESCs and have been transplanted into the spinal cord of chick embryos maintaining their phenotype, but the potential benefits of the neurons stemming from ESCs require further characterization [78]. The poor survival of ESCs in the injured spinal cord can be enhanced by the incorporation of neurotrophic factors, including BDNF [79]. Mouse and human ESCs have also been differentiated *in vitro* into dopaminergic neurons, addressing their potential use for PD treatment [80,81]. These findings have led to the use of dopaminergic neurons derived from

ESCs in preclinical PD transplants, avoiding immune response, resulting in slight motor improvements and a reduction of symptoms [82,83]. The use of ESCs is also a promising approach to address several CNS damages; however, using ESCs requires solving many problems, such as poor cell survival and low efficacy of transplanted cells, ethical aspects because of its embryonic origin and the risk of teratoma formation.

As regards induced pluripotent stem cells (iPSs), they do not pose the ethical problems associated with the use of human ESCs and show a lesser degree of immune response. These cells are obtained from somatic cells by overexpressing four reprogramming factors (Oct4, Sox2, Klf4, and Myc), leading to self-regenerative and pluripotent cells like ESCs. These pluripotent cells may be differentiated into the desirable lineage, including glial cells and neurons, by adequate molecular stimuli, and can be transplanted to address several CNS damages. Nevertheless, the use of iPSs gives rise to problems still to be solved, including teratoma formation, aberrant reprogramming, or presence of transgenes before their clinical application.

In summary, preclinical and clinical trials have shown that implantation of cells in the CNS yields transient benefits based on paracrine effects. However, pure cell supply therapy has brought only modest results lagging behind initial expectations, due probably to a variety of factors such as a poor cell survival in an aggressive environment and an inappropriate axon reconnection and guidance.

5. Cues influencing axon guidance and growth

Biomaterials have been studied in a wide range of tissue engineering fields since decades. They are usually manufactured as scaffolds and their general function is to mimic the specific tissue and cell extracellular matrix (ECM) during the regenerative process. Scaffolds directly affect cells behaviour, in terms of their adhesion, proliferation, migration, differentiation, and maintenance of phenotype. Biomaterials in neural tissue engineering can be tailored to trigger these effects in cell behaviour as well. In addition, scaffolds in neural tissue engineering can be useful to guide and stimulate axon outgrowth, provide a neuroprotective effect in the toxic microenvironment of the injured CNS and support glia migration.

In the last decades, researchers have investigated the cues involved in axon growth and guidance by several *in vitro* studies, which are related to cell substrates (structure, composition, mechanical properties, etc.), physical stimuli (electrical signals and mechanical stretches), cellular cues (glial and other cell types) and permissive and inhibitory molecules of axon growth (ECM components, neurotrophic factors, etc.).

5.1. Molecular cues

Cell behaviour including adhesion, differentiation, migration, etc. and particularly, axonal extension, growth and guiding are regulated by different molecules, which become thus crucial in the tasks to address CNS regeneration. Axonal extension and guidance is initiated by the molecular interaction of microtubules of the growth cone with the surrounding signals (chemical and physical cues). The growth cone development during axonal sprouting is influenced by chemical signals, including ECM and cell adhesion molecules and neurotrophic factors. However, the complex neuronal network within the CNS results from the presence of promoting (growth factors and proteins) and inhibitory molecules of axonal growth (including canonical axon guidance molecules (semaphorins, ephrins, etc.), myelin associated glycoproteins (MAG) and CSPGs). This complex neural network of the CNS begins to form at early stages of embryogenesis and continues through the adulthood. Moreover, some of these molecules may act as promoters or inhibitory ones depending on the stage of the developing CNS.

ECM components-based substrates have been employed to obtain scaffolds with binding moieties for direct cell attachment [84]. The most commonly employed ECM components to address *in vitro* studies of axonal growth and guidance include proteins such as collagen, laminin (LN), fibronectin (FN) and their specific peptides: RGD, GRGDS, and IKVAV, etc. The ability of these molecules to generate bindings to a number of membrane-bound receptors has been exploited by several researchers, which have developed ECM components-based scaffolds to favour neuron attachment and neurite outgrowth including collagen porous scaffolds [85] and fibrin multichannel conduits [86]. However, the majority of the synthetic [87] and natural [88] substrates employed usually lack of protein-binding motifs and other permissive molecules of axon growth such as neurotrophic factors, which may be incorporated post-processing by different methods including physical adsorption, encapsulation into substrates, electrostatic interaction, covalent immobilization and cultures of natural or modified cells with the ability to secrete these molecules. The incorporation of molecules for cell adhesion improves growth cone attachment, neurite growth rate, branching and extension compared with the uncoated biomaterials [89–91], even some aspects of neural and glial cells fate can be enhanced, for example proliferation and migration [87]. The incorporation of cell adhesion molecules on substrates can be accomplished by different techniques, including physical entrapment, chemical linking and physical adsorption. These methodologies combine permissive and limiting pathways for axon guidance, to obtain parallel and preferential pathways or microchannels for axon growth. Gradients of permissive molecules on substrates can also be achieved by different techniques including microcontact printing of proteins, selective detachment of immobilized molecules by laser or electron ablation, among others. These techniques enhance neural attachment and increase neurite lengths, and even can favour the orientation of extended axons in the direction of the gradients and/or along the coated substrates and a bipolar morphology [92,93]. The non-permissive molecules of axon outgrowth play an important role on forming the correct axonal pathways; for instance, axons can be grown and guided into channels surrounded by CSPGs, in a similar way than on cell adhesion molecules [94]. Moreover, parallel axons can be grown in multichannels thanks to a combination of permissive molecules forming the axonal pathways and inhibitory ones forming the limiting regions of axon outgrowth [95].

Peptide motifs can be incorporated to the substrates in similar ways as ECM components, and accomplishing comparable improvements in cell attachment, neurite length, glia migration and orientation of growing axons, since these peptides constitute the functional parts of some relevant proteins implied in cell attachment. Several authors have demonstrated that depleting these functional sites of the cell adhesion proteins leads to an abrupt reduction of neurite lengths and cell migration and the ability of growth cones to turn towards higher peptide densities [96,97]. Some peptides show the ability to form self-assembled nanofibrous gels, with fibres sizes of a few nanometres, by the spontaneous organization of the molecules under physiological conditions, forming non-covalent bonds including hydrogen bonds, Van der Waals forces, electrostatic forces, etc [98]. This self-organization of peptides leads to hydrogels with molecular and topographical cues to enhance differentiation of NSCs into neurons despite astrocytes [98] and neurite extension [99].

Neuronal and glial proliferation differentiation and axonal growth and guidance can be regulated and promoted by growth factors including FGF, insulin-like growth factors (IGF-1, IGF-2), vascular endothelial growth factor (VEGF), and neurotrophic factors, including NGF, BDNF, neurotrophin-3 (NT-3), neurotrophin-4/5 (NT-4/5), glial cell-line derived neurotrophic factor (GDNF), CNTF, etc. These molecules are able to promote the differentiation of embryonic and adult NSCs [100–102] and other adult cells, including MSCs [103] and rat pheochromocytoma (PC-12) cells [104] into neuronal fates. Several studies have shown that solubilized- or substrate-immobilized neurotrophic factors produce higher neurite length, cell attachment and neuron survival [105–107]. Besides, some *in vitro* studies have shown the ability of neurotrophic factors to enhance functional connections showing expression of synapses and different neurotransmitters, even the new-formed synapses presented the ability to respond to external stimuli [105]. Immobilization of neurotrophic factors and growth factors can be accomplished by similar methods to those of proteins or peptides, including physical entrapment, chemical linking and physical adsorption for a controlled release during weeks [106]. Their effects can vary depending on the preparation method; for example, the chemical immobilization of NGF onto substrates has shown an increment of differentiation into neurons compared to physical entrapment. Chemical immobilization has resulted in similar neurite lengths and neuron survival to those studies where the growth factor is daily incorporated into culture media [107]; this result is interesting to improve the effect of biomaterials with neurotrophic or/and growth factors for neural implants. Several immobilization techniques of neurotrophic and growth factors permit to perform gradients thereof on substrates, even into scaffolds, with an important effect in neurite and axon guidance [107,108]; they have shown to elicit turning of growth cones and axons outgrowth in the direction of their gradients. The neuronal behaviour, including neurite outgrowth and length, can be even improved combining several neurotrophic and growth factors and modifying the amount of each one. Moreover, this synergistic effect can be also observed by combining factors and cell adhesion molecules. The combination of factors allows NSCs to differentiate into several neural or glial lineages, and particularly some combinations permit to obtain neuronal cells despite astrocytes, suggesting their potential use to avoid the inhibitory glial scar [109].

ECM components, growth and neurotrophic factors are not the unique molecules employed to address neurite outgrowth; other molecules including immunoglobulin-like adhesion molecules (NCAM and L1), receptors (ephrins), neurotransmitters (acetylcholine), etc. have also been studied to address neural regeneration. Adhesion molecules as L1 and NCAM show similar effects on neurite outgrowth and cell viability than permissive ECM components; however, L1 can induce more the attachment of neurons in detriment of astrocytes compared with LN [89]. Ephrins have an important role in axon growth during CNS development, but in the adulthood, they become inhibitory of axon outgrowth. Curiously enough, on substrates with gradients of ephrins axons grow in the direction of higher gradients but growth stops at higher densities, showing the ambiguous effect of these molecules in axon growth and guidance [110]. Neurotransmitters have an important role in neuron communication and transmission of electrical impulses and, as a consequence, these molecules are vital to obtain a correct synapse formation and function and in neuron survival. Neurotransmitters can be biomimicked by chemical compounds and incorporated to different substrates, showing improvements in neuron survival and neurite extension, even similar to cell adhesion molecules [111].

5.2. Topographical cues

Cell differentiation is a process dependent of external stimuli and is well known that cell-substrate interaction and particularly topographical cues are important in this cellular behaviour. Bellamkonda discussed the anisotropy of both molecular and substrate topography as an important cue to obtain faster and more orientated axon regeneration and to direct growth cones following a unique direction [112]. This idea has resulted in a wide variety of investigations, where researchers have studied axon outgrowth, both *in vitro* and *in vivo*, on anisotropic substrates such as surface patterns, fibres and channels. Different 3D structures with pronounced anisotropy are obtained by different micro and nanofabrication techniques. For example, nanofibres are usually obtained and aligned by electrospinning techniques, while microfibres are fabricated by extrusion with different morphologies [113,114]. Microchannels are principally made by moulding employing fibres templates or other techniques such as laser irradiation, freeze-drying, dipping and ion capillarity [88,115,116]. Nano and micropatterns are fabricated by common techniques such as lithography, photolithography, moulding templates, reactive ion etching, etc. [117], which allow to obtain structures alternating permissive axon pathways (grooves) and high ridges to separate them.

Silicon-based materials such as polydimethyl siloxanes (PDMS) have been commonly employed to develop micropatterns [118]. Researchers have studied exhaustively these systems to optimize their dimensions (groove depth, width and space) to enhance axon growth and guidance. Axon alignment and outgrowth is increased as a response to greater depths of the grooves because cell bodies are restricted to the grooves, avoiding neuronal crossing between grooves. Several studies have stated that an optimal groove width of 20 μm enhances neurite extension and achieves a bipolar structure oriented in the direction of the grooves and with a significant reduction in neurite branching [119]. Other investigations gave similar results using stem cells and neuronal lines, including human neural stem cells [120] and PC-12 cells [121,122]. These studies show a similar behaviour of axons outgrowth in the micropatterned substrates: the channels width must be enough to host the neuronal body but not too much to permit neurite hyperbranching, yielding neuron bipolar and oriented structure. These findings are conditioned to the sizes of different parts of the cells: neuronal bodies are in the range of 10-20 μm, growth cones about 5 μm, while axons and focal adhesions show sizes about 1-3 μm [123]. However some studies have shown unexpected neuron morphologies on microgrooves, in which neurons can bridge between adjacent channels with no underlying support [124].

Some researchers have exploited the idea of obtaining substrates with topographies that mimic the glial scar environment (combination of permissive and inhibitory cues on axon pathways), for example studying the behaviour of co-cultures of astrocytes and neurons. This idea was addressed by Krsk et al. [125], who fabricated silicon wafers with discrete polyethylene glycol (PEG) grafts by electron-beam, which forms an inhibitory region similar to the glial scar. These authors obtained parallel silicon 3 μm-channels separated with PEG lines and silicon arrays with different distances between PEG spacers (1, 2, 3, 4 and 10 μm). They determined that astrocytes and mouse dorsal root ganglia (DRGs) neurons grew parallel to the PEG lines onto the grooves forming elongated structures and oriented actin filaments and neurites, respectively. Furthermore, both neurites and astrocytes grew only onto above distances of 2 μm between PEG spacers, but astrocyte adhesion and spreading were comparable to the unpatterned silicon when the arrays were 10 μm long between PEG spacers.

Nanopatterned substrates (submicron features) provide differences in neurite growth and alignment with respect to the micropatterns, since the effect of nano-size features can be considered as a roughness phenomenon on cell behaviour. The nanostructured materials increase cell-substrate interactions, promoting cell adhesion and hosting neuronal processes, but not cellular bodies. Nanostructured topographies mimic the structure and dimensions of ECM components and chemicals; they cannot host axons and neural bodies, but their sizes are in the range of the ECM components (organized proteins and polysaccharides) involved in cell attachment, spreading and migration, including fibronectin fibrils (about 100 nm), actin filaments and neurofilaments (10 nm) [123]. Several techniques have been developed to obtain them: electron-beam lithography to obtain roughnesses above 3 nm, chemical and electrochemical etching with motifs above 2 nm and multiphoton excitation to address greater structures with 250 nm. Some studies have found that rough surfaces improve cell adhesion compared to the smoothest ones, in which growth cones reach even half their growth [126]. However, nanogrooves lead to less axon guidance than micropatterns because all somas and axons grow along the plateau and only the smallest axons are suitable to be aligned [127].

Nanofibres constitute an interesting nanostructure to address neural cell attachment and axon outgrowth because most of the ECM proteins assemble naturally into fibres with diameters at the nanometric scale in the physiological environment. Nanofibres technologies permit to obtain 3D structures or scaffolds and not simple superficial topography or roughness, which differentiates these substrates from the above mentioned superficial micro and nanopatterns. Researchers have developed different processes to obtain them, including phase separation, electrospinning and self-assembly of peptide nanofibres (SAPNs).

Phase separation allows obtaining fibres between 50 and 500 nm, depending on the process parameters (temperature, polymer concentration, and solvent-polymer system). This method is achieved by several solvent-removal steps with temperature shifts of a polymeric solution, obtaining porous or fibrous structures (both of them interesting structures for different neural tissue engineering problems). These nanofibrous scaffolds provide a substrate for cell differentiation of neural lineages, even neurite outgrowth can be accomplished [128]. However, this nanofibre fabrication method shows several disadvantages because it employs large amounts of solvents and prevents to obtain aligned fibres, which limits their potential uses to address the orientation of glial cells and neurites.

Electrospinning is another current and interesting methodology to obtain nanofibres for neural tissue engineering. It is quite simple and permits to achieve nanofibres with diameters at the nanometric and micrometric scale. In addition, their operation parameters allow manufacturing a wide range of fibrous morphologies including random and aligned fibres and more complex and precise structures. The principle of the electrospinning process consists in the extrusion of a polymer solution through a small orifice (with the help of a pump pushing a syringe piston towards its needle) at high voltages, and the evaporation of the solvent as the jet is projected onto a metallic collector yielding a fibre that superposes with time forming a mat. Size and morphology of the fibres can be modified by varying operation parameters such as polymer concentration, viscosity, polymer-solvents interaction, distance from the collector, the applied voltage and external variables (temperature and humidity). The basic electrospinning process can be easily modified to obtain other fibrous disposition and morphologies: rotating drum collectors or two spaced collectors allow obtaining aligned fibrous tubes, crosslinking agents permit to fabricate polymeric networks, salts may be used to obtain porous fibres, and the most advanced electrospinning processes allow to fabricate fibres structures with a nanometric precision. This technique is able to render fibre mats (membranes) of a large amount of polymeric compositions, both synthetic and natural. The most employed synthetic biomaterials in neural tissue engineering are the family of aliphatic polyesters including poly(L-lactic acid) (PLLA), polyglycolic acid (PGA) and poly-ε-caprolactone (PCL), which are interesting biopolymers to be used in implants for neural tissue engineering for their biodegradability; furthermore, PLLA and PCL are approved by the Food and Drug Administration (FDA) for uses in some clinical methods. However, a wider range of synthetic polymers has been studied as electrospun substrates for axon outgrowth and guidance, such as polyamides, polyacrylates, polyacrylic acids, polydioxanone (PDS), etc. These synthetic polymers can support well glial and neural cells, in general, but they show different effects over cell behaviour due to their discrepant properties. On another hand, some researchers have developed methodologies to obtain electrospun fibres from natural polymers such as collagen, gelatine, chitosan and hyaluronic acid, because they form part of the tissues and, as a consequence, they have similar mechanical properties and cells recognize their domains easier than in the synthetic ones. However, natural polymers generally present poor mechanical properties and the electrospun fibres obtained from them show the limitation of handling for clinical uses. In addition, these materials present high costs, low reproducibility and may contribute to the immunogenic response. In order to meet the good mechanical properties and reproducibility of the synthetic materials and the major biocompatibility of the natural ones, electrospun mats of composites from synthetic and natural polymers (gelatine/PCL, collagen/PCL, PLLA/laminin, etc.) have recently been developed. Another option is to biofunctionalize synthetic fibres with ECM components (collagen, fibrin, etc.) by covalent binding or physical adsorption; these biofunctionalized fibres enhance biocompatibility, neurite extension, and glial cell migration and alignment [91,97].

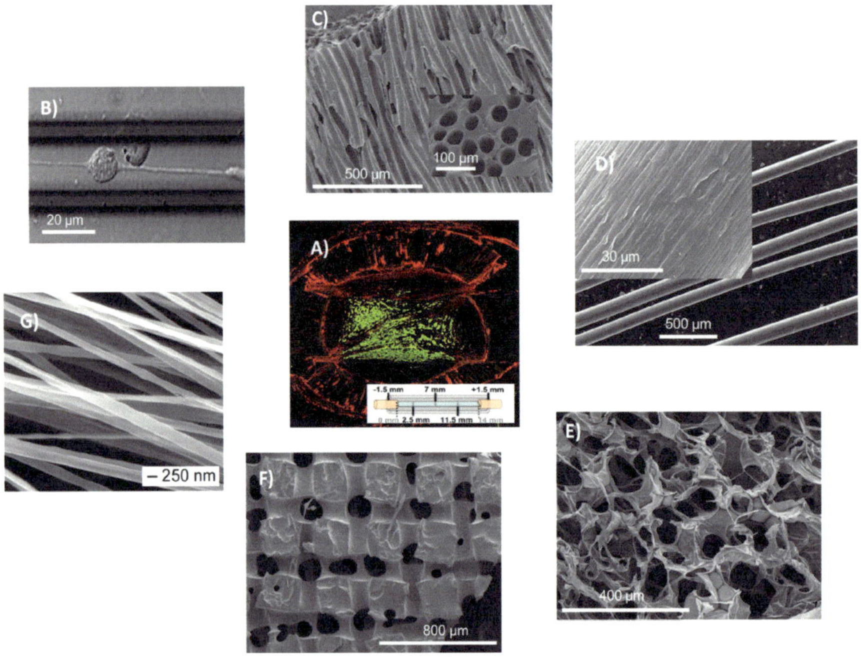

Figure 2. Topographical motifs included in the biomaterials usually employed in CNS regeneration. A) Typical conduits based on an external channel and an inner matrix for peripheral nerve and spinal cord regeneration (from I. P. Clements et al. [116] with authorization). B) PC-12 cell growing along a microopatterned substrate (from M. J. Mahoney et al. [122], with authorization). C) Multichannel poly (ethyl acrylate) scaffold, inset: detail of the multichannel scaffold cross-section (own our group). D) Microfilament of polycaprolactone obtained by micro-extrusion (own our group). E) Crosslinked hyaluronic acid scaffolds with interconnected porous structure (own our group). F) Poly (ethylacrylate)-based scaffolds containing interwoven microchannels (own our group). G) Aligned electrospun fibres of polylactic acid (from J. Xie et al. [113], with permission).

Electrospun scaffolds have been employed as artificial matrices for neurons and glial cells. The most common biodegradable synthetic biomaterials, performed as electrospun scaffolds, show good cytocompatibility and they are even more biocompatible than as flat substrates [129]. Some researchers have investigated cell differentiation in different electrospun fibres and have reported differentiation of NSCs and ESCs to neural lineages [130,131]. More in detail, others have concluded that fibre alignment does not influence the NSCs differentiation to a neuronal lineage, which does depend on the electrospun fibres diameter, because fibres in the nanometric scale enhance NSCs to neural lineage whereas the micrometric ones do not [132]. Similar studies with ESCs have demonstrated that both, random and aligned fibres are suitable for cell differentiation to a neural lineage; however, only the aligned fibres inhibit astrocyte differentiation [131]. This interesting result may be useful in order to employ aligned fibres as substrates for ESCs supplying neurons and inhibiting the glial scar formation in some CNS injuries. Other studies have shown that electrospun scaffolds with aligned fibres enhance a bipolar morphology of glial and neural cells and an extension of axons, even DRG ganglia may be oriented in the direction of electrospun fibres. In addition, neurites grow faster and longer in the aligned fibres than in the random ones [91,133].

5.3. Cellular cues

The role of glial cells in the regeneration of the injured CNS is crucial because they exert several functions over neurons in the healthy and developing CNS. Glial cells can be permissive or inhibitory of axon outgrowth; even they change their roles in the developing, adult or injured CNS. The most studied glial cells to address axon guidance include SCs, astrocytes and olfactory ensheathing cells/olfactory nerve fibroblasts (OECs/ONFs).

SCs promote the axonal cone growth and show a neuroprotective effect in the regenerating axons by the release of some neurotrophic factors in the injured PNS. This property in the PNS environment has been widely explored to address axon outgrowth in the CNS. The principal role of SCs involves myelination and ensheathing the peripheral axons, but they produce MAGs, which are inhibitory of neurite outgrowth and hyperbranching. Some authors have studied the influence of SCs in axonal extension *in vitro*, showing that aligned monolayers of SCs directly affect neurite outgrowth and alignment without the help of other topographical cues [134].

CNS glial cells regulate several functions of neurons; for this reason, these cells have been studied to address their effect in neurite growth. The most widely studied CNS glial cells are the astrocytes, which main function is to provide nutrients to neurons. Astrocytes have been reported to be appropriate glial cells to differentiate neural progenitor cells (NPCs) [135]. Some authors have cultured neurons on monolayers of previously aligned astrocytes by several techniques such as electrical fields [136] or topographical cues [137], demonstrating that the bipolar morphology of astrocytes is able to enhance neurite outgrowth and alignment, even neurite outgrowth can be compared to what happens onto micropatterned proteins. However, co-cultures of neurons and random astrocytes have not demonstrated important changes in neurite outgrowth, because the astrocytes are in a multipolar shape characteristic of the hypertrophied ones, which constitute the glial scar [138].

OECs constitute one of the most promising adult cell types in order to address CNS injuries. These cells present some similarities with SCs and astrocytes (molecular expression, morphology and functions). However, they are the unique glial cells with the ability to cross the PNS-CNS transition zone helping the olfactory axons outgrowth from the nasal epithelium to the olfactory bulb. Their ability to interact with both nervous systems has given rise to the idea that OECs may be able to regenerate axons in the presence of gliotic components. In this sense, authors have demonstrated how OECs enhance neurite extension and increase the amount of axons in gliotic-based substrates such as myelin, even extension and neurite density resulted similar than using permissive substrates for axon growth such as LN [139]. Co-cultures of cortical neurons with aligned OECs have shown higher neuronal lengths and similar alignment of neurites than with astrocytes, but OECs lead to a more bipolar morphology of neurons [140].

Glial cells act on neurite outgrowth and neuron alignment by means of different cues, which include their own topography and molecular cues related with the released inhibitory and permissive molecules. The effect of molecular components has been largely investigated as commented above, but the influence of cell topography is still a rather unexplored phenomenon. A few investigations have addressed the effect of cell topography as an independent factor in neurite alignment and outgrowth by original methods. The methodology employed consists in obtaining reliefs of a substrate that mimic the shape of different cells, employing different techniques such as polymerization and lithography of masks of cellular features obtained by CAD software or impression of cell replicas. Normally, the biomimetic reliefs are formed by bipolar and oriented morphologies to enhance neurite outgrowth and alignment and the most common cells-masks include SCs, astrocytes and endothelial cells morphologies. Aligned cellular motifs of SCs have an influence on neurite alignment and increase their length compared to flat substrates and random motifs, even perpendicular reliefs have an inhibitory effect in axon outgrowth [141]. Although astrocytes and endothelial cells can enhance neurite alignment too, SCs have improved the most these aspects of neurites, higher degrees of alignment and sizes of their dimensions (weight, length and height), in comparison to astrocytes and endothelial cells [142].

5.4. Electrical cues

As is well known, electrical voltages play an important role in cell behaviour. Particularly, the transmission of information between neurons is accompanied by electrical changes in their environment; this phenomenon is known as action potential (AP). Electrical stimuli are produced by the existing voltage gradient across the cell membrane and the complex equilibrium of environmental and cellular ions and ionic channels across the membrane. The steps of the process include depolarization of the neural membrane as a consequence of the stimulus received by dendrites of a neighboring neuron, subsequently the Na^+channels open and if there are sufficient opened channels the entering flux of Na^+modifies the cell membrane potential. The following step includes the repolarization of the membrane by the opening of the K $^+$channels, which act slowly and restore the original action potential. However, before reaching the normal membrane potential, neuron is hyperpolarized preventing it from receiving another stimulus and neuron cannot generate another AP during a brief time, ensuring that signals travel always in a unique direction.

Electrical stimulation of neural cells has led to the development of new frontiers for neural tissue engineering in the recent decades, including investigations of new conductive biomaterials and fabrication of electrodes to address several damages in the CNS. Actually, the efforts have focused on the study of conductive polymers for their electrical conductivity combined with their characteristics as organic polymers. The electrical properties of these polymers are due to the presence of loosely held electrons in their backbones that form conjugated π bonds, and with the help of doping chemicals the polymer can be negatively or positively charged. Traditionally, dopant agents employed have been inorganic compounds such as ionic species (chloride, perclorate, iodine, etc.). However, the introduction of active biological dopants is more attractive from the point of view of biomedical applications [143]. The most commonly studied conductive polymers include polypirroles (PPy), polyaniline (PANI) and polythiophene derivatives, specially, the poly (3, 4-ethylenedioxythiphene) (PEDOT), but other emergent materials as piezoelectric polymers and carbon nanotubes and nanofibres (CNTs and CNFs) as fillers are promising to develop conductive polymers for neural tissue engineering. Several studies have shown that PPy are biocompatible with different cell types [144] and, particularly, with neural cells [145]. PANI and PEDOT as conductive polymers for general tissue engineering applications have been less studied than PPy ones, but in the recent years they are more investigated for these purposes. Recent studies have demonstrated the absence of cytotoxicity of PANI and PEDOT by *in vitro* studies, showing the biocompatibility of these alternative biomaterials [146,147]. Moreover, the tissue response of these three conductive biomaterials has been studied preliminary in some *in vivo* studies, resulting in non-toxic effects and with an acceptable inflammatory response [148–150].

External electrical stimuli can be applied to cells or tissues to produce an artificial AP leading to electrical changes in cells; this phenomenon is termed as electrical stimulation. Electrical stimulation can influence cellular activities such as proliferation, migration, differentiation, extension of processes and protein expression [151–153]. Just as other cells, neuron activity can be affected by electrical fields, both direct and alternate currents (DC and AC, respectively). Some works have shown how an electrical stimulation influences the direction and extension of neurite outgrowth [148,153] and the alignment of other cellular types such as astrocytes as well [154]. In other studies, the electrical stimulation has been found to increase neurite extension in differentiated PC-12 cells cultured on PPy films [155]. Most recent studies have corroborated the enhancement of neurite outgrowth with electrical stimulation in other cell types and different conductive biomaterials, for example NSCs on PANI [156], cochlear neural explants on PPy [157] and DRG explants on PPy [158]. In addition, these works have demonstrated, as Patel et al. [159] suggested, that one explanation for the increment in neurites length observed is the electrophoretic accumulation of proteins on the surface of the substrate.

As stated previously, CNTs and CNFs are an emergent and promising group of conducting materials for neural tissue engineering. CNTs are composed of rolled structures of sheets of graphene, either single-(SWCNTs) or multi-walled (MWCNTs). The most relevant properties of these materials are their ability to conduct electrical stimuli, form structures with a very large exposed surface per volume unit, and good mechanical properties (strength, flexibility, ductility, etc.). The first experiments with CNTs showed that they are cytotoxic when they are in suspension in a cell culture; nevertheless, posterior studies did not show any cytotoxicity when they are immobilized in the culture [160]. This fact meant the beginning of their use in tissue engineering, and for this purpose, it was necessary to obtain CNTs with higher purity and fabricate more soluble CNTs by functionalization of their surfaces. Other novel conducting polymers are the piezoelectric ones, which have the ability to generate electrical stimuli by the help of small mechanical deformations, being the application of external stimuli unnecessary. The most common piezoelectric polymer is the poly (vinylidene fluoride) (PVDF). This material has demonstrated good attachment and proliferation of neural cells when it is combined with l-lysine [161]; even neurite outgrowth can be enhanced compared with non-piezoelectric materials [162]. Unfortunately, piezoelectric biomaterials have the drawback of not allowing the external control of the electrical stimulation, for what it seems preferable to employ conducting polymers in neural tissue engineering.

Conducting polymers possess a high degree of conjugated bonds providing them rigidity, insolubility and difficult processability, for what these polymers may be transformed or employed in composites with others. Besides, they are usually non-biodegradable and usually offer poor biocompatibility; for these reasons some researchers have investigated the possibility of obtaining conducting biodegradable biomaterials by different techniques including polymerizations that form degradable linkages, blending conductive polymeric segments with biodegradable polymers, and fabrication of copolymers of biodegradable monomers and conductive biostable ones [163,164]. With the aim of improving neurite outgrowth and alignment of cells, bioactive molecules such as neurotrophic factors have been immobilized on conducting polymers [165], their topography has been modified by surface treatments to generate microchannels [166] or they have been electrospun to obtain nanofibrous matrices [158].

5.5. Mechanical cues

The CNS is one of the softest tissues in the body, with elastic modulus between 0.1 and 10 kPa [167]; others, more rigid as bone and connective tissues, present higher elastic moduli ranging 15-30 MPa and 100-1000 kPa, respectively [168]. Rigidity has an important role at early stages of embryogenesis and development of tissues, and particularly in the CNS, since cells regulate the formation of ECM components and their migration by mechanical cues. The rigidity of the brain increases with aging, and the highest elastic modulus of the nervous system tissues corresponds to the spinal cord [169]. This accounts for the modulation of the differentiation of neural precursors into glia or neurons by changing the substrate's rigidity; for example, ESCs differentiate into glia in soft substrates whereas they do to neurons in more rigid ones [170], and NPCs and NSCs show a similar behaviour, differentiating preferably into neurons on softer substrates with moduli between those of the neonatal and the adult brain [169,171].

In general, axon extension increases in softer substrates, so elastic biomaterials are more suitable to address axon regeneration. The most commonly employed substrates with modulated stiffness in neural regeneration are gels as hyaluronic acid, fibrin, polyacrylamide and PEG due to their mechanical properties similar to the CNS tissue. Similar trends in axonal extension and branching have been observed in different types of neurons: the softer or more elastic substrates generally increasing neurite length, neuron attachment and branching compared to harder ones [170,172,173]. However, axonal extension decreases in very soft substrates, when they present elastic moduli much lower than that of the CNS tissue [174,175]. Co-cultures of astrocytes and neurons on hydrogels with tuneable mechanical properties have shown that, unlike neurons, astrocytes attachment decreases on soft substrates. This interesting result has highlighted the beneficial uses of soft substrates to address CNS injuries with presence of glial scar [170]. However, neurons have shown good attachment and large neurite extension and branching cultured on substrates, such as PDMS, which possesses higher elastic modulus than CNS tissue, suggesting that neuronal attachment depends on substrate chemistry too [176].

In the latest stage of embryogenesis and during the maturation up to the adult CNS, axons are submitted to great strains resulting in abrupt changes in their extension. The knowledge of this natural phenomenon has led to investigate how strain can influence axon extension. As an example, axon outgrowth has been studied in engineered bioreactors with the ability to stretch axons on permissive and flexible substrates such as collagen. The normal axon outgrowth is limited to 1 mm/day by the synthesis and transport of proteins and structural components to permit the extension of growth cones. However, cultures of DRG on collagen in a stretch-grown bioreactor and applying a constant strain of 1 cm/day have demonstrated that axons can grow up to 5 cm in only 8 days, which is a much longer extension than that reached by conventional therapies in the regeneration of CNS tracts (2-3 cm). Moreover, the stretched axons maintained their typical cytoskeletal proteins along them and an appropriate structure in their cross section [177], even the electrical signals showed normal action potentials due to the increment of the $K^+$and Na^+channels density [178]. Cyclic strains can influence neurite outgrowth just as constant ones. Cultures of neural cells with cyclic strains have demonstrated a synergistic effect of applied strain and cycles frequencies. This effect has been studied in cultures of PC-12 cells on PDMS substrates with cyclic stretch and strain levels of 4, 8 and 16% and strain rates of 0.1, 0.5 and 1 Hz, showing that only two operation conditions enhance neurite density, length and orientation [179]. In contrast, other operation conditions seem to inhibit neurite length, density and orientation compared to static substrates.

5.6. Physical and chemical cues

Some investigations are related to the study of the effect on neural cells of surface properties of biomaterials such as charge, functional groups density and wettability. Surface wettability can be modified by several techniques such as plasma polymerization, even surfaces with gradients of wettability can be obtained. In works on the effect of surfaces with uniform or graded wettability on hippocampal neurons it can be observed that graded substrates and particularly their most hydrophilic regions show an increased cell density, longer and faster development of the processes than uniform substrates and hydrophobic regions [180]. The surface gradients seem to promote interactions between neighbouring cells and the development of neuronal networks. Similar results were found by culturing PC-12 cells onto polyethylene surfaces with wettability gradients prepared by a corona discharge treatment [181]: cell density was the highest at intermediate wettability of 55º but the length of neurite processes increased on more hydrophilic regions, suggesting an important role of surface hydrophilicity on the differentiation of cells into neurons. Different gradients of wettability on the surfaces of a substrate can be obtained by self-assembled monolayers (SAMs), by the addition of different siloxanes with methyl-terminated molecules. On these substrates, PC-12 cells can differentiate to neurons with neurotrophic factor-free medium, and the differentiation is enhanced on substrates with higher surface wettability. These results suggest the importance of mechanical, chemical and spatial cues in neural fate [182].

6. Biomaterials for traumatic brain injury and similar neuronal damages

Traumatic brain injury and similar brain damages including hypoxia or stroke produce an abrupt loss of neuronal and associated glial population at a primary stage. This initial damage is followed by inflammatory processes leading to a secondary loss of neurons and glial cells and the astrocyte activation forming the inhibitory glial scar for axon outgrowth. Brain tissue damage constitutes, thus, a multistep process leading to the formation of a lesion cavity and necrosis, surrounded by glial scar avoiding axonal regrowth and, as a consequence, the successful healing of the damaged tissue. The cavity generates a loss of brain tissue forming a hostile region to support cells that, together with the inhibitory glial scar, avoid its regeneration leading to psychiatric and neurological symptoms. The inhibitory nature of the CNS due to the presence of astrocytes and oligodendrocytes, the lack of enough neurogenesis niches and the toxic microenvironment after brain injury avoid the regeneration in the brain.

The justification for employing biomaterials after brain damages is principally this lack of cell support, and biomaterial scaffolds should be able to mimic the natural environment of cells promoting their attachment and/or entrapment aiding the regenerative process. Scaffolds can lodge cells in a tridimensional context, allowing diffusion of nutrients and acting as substrates to enhance cell survival, neurite extension, axon regeneration and cell infiltration. The selection of the biomaterials in brain regeneration is based on the principles to achieve neuroprotection of the surrounding native tissue avoiding the secondary cell death, and neuroregeneration of the damaged tissue to restore functions.

Normally, the use of natural polymers is preferable rather than synthetic ones because most of these biomaterials are present in the tissues and their effects are well known; however, many synthetic biomaterials have though a good integration with brain tissue and present some advantages respect to the natural ones, including lower cost, enhancement of mechanical properties, handling, etc. The functional groups within the employed biomaterials can affect cell attachment, since surface charge, hydrophilicity and hydrophobicity are determinant on the biocompatibility of the biomaterials. Cells prefer to attach on positive surfaces because their membrane contains residues of sialic acid, which provide a net negative charge. On the other hand, substrate hydrophilicity and hydrophobicity govern the formation of non-specific protein layers, which can cause inflammation, and the conformation of proteins involved in cell attachment.

The device size is an important parameter in order to cause the minimal damage due to the invasive surgery. Besides, the biomaterial morphology is crucial to ensure nutrient transport, vascularization, cell attachment and colonization in the inner structure. Moreover, ideal structures should provide high surface/volume proportions (interconnected porous scaffolds, microspheres, nanofibres, etc). The surface of substrates plays also an important role in cell attachment and behaviour (view previous section). Mechanical properties such as elasticity influence cell spreading; thus, softer substrates (mimicking the mechanical properties of the brain) can favour the differentiation of neural precursors and NSCs into neural lineages. From the biological point of view, a basic requirement of any material used is obviously its biocompatibility, avoiding adverse cell response, which leads to the formation of gliotic scar and/or acute inflammation. Thus, researchers look for materials not inducing astrocyte reaction and showing a modest inflammatory process. Besides, the biomaterial should preferably biodegrade without leaving any foreign residue as the regenerating tissue replaces it.

Preclinical strategies using biomaterials to address brain damages should satisfy these requirements and provide brain neuroprotection and neuroregeneration. Most of the studied biomaterials for brain regeneration include *in situ* formed gels [183] and porous interconnected scaffolds [184]. In addition, other types of structures such as fibres [185], aligned microchannels and interwoven microchannels have been also proposed [186].

6.1. Natural biomaterials

Natural polymers are similar to some substances present in tissues (polysaccharides, proteins and glycosaminoglycans), even many natural biomaterials form part of the ECM compounds such as collagen and hyaluronic acid. Therefore, implant devices based on natural polymers have in general high biocompatibility. Moreover, in some cases, they may provide signals to cells improving cell-substrates interactions and modeling cell behaviour. They are usually biodegradable by enzymatic degradation, making them good candidates for implantable devices in brain.

A commonly employed biomaterial for tissue engineering is collagen, which is a fibrillar protein forming part of the ECM compounds. Collagen is a natural biodegradable polymer and shows good biocompatibility when it is employed for brain regeneration purposes. This natural polymer has been implanted in the brain forming interconnected porous scaffolds or *in situ* gels. Several authors have employed commercial scaffolds from decellularized and dried tissues from human donors, maintaining the natural structure of collagen tissues [187–189]. Moreover, collagen may be tailored by different processes, such as freeze-drying, in order to obtain porous scaffolds, which are more suitable for cell ingrowth [190]. However, this biomaterial alone does not reduce the size of the lesion cavity nor produces significant functional recovery [187]. The implantation of crosslinked scaffolds based on mixtures of collagen and glycosaminoglycans (GAGs) such as hyaluronan, chondroitin sulphate and heparan sulphate seems to lead to some functional recovery [188]. Freeze-dried collagen-GAG scaffolds with 140 μm-pore diameters and porosity around 96% remained in rat brains 28 days after implantation, reduced the lesion volume, increased the amount of cells in the surrounding tissue, promoted the secretion of neurotrophic factors such as GDNF and BDNF by the host cells and stimulated neurogenesis of neural precursors from the SVZ. Consequently, the inflammatory response of brain tissue was not excessive, promoted cell proliferation and differentiation of migratory neuroblasts, suggesting certain degree of regeneration. Moreover, the results from different tests regarding behaviour, motor and reflex response in animals (adhesive-removal tests and neurological severity scores-NSS-) suggest that animals with biomaterial implants show less neurological deficit than the untreated groups; therefore, collagen-GAG scaffolds achieve some functional recovery without the need to combine with other therapies including drug release or cell entrapment. Some brain functional recovery can be accomplished combining collagen-based interconnected porous scaffolds with seeded and expanded MSCs. MSCs have the ability to release neurotrophic factors and migrate into the lesion boundary zone, remodelling the injured brain tissue and reducing the lesion cavity [188–190]. The implantation of these hybrids in the brain of rats and mice promoted angiogenesis by the formation of new vessels, and increased the neural fibre length in the scaffolds. Some neuroregeneration was assessed by sensomotor functional recovery and improvements in spatial learning, which are higher compared with separated treatments, due to the beneficial effects of the collagen substrates to support and localize cells and the molecular signals produced by MSCs [187]. The use of this biohybrid system implies an important limitation that is the need for open-surgery to implant it in the brain. To avoid such invasive process, an *in situ* gelation can be followed, since collagen is able to form gels at physiological temperatures. For example, mixtures of collagen with NSCs and proteins (fibronectin and laminin) have been gelled *in situ* and employed in the brain. This post-implantation gelation of collagen with NSCs allows cells to distribute widely in the lesion site, oppositely to what happens when NSCs are implanted alone. The incorporation of proteins improves transplanted cell survival during weeks, and the use of laminin shows the highest levels of NSCs survival, which is related with the improvements of cognitive functions suggesting a higher grade of neuroregeneration [183].

Hyaluronic acid (HA) is a negatively charged GAG with a high molecular weight and an abundant ECM component found in the CNS tissue. As biomaterial, it presents interesting properties for neuroregeneration purposes, including biocompatibility, biodegradability and a high degree of swelling. HA products from the degradation process stimulate endothelial cell proliferation and migration promoting angiogenesis. The high degree of swelling permits to accommodate and cover entirely the lesion cavity, conferring it mechanical properties similar to those of the brain tissue and allowing nutrients, ions and solutes transport. Besides, HA of high molecular weight has an anti-inflammatory effect, allowing the control over microglia activation and avoiding foreign body reaction. The ability of HA to absorb large amounts of water allows it to be cross-linked under diluted solutions, obtaining hydrogels. HA cross-linked hydrogels and a successive step of freeze-drying allow obtaining porous structures, which can be from nanoporous to interconnected macroporous (scaffolds) structures. The formation of each structure depends on the crosslinking conditions and is due to the freeze-drying of the water absorbed by the molecule. Thus, the principal advantage of porous scaffolds is to increase the surface area for cell attachment and ingrowth. HA-based hydrogels containing pores up 60 μm implanted in the rat brain have demonstrated their angiogenic activity, inhibition of formation of on the one hand a glial scar and on the other of a secondary cavity or an envelope by hyperplasia or connective tissue and cell aggregation [187]. The major limitation of HA is its poor cell adhesion; subsequently, it is usually employed with adhesion molecules or in combination with other polymers. For example, HA-based scaffolds with immobilized LN have demonstrated improvements in the astrocyte response and the stimulation of neurite ingrowth, suggesting that HA-based scaffolds with these proteins promote neuroregeneration [191]. HA-based scaffolds with peptides such as IKVAV show similar effects than with laminin immobilization, that is, a limited and localized inflammatory response, angiogenesis, neurite ingrowth and the complete integration of the scaffold within the host tissue [184]. In order to obtain interconnected porous scaffolds, HA may be copolymerized with adhesive molecules. Therefore, scaffolds based on copolymers of HA and poly-d-lysine enhance cell migration and attachment without increasing inflammatory

http://dx.doi.org/10.5772/59339

response when they are implanted in rat brains [192]. Poly-d-lysine in combination with HA grants scaffolds with a more positive net charge, which improves cell attachment as mentioned above. The limitation to implant HA scaffolds is again the invasive surgery required, but HA can be combined with other polymers to form gels *in situ*. For example, combining methylcellulose with HA thermoresponsive gels can be obtained, which, similarly to HA scaffolds, reduce the lesion cavity and attenuate microglia activation [108]. The combination of HA hydrogels with neuroprotective molecules, such as erythropoietin (EPO), promotes migration and proliferation of NSCs and NPCs from the SVZ, reduces apoptosis and increases the amount of neurites, suggesting some neuroregeneration in the injured cortex [193]. The neuroregenerative effect of HA biomaterials can also be promoted by attaching to the hydrogel receptor antibodies of inhibitory molecules for neural growth, such as NoGo [194].

Chitosan is derived from the alkaline deacetylation of chitin, a natural polysaccharide found in crab shells and many shellfish. The degree of deacetylation produces a positive charge and, consequently, increases cell attachment, reduces inflammation and enhances biocompatibility. Chitosan is naturally degraded by the effect of lysozyme and its biodegradability can be controlled by the degree of deacetylation. It has shown good biocompatibility with neurons and glial cells *in vitro*, although the viability of neural cells should be improved [195]. Chitosan becomes a thermally responsive gel with glycerolphosphate salt at physiological temperatures, and the gelation process forms a hydrogel able to attach cells and transport nutrients, even the degradation process allows cell infiltration. This property has been exploited to obtain gels *in situ* for brain repair avoiding invasive surgeries. However, the implantation of chitosan gels into rat brain has been found to produce a high inflammatory response by activation of macrophages, leading to the complete phagocytation of the biomaterial in a few days [196]. The activation of macrophages is due the receptor-mediated binding of N-acetylglucosamine, and this effect should be controlled for its future application in the human brain.

Methylcellulose is a cellulose derivative obtained by methyl substitution, forming a water-soluble compound and a thermoresponsive gel. The gelation of this compound can be performed at physiological temperature by altering salt concentration and composition, allowing it to be injected into brain lesions by a minimal invasive surgery. Moreover, its potential use for brain regeneration comes motivated by the approbation of a methylcellulose system (Methocel A®) in nerve repair by the FDA. Methylcellulose gels injected in rat brain have demonstrated reduction of cavity area and glial scar thickness and a similar inflammation area compared with control groups (lesion without gel injection), concluding that methylcellulose is an appropriate biocompatible biomaterial for brain regeneration [197]. However, the principal limitation of methylcellulose gels is that they are not biodegradable; for this reason they have been mixed with biodegradable polymers such as HA to obtain semi-biodegradable gels for their use in the brain [193].

Other ECM compounds, including LN and fibrin, have been employed as permissive substrates for brain regeneration; they are usually employed as injectable *in situ* forming gels or by mixing with other polymers to improve cell attachment. These substrates are able to immobilize neurotrophic factors such as BDNF yielding a permissive substrate for neurite growth and a neuroprotective effect by the incorporation of this molecule. The use of these systems in ischemic models has resulted in a reduction of the infarct volume and in the stimulation of neurogenesis in the dentate gyrus of hippocampus of rats [197]. Hybrid systems employing fibrin gels and MSCs are improved by the incorporation of a growth factor (bFGF), attaining advances in cell survival by the reduction of the apoptotic activity. These systems implanted into rat brain provided more significant improvements in morphological and functional recovery, including reduction of the cavity lesion and neurite and astrocyte infiltration within the gels [198]. The use of SAPNs in brain regeneration has resulted promising due to the ability of these peptides to form nanofibres percolating in a continuous network under physiological conditions. The peptide solution can be injected with a non-invasive surgery and the self-assembled peptides form nanofibrous gels by the presence of ions, filling entirely the cavity lesion and yielding a natural substrate with molecular and topographical cues for axon ingrowth. The gelled SAPNs are integrated within the host tissue and reduce the secondary injury that causes further cell loss, decreasing the lesion volume. Moreover, the astrocyte activation is lower, which reduces the thickness of the glial scar, and the microglia activation decreases during the acute inflammation, suggesting that SAPNs can be successful employed in brain regeneration [199].

6.2. Synthetic biomaterials

Unlike natural polymers, synthetic ones show lower costs, may be tailored in multiple fashions with different properties, are easier to be obtained and the batch processes are more repetitive than natural ones. The fabrication methodologies and modifications of synthetic polymers allow obtaining a wide range of properties and structures for the purposes of brain regeneration. Thus, substrates with properties similar to those of neural tissue can be produced, either biodegradable or bioestable in the shape of structures mimicking the natural ECM.

Acrylate-based biomaterials have been widely proposed for peripheral nerve repair. Poly (2-hydroxyethyl methacrylate) (PHEMA) and poly (hydroxypropyl methacrylate) (PHPMA) are the most common acrylate substrates employed in brain regeneration, for its ability to form hydrogels, yielding substrates with similar properties to the natural tissue. Both polymers show a good biocompatibility and integration with the host tissue adhering firmly to the surrounding brain tissue [200]. The implantation of crosslinked PHEMA and PHPMA materials owning interconnected porous structures led to a dense glial scar and CSPGs deposition surrounding the interface material-host tissue. However, both polymers allowed the infiltration of astrocytes, tissue compounds (fibroblasts, collagen and blood vessels) and neurites within the porous structure. PHPMA scaffolds showed improvements towards brain regeneration in terms of neurite infiltration and presence of larger amounts of neurites and connective tissue within their pores. However, these scaffolds presented a low cell adhesion, which can be enhanced by the incorporation of adhesive peptides. Thus, the immobilization of IKVAV and RGD peptides on PHPMA led to the structural continuity across the cavity, resulting in a more adequate substrate for axonal ingrowth and reduction of glial scar [201,202]. PHPMA has been combined with RGD motifs and engineered fibroblasts to express CNTF and/or BDNF, resulting in larger amounts of regrowing axons, less astrocyte proliferation and similar inflammatory response than hydrogels containing only engineered fibroblasts [202].

Aliphatic polyesters are a family of synthetic polymers widely employed for CNS regeneration; they are easily obtained by ring opening polymerization. The most common aliphatic polyesters are PCL, PLLA, PGA and PLLA-PGA co-polymers (PLGA). The FDA has approved the use of these polymers for several tissue engineering purposes, which makes them attractive in the regeneration of brain damages. A large variety of structures can be achieved with these polymers due to their easy processing (porous scaffolds, electrospun fibres, etc.). They show different mechanical properties and biodegradation rates, the development of a wide range of intermediate ones by the possibility of forming co-polymers. PCL and PLGA have been designed as porous scaffolds with a spongy structure for TBI models and have demonstrated a good biocompatibility with rat brain tissue [203]. Both polymers have shown the ability to maintain the volume of the lesion cavity rather than control groups, preserving the healthy tissue of the primary injury and attenuating the secondary tissue loss. PCL and PLGA reduce cell death, induce neurite ingrowth and do not cause severe inflammation. However, PCL has some advantages over PLGA scaffolds because it provokes a lower microglia and astrocyte activation, and promotes higher levels of neurite ingrowth [203]. PCL with different architectures has been tested for brain regeneration. As an example, PCL nanofibres, both random and aligned, implanted in rat brain have shown a good biocompatibility through the absence of encapsulation and foreign body response, and they have revealed unexpected results. The random nanofibrous scaffolds allow neurite infiltration in the fibrous mesh, whereas the aligned ones show no neurite ingrowth and they are preferentially localized on the surface of the mesh after 7 days [185]. This phenomenon reveals that aligned fibres constitute a good substrate for neurite alignment and growth on its surface by the phenomenon of perpendicular contact guidance without infiltration into the inner network. PCL channel-based architectures with different internal structure (single channel, parallel channels and perpendicular channels) have been employed to address brain damages and guide axon growth in rats. All channel-based structures have demonstrated to be useful for neural ingrowth and they reduce glial scar formation. Particularly, orthogonal channels resulted in the optimal structures for neurite alignment constituting a more permissive environment for cell ingrowth and tissue integration, while parallel channels favour the re-growth of neurites along the surface of the channel [186].

The most common materials employed for brain regeneration purposes are the mentioned polyacrylates and aliphatic polyesters, but other families of synthetic polyesters have also been tested, such as poly (trimethylene carbonate) (PTMC). PTMC can be obtained with different biodegradabilities, the PTMC with fast degradation and slow resorption showing a similar degree of brain regeneration than PLGA. Oppositely, PTMC with a fast rate of degradation and resorption presents a faster and more efficient tissue recovery than common PLGA [204]. Silicon substrates are components also employed in neural interfaces such as electrodes and their functionalization with the protein of neuron adhesion, L1, has shown reduction in astrocyte reaction, fibroblast adhesion, and microglia and macrophages reaction. Moreover, L1-functionalized silicon increases neurons density at the implant-tissue interface [205].

CNTs have also a great interest in neural tissue engineering due to their electrical properties, and their problems of biocompatibility can be elicited by chemical modifications. Indeed, both hydrophobic and hydrophilic modified CNTs have been implanted in the damaged brain of rats resulting in functional recovery. Their benefits in brain regeneration can be enhanced by combining NPCs and CNTs; these systems reduce the gliogenesis of NPCs promoting the differentiation into functional neurons, which restor electrical activity and lead to synaptic formation [206].

7. Biomaterials for spinal cord regeneration

SCI is usually caused by traumatic injuries, including traffic accidents, acts of violence, falls and sports injuries. Damages in spinal cord are due to a mechanical insult following a fractured vertebra or disk intruding into the spinal canal, and can lead to a partial or complete spinal cord transection. The majority of patients die for respiratory complications, because the injury at any level of the spinal cord implies difficulties in respiratory function due to the destruction of the descending motor tracts, which regulate muscles, such as diaphragmatic, thoracic, and abdominal, implied in the breathing process. Consequently, the patients with SCI require assisted breathing, and the clinical treatments are limited to reduce the pain and prevent secondary injuries, including the administration of anti-inflammatory drugs. However, the classical clinical treatments do not achieve tissue regeneration nor recovery of the lost functions.

In early stages, the damaged tissue presents necrosis due to physical trauma, hemorrhage and disruption of the BSCB leading to ischemic processes. The disruption of the BSCB produces the infiltration of foreign components, including fibroblasts, macrophages and cytokines, which are involved in the activation of astrocytes leading to the formation of the gliotic scar and inflammatory processes. After a few days, the apoptotic pathways are activated by the presence of a toxic microenvironment, producing glial and neurons loss without the possibility of replacement by the absence of sufficient endogenous progenitors. The primary axonal loss is followed by a continuous axonal degeneration due to demyelination by oligodendrocytes loss, the block of signalling and the lack of a permissive microenvironment for axon regrowth. Moreover, the axon degeneration progresses over months and years, even it can affect the distal ends of several axons (Wallerian degeneration) leading to the loss of entire spinal tracts.

The axons in the spinal cord form longitudinal bundles, tracts, and they are similar to the nerve fascicles that compose the peripheral nerves. Thus, several researchers have proposed the use of nerve conduits for this pathology due to their success for peripheral nerve repair. However, the spinal cord presents some physiological differences and the regeneration is more complex due to the limitation of regeneration in the CNS. Owing to the complex geometries, partial transection founded at spinal cord lesions and the need of invasive surgeries for the implantation of guidance conduits, many researchers have considered more convenient the use of injectable hydrogels.

As stated, the regeneration of the spinal cord requires guidance structures for a correct axon regrowth, since the tracts in the spinal cord are formed by aligned bundles of axons. Particularly, tracts in the spinal cord own diameters about 100-1000 μm, so the implanted scaffolds in the spinal cord must be permissive to the formation of axon bundles around these diameters. Moreover, the myelinated axons in the spinal cord environment show diameters around 20 μm, thus the SCI scaffolds should be able to guide individual neurite outgrowth as well. The degree of swelling of a material under physiological conditions is important to maintain an appropriate alignment and avoid the compression of the regenerating tracts. Other properties, similarly to what happens with brain regeneration, are important to design scaffolds for spinal cord regeneration; they include biocompatibility, controlled biodegradability, high surface/volume to permit cell attachment and the correct nutrient transport, similar mechanical properties to the host tissue, etc. Guidance structures can be fabricated by different techniques; the most common ones include rolled sheets of aligned electrospinning, scaffolds obtained by freeze-drying of polymeric solutions or moulding combined with particle leaching to generate porosity, or printing the polymers previously designed by CAD software.

7.1. Natural biomaterials

The natural polymers employed in spinal cord regeneration are similar to the ones proposed for brain regeneration, and include fibrillar proteins, polysaccharides, glycosaminoglycans and carbohydrates. The most common are the ECM components collagen and hyaluronic acid, polymers derived from algae such as agarose and alginate, and proteins derived from blood plasma such as fibrin and fibronectin. They are potential materials for the purpose of spinal cord regeneration for their possibility of forming hydrogels *in situ*, avoiding invasive surgeries. However, they are commonly fabricated as porous scaffolds and guidance conduits to be implanted in the spinal cord lesions and to permit axon regeneration in an aligned manner.

Collagen has been successfully employed in peripheral nerve guidance channels, resulting in conduits approved by the FDA as Neuromatrix™ and Neuroflex™. Collagen shows mechanical properties similar to spinal cord tissue, and possesses peptide-motifs for cell adhesion, migration, proliferation, etc. It has a good biocompatibility and integration with the host tissue of the spinal cord, and can be obtained in different shapes, including the aforementioned guidance conduits [207,208]. However, this material lacks ability to support and promote axon growth, and consequently its injection to obtain a gel *in situ* has resulted in dense inclusions within the gel and cavities at the interface biomaterial/host tissue which impede axon growth [207]. In most cases, axon attachment and guidance can be promoted by topographical cues. For this reason, aligned and random collagen electrospun fibres have been developed forming conduits for SCI models, which have shown good results concerning the integration with the spinal cord tissue: reduction of the acute inflammatory response, non-formation of glial scar at the interface between the lesion site and the scaffold, cellular infiltration into the scaffold and axonal sprouting [208]. However, this system has resulted in low axonal sprouting and the aligned conduit does not promote neurite alignment. That is, collagen alone does not promote axon regeneration or alignment, which is a fundamental task in spinal cord repair, and its use requires functionalization to incorporate molecular cues, or to combine it with other permissive substrates of axon guidance and growth. Indeed, collagen has been employed as filling of other polymers' conduits, resulting in the absence of axonal ingrowth and regrowth, but when collagen includes growth factors (FGF or NT3), the axons grow into the channels [209]. The immobilization of peptides such as laminin improves axonal regeneration in the SCI; thus, laminin gradients in a collagen gel filling have shown to increase axonal length and provide directional guidance to the collagen fibres [210].

HA, like collagen, shows a good compatibility with spinal cord tissue but does not promote axon attachment and growth, impeding regeneration. However, the implantation of HA gels in spinal cord lesions is able to reduce the lesion cavity, reduces inflammation and gliotic scar in the surrounded tissue. The reduction of inflammation and gliosis is due to the own nature of HA: the negative charge of the HA chain inhibits cell attachment and their binding site interactions with the cell membrane receptor CD44 of macrophages, limiting the inflammatory effects [211]. HA can be functionalized or combined with other polymers to improve axon attachment and regeneration. For example, the incorporation of poly-l-lysine into HA gels can promote neurite attachment, while the incorporation of the nogo66 receptor antibody, which has an agonist interaction with the inhibitory molecule nogo66, promotes axon regrowth. This system has yielded to the reduction of the glial scar at injury sites promoting wound healing and has facilitated angiogenesis; it has even shown improvements in axon regeneration, leading to both myelinated and unmyelinated axons [212]. The combination of HA with methylcellulose (HAMC) produces gels *in situ*, which are employed in brain damages and SCI, too. This gel has the advantage of covering entirely the lesion cavity, and the possibility of incorporating factors or other neuroprotective compounds enhancing SCI regeneration. For instance, the incorporation of EGF and EPO in the HAMC gels has improved neuroprotection at the implant site and increased neuron density [213,214].

Agarose hydrogels, like collagen and hyaluronic acid, is not adequate for axon infiltration, but is a good substrate when it is functionalized with protein gradients, of laminin for example. Axons grow, preferably, up the laminin gradient *in vitro*, which means promotion of axon extension and guidance [215]. Agarose permits to obtain conduits with an internal structure of aligned linear pores, by means of a freeze-drying process with cooling gradients [216]. These materials show a good biocompatibility and integration within the host tissue, because they do not cause fibrous encapsulation, nor foreign response, the scaffold-host tissue interface is continuous and the cystic cavity is rarely observed. The immobilization of BDNF into the channels promotes blood vessels formation and alignment along the longitudinal orientation of the channels and enhances axonal infiltration into agarose scaffolds [115]. Agarose can also be employed as an *in situ* forming gel, because it is a thermoresponsive hydrogel at temperatures over 17ºC, but gelation upon cooling is a drawback of injectable hydrogels. However, Kaoka et al. employed a customized device for quickly cooling; they implanted agarose gelling *in situ* with steroids, for SCI models, resulting in a reduction of macrophage infiltration and lesion cavity [217].

http://dx.doi.org/10.5772/59339

Alginate is another polymer derived from algae employed for spinal cord tissue engineering. This material has the ability of forming gels by the interaction of multivalent ions, especially Ca^{+2} ions. It has been employed to obtain conduits for spinal cord repair by an ionic diffusion method, where Ca^{+2} ions in an alginate solution diffuse by capillarity forming parallel and aligned conduits with adjustable pore diameters [218]. Alginate gelation *in situ* under physiological conditions can be performed by Ca^{+2} ions forming hydrogels, thus this material can be employed to form gels in the spinal cord cavities. These gels permit axonal support and extension and inhibit the gliotic scar formation in the injured spinal cord [219]. The use of alginate for spinal cord tissue engineering instead of agarose is due to its biodegradability.

Fibrin and fibronectin are two glycoproteins derived from plasma: fibrin is a fibrillary acidic protein glycoprotein polymer and fibronectin is a plasma glycoprotein dimer. Fibronectin gels can be obtained with linearly aligned fibres, which orient axonal growth in spinal cord implants [220] and are more permissive for axon ingrowth than collagen or fibrinogen [210]. However, fibronectin gels implanted alone in the spinal cord generate cavities due to their slow gelation ability. Fibrin has been employed to fill synthetic conduits, made of poly(2-hydroxyethyl methacrylate-co-methyl methacrylate) P(HEMA-MMA) for example, resulting in a greater axonal regeneration specially from vestibular neurons, even obtaining improvements in locomotor function after complete spinal cord transaction at T8 in rats [209]. Fibrin can be modified with heparin to incorporate neurotrophic factors, such as NT-3 [221]. The incorporation of NT-3 into the fibrin matrix reduces the activation of astrocytes inhibiting the glial scar formation. The reduction of the inhibitory components for axon regeneration can also be performed by incorporating the enzyme chondroitinase ABC (ChABC), which cleaves the CSPGs chains generating a more permissive environment for axon regeneration [222]. Fibrin gels are an adequate vehicle to encapsulate embryonic stem cells; its implantation along with pluripotent cells and growth factors, including NT-3 and PDGF in subacute rat model of SCI has been found to increase the survival of transplanted cells and resulted in more differentiated neurons [223]. However, the use of fibrin gels for these applications is falling into decline for its fast degradation. Because of this drawback, fibrinogen has been combined with fibronectin to obtain more durable and efficient gels [207]. The fibronectin/fibrinogen gels have demonstrated a good integration with the host tissue, axonal ingrowth associated with SCs infiltration and deposition of laminin, abundant vascularization and non-formation of cavities. That is, fibronectin/fibrinogen gels are more suitable for SCI regeneration than any of them separately, even than collagen gels.

SAPNs are an interesting strategy to address SCI, due to their ability to be injected and form nanofibrous gels *in situ*, acting as contact guidance substrates for axon outgrowth without employing any invasive surgery. As an example, a negatively charged peptide amphiphile incorporating the neuroactive pentapeptide epitope from laminin, IKVAV, has been injected after laminectomy of spinal cord of mouse at the T10 vertebral segment, resulting in a reduction of astrogliosis and oligodendrocyte apoptosis. Moreover, functional recovery was assessed by the Basso, Beattie and Bresnahan (BBB) locomotor scale modified for mouse at 9 weeks after SAPN injection [224]. However, the stability of many SAPNs is poor and need structural modifications. In this sense, the RAD16-I (Puramatrix™) ionic peptide (multiple of the RADA amino acid sequence) has been modified with bone marrow homing motifs to enhance its stability and bioactivity in the injured spinal cord environment after its gelation [225]. The injection of this modified RAD16-I has shown to promote the upregulation of trophic factors and ECM remodelling proteins, an increment in cellular infiltration, together with laminin and collagen, axon regeneration after 7 days and some improvements in the motor and coordination function 8 weeks after implantation. RAD16-I has been implanted in the transected dorsal column of spinal cord of rats, carrying SCs and embryonic NPCs to replace dead cells and promote neuroprotection [226]. The implanted cells have survived after 6 weeks, SCs mature with a tubular morphology and NPCs differentiate into neurons, astrocytes and oligodendrocytes. This system has shown a good integration with the host tissue without gaps between the implants and the injured tissue at 6 weeks. In addition, host cells have been found to migrate into the gel and extensive blood vessels have been formed.

7.2. Synthetic biomaterials

The synthetic polymers permit to obtain a wide range of structural designs for spinal cord repair, including conduits to form bridges between the rostral and caudal ends of the injured spinal cord, and their properties are more controllable than those of natural polymers. The synthetic polymers most employed for spinal cord tissue engineering are either biodegradable materials, mainly the aliphatic polyesters (PLA, PGA, PLGA, PCL), or bioestable ones, such as acrylate-based hydrogels (PHEMA and PHPMA).

PLA is a good choice for spinal cord regeneration because the FDA has approved its use for several clinical applications, including absorbable sutures and fabrication of peripheral nerve grafts. PLA scaffolds have been designed as conduits for axon guidance in the environment of the spinal cord by a freeze-drying technique, which allows obtaining aligned pores to guide axon growth. However, the implantation of these conduits with entrapped BDNF in the transected adult rat thoracic spinal cord does not improve axon regeneration [227]. Further studies have incorporated engineered SCs expressing BDNF and NT-3 into porous PLA scaffolds, resulting in modest axon regeneration after 6 weeks, but only a few implanted cells survived after one week [228]. Regeneration employing PLA can be improved incorporating topographical cues for axon growth and guidance. For example, PLA conduits have been developed by rolling random and aligned PLA electrospun fibres [229]. These conduits, implanted into a 3 mm gap created after complete transaction of thoracic rat spinal cord, increased cell proliferation and allowed host cells and tissue infiltrate into the lumen of the conduit, filling the scaffold 4 weeks after implantation, closing the gap and forming a continuous tissue. Moreover, the PLA fibres promote the regeneration of supraspinal and propriospinal neurons and glial cells, and aligned fibres promote longer bridge regeneration without adding other cues for axon regeneration, pointing out the importance of topographical cues for axon growth and guidance. However, the electrospun PLA conduits are structurally unstable, easily fragmenting and collapsing.

Lactic acid is normally copolymerized with another aliphatic polyester, glycolic acid (GA), to obtain the copolymer PLGA, which enhances the properties of implantable conduits and allows a better control over the degradation rate. Such as PLA, PLGA can be employed to fabricate conduits for spinal cord regeneration. Thus, aligned porous conduits of PLGA with interconnected walls have been obtained by a gas foaming/particulate leaching process for spinal cord regeneration of rats [230]. These conduits allow the infiltration and longitudinally alignment of different cells including fibroblasts, macrophages, glia, endothelial cells and neurofilaments, which grow up to 2.5 mm from the proximal end of the injured spinal cord. Macrophages are activated in the acute inflammation step but their activity decreases over time, while astrocytes are localized around the structure and the CSPGs levels surrounding the bridge decrease. These conduits can be fabricated with different porous structures, either with small pores (less than 38 μm) or larger ones (63-108 μm), showing differences in the distribution of cell ingrowth; small pores lead to a less uniform cell ingrowth but an increment of endothelial cell infiltration and formation of vessel-like structures. However, conduits of both sizes provided some functional recovery in rats after 14 days, indicating the implant stability and the plastic organization.

The PLGA copolymer has been employed in combination with neurotrophic factors to promote neuroprotection and enhance regeneration in mice models of SCI [231]. As an example, PLGA was shaped as microspheres containing sonic hedgehog (Shh), which is a multifactorial growth factor related to the spinal cord development, and then it was injected in spinal cord lesions. This system reduces the glial scar at the injury site, not affecting the inflammatory response, and provokes an increment in the number of proliferative cells, including NPCs, which provide a permissive environment for regeneration. These benefits result in improvements in the locomotor function recovery, that is, some kind of regeneration. PLGA conduits have also been loaded with neurotrophins such as NT-3 and BDNF [232]. These PLGA conduits were fabricated by a gas foaming/particulate leaching method, injecting a mixture of PLGA microspheres and salt particles into an aluminium mould with oriented wires and subjecting the mixture under pression to obtain a continuous and porous PLGA multichannel. Then the microchannels were filled with hydroxyapatite complexed with a lentivirus encoding luciferase. Once implanted into a rat model of hemisectioned spinal cord, these conduits allowed fibroblasts and macrophages infiltration, but astrocytes were only observed in the surrounding tissue. The axon density per channel is increased by releasing neurotrophins in the rostral position, suggesting the beneficial effect of NT-3 and BDNF in neuroprotection and regeneration. Moreover, the myelinization was enhanced in the channels with lentivirus encoding both neurotrophins. The neuroprotective effect can also be achieved by incorporating cells secreting neurotrophic and growth factors into the PLGA conduits. The incorporation of SCs into PLGA multichannel conduits has demonstrated robust axonal regeneration, but not functional improvements though [233]. The effect of different channels diameters in these structures has been studied [234], and those with diameters of 450 μm and 660 μm demonstrated axon regeneration after 3 months post-implantation in the transected rat spinal cord, but the longest axon fibres were found in the smallest channels. Other cells have been seeded in PLGA conduits and implanted in rat model of transected spinal cord, including NSCs [235]. In this work, PLGA conduits fabricated by a moulding-solvent evaporation process seeded with SCs or NSCs were implanted after an SCI, resulting in similar axonal regeneration in both seeded conduits, but relatively poorer than in the control healthy group, and none functional recovery. However, the system with NSCs seeded in the conduits led to an interesting result, since NSCs preferentially differentiated into neurons with some cells with astrocyte phenotype.

PCL has also been proposed for spinal cord regeneration, due to its good mechanical properties and its slower degradation rate compared with the other aliphatic polyesters. Porous PCL conduits have been fabricated by a salt-leaching and an injection-moulding process, obtaining different morphologies for spinal cord regeneration purposes, including cylinders, tubes, multichannels, open-path architectures with a central core and open-path architectures without such central core [236]. The different morphologies influence the regenerative process of the spinal cord and provide different results relative to cell infiltration and tissue response. In the cylindrical, tubular and multichannel architectures, cell growth has been found to occur along the outside of the implants, the interior walls being filled with cellular debris, macrophages, and fibroblast-like cells after 1 month. However, the result most discourageous of these three structures was the secondary cell death within and around the implants after 3 months. Besides, both open-path structures produced growth of tissue into and along the entire length of the conduits and less cysts and fibrotic tissue around the implants and the stumps. In addition, the open-path structures promoted astrocyte infiltration and sufficient contact guidance to allow the extension of axons across the entire defect length, even many fibres accomplished to be myelinated.

PCL can be chemically modified to obtain a biodegradable crosslinkable polymer by the addition of fumaryl chloride and PCL diol under ultraviolet light. This crosslinkable polymer, poly(ε-caprolactone fumarate) (PCLF), has been employed to fabricate multichannel conduits for spinal cord regeneration [237]. The benefits in spinal cord injuries of PCLF conduits have been compared with those of PLGA ones, resulting in a major number of axons in the first ones. PCL in combination with PLGA has been employed to obtain electrospun fibres conduits containing a filling of the RAD16-I gel to promote axon attachment and regeneration, and growth factors and neurotrophic factors (BDNF, CNTF, chABC and VEGF) to be used in a chronic SCI rat model [238]. The incorporation of RAD16-I into the lumen of the conduits led to a reduction of cavities, the gliotic scar occasionally appeared surrounding the implants but less frequently inside the lumen, promoted tissue infiltration, neurite regeneration and blood vessel formation and alignment. The incorporation of cytokines enhanced neuron maturation and the formation of axon bundles in all the microchannels, and the regeneration even continued after 6 months, measured by electrophysiological activity and locomotor improvements.

Poly-β-hydroxybutyrate (PHB) is a polymer produced by a wide variety of bacteria, it degrades slowly at physiological temperatures and the metabolites are secreted in urine. In contrast to aliphatic polyesters, the slow degradation rate and continuous elimination of the PHB avoid the formation of acidic debris, which can cause inflammation. PHB has been employed to fabricate conduits composed of two perpendicular layers containing parallel fibres for spinal cord regeneration [239]. These conduits have been coated with alginate, seeded with SCs and implanted into the gap generated after cervical SCI in rats, resulting in a reduction of the spinal cord cavity and the retrograde degeneration of the injured spinal tract. Further studies with PHB fibres conduits coated with alginate and seeded with SCs resulted in improvements in cell attachment, proliferation and survival of SCs and promotion of axonal regeneration within the conduit [240]. However, only the descending raphaespinal and sensorial neurons grew within and along the conduit, but not the rubrospinal ones.

PHEMA-and PHMA-based hydrogels are bioestable materials and due to their swelling rate, they show similar properties to those of the spinal cord tissue. The regenerative ability of both types of hydrogels has been compared in studies of adult and developing rat spinal cords [241]. Both hydrogels showed a good integration with the host tissue, presenting a minimal glial scar and cystic formation and the infiltration of connective tissue. They promoted the infiltration of blood vessels and axons into the pores, but axons grew into the central part of the hydrogels after eight weeks. Moreover, SCs infiltrated into the hydrogels and along the axons, suggesting the possibility of obtaining myelinated axons. These hydrogels can be modified with peptides and neurotrophic factors to improve axon regeneration and generate neuroprotection. Thus, the incorporation of the cell adhesive site of fibronectin RGD (NeuroGel™) into PHMA hydrogels with a porous structure obtained by freeze-drying has been found to promote axon regeneration in injured spinal cord [241]. These systems provoked an early inflammation a few days following implantation, which decreased gradually. One week after implantation, the proliferation of capillary sprouts was observed, along with a glial scar surrounding the implant and a few number of fibroblasts, fibrocytes and macrophages. Over the next weeks, the implants showed ingrowth of capillaries, infiltration of connective tissue, glial processes and axons. Moreover, the implants presented neither cystic formation nor secondary axon degeneration, and the axonal pathways were altered into the rostral stump, showing signs of morphological changes and regeneration indicating growth cone formation. Similar studies on PHPMA hydrogels with immobilized RGD and neurotrophic factors (BDNF and CNTF) resulted in a major number of axons growing into the hydrogels compared with unmodified ones [202].

Conduits of PHEMA-co-methyl methacrylate with different internal matrices –fillings- (collagen, fibrin, matrigel and methylcellulose) and growth factors (FGF and NT-3) have been compared with multichannel conduits and have demonstrated improvements in the regeneration of selective tracts after complete spinal cord transection at T8 in rats [209]. This study shows that all the internal matrices increase axon density within the channels except the collagen matrix without growth factors and the multichannel conduits without fillings. The regeneration of sensory axons was not influenced by the internal matrix and axon growth factors even hinder it. The different matrices and growth factors affect selectively to different tracts: fibrin promotes the greatest axonal regeneration of reticular neurons, whereas methylcellulose does of vestibular and red nucleus neurons, FGF increases the axonal regeneration of vestibular neurons, and NT-3 decreases the total amount of axons regenerating from brainstem neurons. Besides, fibrin filling and multichannels accomplish improvements in the locomotor function. With the aim of improving axon regeneration in the injured spinal cord, HPMA-and HEMA-based hydrogels have been combined with MSCs so as to deliver neurotrophic factors. In a preliminary study, the survival of MSCs within HPMA and PHEMA implants has been compared, being it higher in the HPMA hydrogels [242]. Besides, this work also demonstrated that most of the cells locate in the pores but a few cells migrate to the border zones of the spinal cord, pointing out that the use of these hydrogels can be helpful as carriers of stem cells and their delivery in the lesion site of the spinal cord.

8. Biomaterials in neurodegenerative diseases affecting the striatum

Neurodegenerative disorders are characterized by a progressive loss of neuronal subtypes, and many of these diseases ultimately lead to a loss of the patient autonomy and even death. In this chapter, two neurodegenerative disorders that affect the striatum in the brain are dealt with PD and HD. Both show similar pathology, with many differences, affecting localized zones of the striatum in the earlier stages and causing the loss of neural subtypes: dopaminergic neurons in PD and GABAergic ones in HD. The experimental therapies are based on cell therapy and drug delivery, and the role of biomaterials is to incorporate cells or drugs to immobilize them into the site of interest.

PD is a neurodegenerative disease affecting around 1% of the population older than 60 years. It is caused by several factors, but researchers have identified six gene mutations associated with this incidence, including α-synuclein, parkin, UCH-LI, PINK1, DJ-1, and LRRK2/ dardarin. This pathology is characterized by the loss of dopaminergic neurons in the substantia nigra of the nigrostriatal tract. For this reason, the striatum reduces dopamine levels and neuronal projections. Other aspect of PD is the generation of abnormal intraneuronal aggregates of protein, especially α-synuclein, called Lewy bodies. Further stages of this disease involve degeneration in the cerebral cortex and autonomic nervous system. The manifestations of PD include tremor, bradykinesia, rigidity, cognitive decline and finally death. The animal model commonly employed to study the PD is obtained by administration of 6-hydroxidopamine, which causes the lesion of dopaminergic pathways. In the initial stages of disease, L-DOPA administration is the most potent therapy to improve motor functions. However, this treatment becomes less effective in further stages, making necessary to find new therapies.

HD is a dominantly inherited neurodegenerative disease that results from the polyglutamine expansion in the gene encoding the huntingtin protein. This disease is characterized by the loss of brain striatal GABAergic neurons, and the degenerative process progressively extends to other brain regions including cortex. The pathogenesis of HD and the function of the huntingtin protein are not well understood, but this pathology is characterized by the loss of medium spiny neurons in the striatum and loss of neurons in the cortex leading to movement disorders (chorea), dementia, and eventually death. The early stages of HD show a localized damage in the striatum, which leads this disease to be a potential target for tissue engineering employing biomaterials. The animal models of HD are obtained by exposing neurons from striatum to drugs or poisons and by using transgenic animal models overexpressing polyglutamine expansions.

Biomaterials for drug delivery are promising since they can be easily modified to adjust the release kinetics to the target organ, tissues or cells, they can selectively cross biological membranes including the impermeable BBB, and the release can be controlled by modulating the biodegradable properties of the biomaterial carriers. The most common structures employed for drug delivery in these neurodegenerative diseases are nanoparticles, liposomes and hydrogels. The principal problem of the oral administration of $_L$-DOPA is that its prolonged use entails tolerance, and its efficacy is limited by the permeability of the BBB. The implantation of dopamine-loaded biomaterials into the striatum would be a more efficient method of controlled release to address PD. The incorporation of dopamine in hydrogels has demonstrated good results, employing non-cytotoxic biomaterials such as dextran crosslinked with gelatine for models of PD [243]. The delivery of dopamine can be accomplished during weeks and produces motor functional recovery in models of hemiparkinsonian rats. However, the release of dopamine occurs during short times and the behavioural bias reappears. GDNF is another molecule employed in PD studies, since it has a trophic effect in the dopaminergic system increasing the production of dopamine, stimulating the regeneration of dopaminergic neurites [244]. Therefore, the incorporation of GDNF in PLGA microparticles and their posterior implantation into the striatum of parkinsonian rats has demonstrated improvements in motor function and increments of fiber density in the striatum [245]. The delivery of GDNF in the striatum in PD models has also been achieved by transfected fibroblasts expressing this trophic factor immobilized into alginate-poly-$_L$-lysine microparticles [246]; this implantable device has shown to increase cell survival and release GDNF during around six months in parkinsonized rats, resulting in improvements of behaviour.

The immobilization of cells within biomaterials is usually carried out by encapsulating cells into polymeric microparticles or by attaching them on the biomaterials' surfaces. The use of biomaterials pursues to increase cell survival of transplanted cells, control cell fate, maintain cell phenotype and promote cell engraftment into the interest site. The deficit of dopamine and dopaminergic neurons loss typical of PD disease could be improved by the incorporation of cells expressing this neurotransmitter or by stromal cells with the ability to differentiate into dopaminergic neurons. In both cases, cell survival would be improved by combining cells with biomaterials. For example, collagen-coated dextran microparticles (Cytodex 3®) together with chromaffin cells have been implanted into the striatum of rats with 6-hydroxydopamine-induced injuries of the substantia nigra [247]. The implantable biohybrids did not cause any inflammation response, increased cell survival and produced functional benefits after 8 months. Similar results have been obtained by incorporating foetal ventral mesencephalon (FVM) cells into the same Cytodex3® microcarriers. Another system to incorporate chromaffin cells into the striatum of PD animal models has been developed by employing glass bead microparticles, which has led to similar results, even long-term functional recovery after implantation in hemiparkinsonian rats [248]. Alginate is another natural polymer employed to encapsulate cells for the treatment of neurodegenerative diseases affecting the striatum. Its microparticles have been combined with choroid plexus cells, which secrete several neurotrophic factors, to study their effect in HD rat models [249]. This device had a neuroprotective effect by the neurotrophic factor release, a reduction of the lesion cavity and improvements in the motor function. Synthetic polymers have also been employed in the treatment of neurodegenerative diseases affecting the striatum. As an example, NT-3 immobilized into PLGA microparticles with LN in their surface have been employed to attach and expand MSCs, and implanted in hemiparkinsonian rats [250]. LN improves the attachment of MSCs to the surface, while the neurotrophic factor NT-3 promotes the differentiation of MSCs into neurons. This device led to improvements of neuron differentiation and the regeneration of the nigrostriatal pathway, and subsequently to modest improvements in the behaviour.

9. Concluding remarks

There is a high interest in the use of biomaterials to repair damage in the brain in many of the most prevalent and burden causing neurologic diseases. Biomaterials can provide a better environment for the homing and survival of neurons, glia and brain blood vessels, act as guidance cues for promoting axonal growth, and provide a proper stimulus for the formation of new synapses and to promote neural plasticity.

Although there is still no clinical application of this research, the wealth of data obtained may foresee a successful solution in the near future, which still will have to go through all the regulatory and clinical essays process.

If we are able to provide a way to regenerate and reconstruct the damaged brain, we will turn the available concept of clinical neurosciences, from procedures that are directed to protect the damaged brain from further injuries into truly reconstructive neurosciences.

Acknowledgements

The authors acknowledge funding through the Spanish Ministerio de Ciencia e Innovación through project MAT2011-28791-C01-02-03.

Author details

Manuel Pérez-Garnes[1], Juan A Barcia[2], Ulises Gómez-Pinedo[3], Manuel Monleón Pradas[1,4] and Ana Vallés-Lluch[1*]

*Address all correspondence to: avalles@ter.upv.es

1 Centro de Biomateriales e Ingeniería Tisular, Universitat Politècnica de València, Cno. de Vera s/n, Valencia, Spain

2 Servicio de Neurocirugía, Hospital Clínico San Carlos, c/Profesor Martín Lagos, s/n, Madrid, Spain

3 Instituto de Neurociencias, Hospital Clínico San Carlos, c/Profesor Martín Lagos, s/n, Madrid, Spain

4 Networking Research Center on Bioengineering, Biomaterials and Nanomedicine (CIBER-BBN), Valencia, Spain

References

[1] Alvarez-Buylla A, Garcia-Verdugo JM. Neurogenesis in adult subventricular zone. The Journal of Neuroscience 2002; 22(3) 629–34.

[2] Sanai N, Tramontin AD, Quiñones-Hinojosa A, Barbaro NM, Gupta N, Kunwar S, et al. Unique astrocyte ribbon in adult human brain contains neural stem cells but lacks chain migration. Nature 2004; 427(6976) 740–4.

[3] Arvidsson A, Collin T, Kirik D, Kokaia Z, Lindvall O. Neuronal replacement from endogenous precursors in the adult brain after stroke. Nature Medicine 2002; 8(9) 963–70.

[4] Galan L, Gomez-Pinedo U, Vela-Souto A, Guerrero-Sola A, Barcia JA, Gutierrez AR, et al. Subventricular zone in motor neuron disease with frontotemporal dementia. Neuroscience Letters 2011; 499(1) 9–13.

[5] Martínez-Ramos C, Vallés-Lluch A, Verdugo JMG, Ribelles JLG, Barcia Albacar JA, Orts AB, et al. Channeled scaffolds implanted in adult rat brain. Journal of Biomedical Materials Research Part A 2012; 100(12) 3276–86.

[6] Lois C, García-Verdugo JM, Alvarez-Buylla A. Chain migration of neuronal precursors. Science 1996; 271(5251) 978–81.

[7] Menn B, Garcia-Verdugo JM, Yaschine C, Gonzalez-Perez O, Rowitch D, Alvarez-Buylla A. Origin of oligodendrocytes in the subventricular zone of the adult brain. The Journal of Neuroscience 2006; 26(30) 7907–18.

[8] Bellenchi GC, Volpicelli F, Piscopo V, Perrone-Capano C, di Porzio U. Adult neural stem cells: an endogenous tool to repair brain injury? Journal of Neurochemistry 2013; 124(2) 159–67.

[9] Doetsch F, Caillé I, Lim DA, García-Verdugo JM, Alvarez-Buylla A. Subventricular zone astrocytes are neural stem cells in the adult mammalian brain. Cell 1999; 97(6) 703–16.

[10] Ricard J, Salinas J, Garcia L, Liebl DJ. EphrinB3 regulates cell proliferation and survival in adult neurogenesis. Molecular and Cellular Neurosciences 2006; 31(4) 713–22.

[11] Doetsch F, Petreanu L, Caille I, Garcia-Verdugo JM, Alvarez-Buylla A. EGF converts transit-amplifying neurogenic precursors in the adult brain into multipotent stem cells. Neuron 2002; 36(6) 1021–34.

[12] Zhang RL, LeTourneau Y, Gregg SR, Wang Y, Toh Y, Robin AM, et al. Neuroblast division during migration toward the ischemic striatum: a study of dynamic migratory and proliferative characteristics of neuroblasts from the subventricular zone. The Journal of Neuroscience 2007; 27(12) 3157–62.

[13] Gotts JE, Chesselet M-F. Vascular changes in the subventricular zone after distal cortical lesions. Experimental Neurology 2005; 194(1) 139–50.

[14] Seri B, García-Verdugo JM, McEwen BS, Alvarez-Buylla A. Astrocytes give rise to new neurons in the adult mammalian hippocampus. The Journal of Neuroscience: The Official Journal of the Society for Neuroscience 2001; 21(18) 7153–60.

[15] Rozental R, Morales M, Mehler MF, Urban M, Kremer M, Dermietzel R, et al. Changes in the properties of gap junctions during neuronal differentiation of hippocampal progenitor cells. The Journal of Neuroscience: The Official Journal of the Society for Neuroscience 1998; 18(5) 1753–62.

[16] Palmer TD, Takahashi J, Gage FH. The adult rat hippocampus contains primordial neural stem cells. Molecular and Cellular Neurosciences 1997; 8(6) 389–404.

[17] Ohori Y, Yamamoto S, Nagao M, Sugimori M, Yamamoto N, Nakamura K, et al. Growth factor treatment and genetic manipulation stimulate neurogenesis and oligodendrogenesis by endogenous neural progenitors in the injured adult spinal cord. The Journal of Neuroscience: The Official Journal of the Society for Neuroscience 2006; 26(46) 11948–60.

[18] Eriksson PS, Perfilieva E, Björk-Eriksson T, Alborn AM, Nordborg C, Peterson DA, et al. Neurogenesis in the adult human hippocampus. Nature Medicine 1998; 4(11) 1313–7.

[19] Jackson EL, Garcia-Verdugo JM, Gil-Perotin S, Roy M, Quinones-Hinojosa A, VandenBerg S, et al. PDGFR alpha-positive B cells are neural stem cells in the adult SVZ that form glioma-like growths in response to increased PDGF signaling. Neuron 2006; 51(2) 187–99.

[20] Quiñones-Hinojosa A, Sanai N, Soriano-Navarro M, Gonzalez-Perez O, Mirzadeh Z, Gil-Perotin S, et al. Cellular composition and cytoarchitecture of the adult human subventricular zone: a niche of neural stem cells. The Journal of Comparative Neurology 2006; 494(3) 415–34.

[21] Zheng W, Nowakowski RS, Vaccarino FM. Fibroblast growth factor 2 is required for maintaining the neural stem cell pool in the mouse brain subventricular zone. Developmental Neuroscience 2004; 26(2-4) 181–96.

[22] Mohapel P, Frielingsdorf H, Häggblad J, Zachrisson O, Brundin P. Platelet-derived growth factor (PDGF-BB) and brain-derived neurotrophic factor (BDNF) induce striatal neurogenesis in adult rats with 6-hydroxydopamine lesions. Neuroscience 2005; 132(3) 767–76.

[23] Jin K, Zhu Y, Sun Y, Mao XO, Xie L, Greenberg DA. Vascular endothelial growth factor (VEGF) stimulates neurogenesis in vitro and in vivo. Proceedings of the National Academy of Sciences of the United States of America 2002; 99(18) 11946–50.

[24] Chen K, Henry RA, Hughes SM, Connor B. Creating a neurogenic environment: the role of BDNF and FGF2. Molecular and Cellular Neurosciences 2007; 36(1) 108–20.

[25] Baldauf K, Reymann KG. Influence of EGF/bFGF treatment on proliferation, early neurogenesis and infarct volume after transient focal ischemia. Brain Research 2005; 1056(2) 158–67.

[26] Xu Y, Tamamaki N, Noda T, Kimura K, Itokazu Y, Matsumoto N, et al. Neurogenesis in the ependymal layer of the adult rat 3rd ventricle. Experimental Neurology 2005; 192(2) 251–64.

[27] Pencea V, Bingaman KD, Wiegand SJ, Luskin MB. Infusion of brain-derived neurotrophic factor into the lateral ventricle of the adult rat leads to new neurons in the parenchyma of the striatum, septum, thalamus, and hypothalamus. The Journal of Neuroscience: The Official Journal of the Society for Neuroscience 2001; 21(17) 6706–17.

[28] Charrier C, Coronas V, Fombonne J, Roger M, Jean A, Krantic S, et al. Characterization of neural stem cells in the dorsal vagal complex of adult rat by in vivo proliferation labeling and in vitro neurosphere assay. Neuroscience 2006; 138(1) 5–16.

[29] Magavi SS, Leavitt BR, Macklis JD. Induction of neurogenesis in the neocortex of adult mice. Nature 2000; 405(6789) 951–5.

[30] Couillard-Despres S, Winner B, Karl C, Lindemann G, Schmid P, Aigner R, et al. Targeted transgene expression in neuronal precursors: watching young neurons in the old brain. The European Journal of Neuroscience 2006; 24(6) 1535–45.

[31] Snyder EY, Yoon C, Flax JD, Macklis JD. Multipotent neural precursors can differentiate toward replacement of neurons undergoing targeted apoptotic degeneration in adult mouse neocortex. Proceedings of the National Academy of Sciences of the United States of America 1997; 94(21) 11663–8.

[32] Rakic P. Neurogenesis in adult primate neocortex: an evaluation of the evidence. Nature Reviews Neuroscience 2002; 3(1) 65–71.

[33] Gould E. How widespread is adult neurogenesis in mammals? Nature Reviews Neuroscience 2007; 8(6) 481–8.

[34] Goldman JE. Lineage, migration, and fate determination of postnatal subventricular zone cells in the mammalian CNS. Journal of Neuro-Oncology 1995; 24(1) 61–4.

[35] Taupin P. Adult neural stem cells, neurogenic niches, and cellular therapy. Stem Cell Reviews 2006; 2(3) 213–9.

[36] Goings GE, Kozlowski DA, Szele FG. Differential activation of microglia in neurogenic versus non-neurogenic regions of the forebrain. Glia 2006; 54(4) 329–42.

[37] Sun X, Zhang Q-W, Xu M, Guo J-J, Shen S-W, Wang Y-Q, et al. New striatal neurons form projections to substantia nigra in adult rat brain after stroke. Neurobiology of Disease 2012; 45(1) 601–9.

[38] Alfaro-Cervello C, Cebrian-Silla A, Soriano-Navarro M, Garcia-Tarraga P, Matías-Guiu J, Gomez-Pinedo U, et al. The adult macaque spinal cord central canal zone contains proliferative cells and closely resembles the human. The Journal of Comparative Neurology 2014; 522(8) 1800–17.

[39] Horner PJ, Gage FH. Regenerating the damaged central nervous system. Nature 2000; 407(6807) 963–70.

[40] Willerth SM, Sakiyama-Elbert SE. Cell therapy for spinal cord regeneration. Advanced Drug Delivery Reviews 2008; 60(2) 263–76.

[41] Blakemore WF. Remyelination of CNS axons by Schwann cells transplanted from the sciatic nerve. Nature 1977; 266(5597) 68–9.

[42] Popovich PG, Guan Z, Wei P, Huitinga I, van Rooijen N, Stokes BT. Depletion of hematogenous macrophages promotes partial hindlimb recovery and neuroanatomical repair after experimental spinal cord injury. Experimental Neurology 1999; 158(2) 351–65.

[43] Tetzlaff W, Okon EB, Karimi-Abdolrezaee S, Hill CE, Sparling JS, Plemel JR, et al. A systematic review of cellular transplantation therapies for spinal cord injury. Journal of Neurotrauma 2011; 28(8) 1611–82.

[44] Shyu W-C, Liu DD, Lin S-Z, Li W-W, Su C-Y, Chang Y-C, et al. Implantation of olfactory ensheathing cells promotes neuroplasticity in murine models of stroke. The Journal of Clinical Investigation 2008; 118(7) 2482–95.

[45] Lakatos A, Barnett SC, Franklin RJM. Olfactory ensheathing cells induce less host astrocyte response and chondroitin sulphate proteoglycan expression than Schwann cells following transplantation into adult CNS white matter. Experimental Neurology 2003; 184(1) 237–46.

[46] López-Vales R, Forés J, Navarro X, Verdú E. Chronic transplantation of olfactory ensheathing cells promotes partial recovery after complete spinal cord transection in the rat. Glia 2007; 55(3) 303–11.

[47] Guest JD, Herrera L, Margitich I, Oliveria M, Marcillo A, Casas CE. Xenografts of expanded primate olfactory ensheathing glia support transient behavioral recovery that is independent of serotonergic or corticospinal axonal regeneration in nude rats following spinal cord transection. Experimental Neurology 2008; 212(2) 261–74.

[48] Kalincik T, Jozefcikova K, Sutharsan R, Mackay-Sim A, Carrive P, Waite PME. Selected changes in spinal cord morphology after T4 transection and olfactory ensheathing cell transplantation. Autonomic Neuroscience: Basic & Clinical 2010; 158(1-2) 31–8.

[49] Shi X, Kang Y, Hu Q, Chen C, Yang L, Wang K, et al. A long-term observation of olfactory ensheathing cells transplantation to repair white matter and functional recovery in a focal ischemia model in rat. Brain Research 2010; 1317 257–67.

[50] Dewar D, Bentley D, Barnett SC. Implantation of pure cultured olfactory ensheathing cells in an animal model of parkinsonism. Acta Neurochirurgica 2007; 149(4) 407–14.

[51] Gómez-Pinedo U, Vidueira S, Sancho FJ, García-Verdugo JM, Matías-Guiu J, Barcia JA. Olfactory ensheathing glia enhances reentry of axons into the brain from peripheral nerve grafts bridging the substantia nigra with the striatum. Neuroscience Letters 2011; 494(2) 104–8.

[52] Snyder BJ, Olanow CW. Stem cell treatment for Parkinson's disease: an update for 2005. Current Opinion in Neurology 2005; 18(4) 376–85.

[53] Ma Y-H, Zhang Y, Cao L, Su J-C, Wang Z-W, Xu A-B, et al. Effect of neurotrophin-3 genetically modified olfactory ensheathing cells transplantation on spinal cord injury. Cell Transplantation 2010; 19(2) 167–77.

[54] Radtke C, Akiyama Y, Brokaw J, Lankford KL, Wewetzer K, Fodor WL, et al. Remyelination of the nonhuman primate spinal cord by transplantation of H-transferase transgenic adult pig olfactory ensheathing cells. FASEB Journal: Official Publication of the Federation of American Societies for Experimental Biology 2004; 18(2) 335–7.

[55] Huang H, Chen L, Wang H, Xiu B, Li B, Wang R, et al. Influence of patients' age on functional recovery after transplantation of olfactory ensheathing cells into injured spinal cord injury. Chinese Medical Journal 2003; 116(10) 1488–91.

[56] Su Z, He C. Olfactory ensheathing cells: biology in neural development and regeneration. Progress in Neurobiology 2010; 92(4) 517–32.

[57] Honmou O, Onodera R, Sasaki M, Waxman SG, Kocsis JD. Mesenchymal stem cells: therapeutic outlook for stroke. Trends in Molecular Medicine 2012; 18(5) 292–7.

[58] Martínez-Ramos C, Lainez S, Sancho F, García Esparza MA, Planells-Cases R, García Verdugo JM, et al. Differentiation of postnatal neural stem cells into glia and functional neurons on laminin-coated polymeric substrates. Tissue Engineering Part A 2008; 14(8) 1365–75.

[59] Soria JM, Sancho-Tello M, Esparza MAG, Mirabet V, Bagan JV, Monleón M, et al. Biomaterials coated by dental pulp cells as substrate for neural stem cell differentiation. Journal of Biomedical Materials Research Part A 2011;

[60] Hofstetter CP, Schwarz EJ, Hess D, Widenfalk J, El Manira A, Prockop DJ, et al. Marrow stromal cells form guiding strands in the injured spinal cord and promote recovery. Proceedings of the National Academy of Sciences of the United States of America 2002; 99(4) 2199–204.

[61] Cao QL, Zhang YP, Howard RM, Walters WM, Tsoulfas P, Whittemore SR. Pluripotent stem cells engrafted into the normal or lesioned adult rat spinal cord are restricted to a glial lineage. Experimental Neurology 2001; 167(1) 48–58.

[62] Setoguchi T, Nakashima K, Takizawa T, Yanagisawa M, Ochiai W, Okabe M, et al. Treatment of spinal cord injury by transplantation of fetal neural precursor cells engineered to express BMP inhibitor. Experimental Neurology 2004; 189(1) 33–44.

[63] Esneault E, Pacary E, Eddi D, Freret T, Tixier E, Toutain J, et al. Combined therapeutic strategy using erythropoietin and mesenchymal stem cells potentiates neurogenesis after transient focal cerebral ischemia in rats. Journal of Cerebral Blood Flow and Metabolism: Official Journal of the International Society of Cerebral Blood Flow and Metabolism 2008; 28(9) 1552–63.

[64] Chen J, Li Y, Katakowski M, Chen X, Wang L, Lu D, et al. Intravenous bone marrow stromal cell therapy reduces apoptosis and promotes endogenous cell proliferation after stroke in female rat. Journal of Neuroscience Research 2003; 73(6) 778–86.

[65] Yang K-L, Chen M-F, Liao C-H, Pang C-Y, Lin P-Y. A simple and efficient method for generating Nurr1-positive neuronal stem cells from human wisdom teeth (tNSC) and the potential of tNSC for stroke therapy. Cytotherapy 2009; 11(5) 606–17.

[66] Bang OY, Lee JS, Lee PH, Lee G. Autologous mesenchymal stem cell transplantation in stroke patients. Annals of Neurology 2005; 57(6) 874–82.

[67] Amin EM, Reza BA, Morteza BR, Maryam MM, Ali M, Zeinab N. Microanatomical evidences for potential of mesenchymal stem cells in amelioration of striatal degeneration. Neurological Research 2008; 30(10) 1086–90.

[68] Edalatmanesh M-A, Matin MM, Neshati Z, Bahrami A-R, Kheirabadi M. Systemic transplantation of mesenchymal stem cells can reduce cognitive and motor deficits in rats with unilateral lesions of the neostriatum. Neurological Research 2010; 32(2) 166–72.

[69] Sadan O, Bahat-Stromza M, Barhum Y, Levy YS, Pisnevsky A, Peretz H, et al. Protective effects of neurotrophic factor-secreting cells in a 6-OHDA rat model of Parkinson disease. Stem Cells and Development 2009; 18(8) 1179–90.

[70] Levy YS, Bahat-Stroomza M, Barzilay R, Burshtein A, Bulvik S, Barhum Y, et al. Regenerative effect of neural-induced human mesenchymal stromal cells in rat models of Parkinson's disease. Cytotherapy 2008; 10(4) 340–52.

[71] Vazey EM, Chen K, Hughes SM, Connor B. Transplanted adult neural progenitor cells survive, differentiate and reduce motor function impairment in a rodent model of Huntington's disease. Experimental Neurology 2006; 199(2) 384–96.

[72] F LM, Neuman T, Rezak M. Therapeutic Microinjection of Autologous Adult Human Neural Stem Cells and Differentiated Neurons for Parkinson's Disease: Five-Year Post-Operative Outcome. Open Stem Cell Journal 2009; 1 20–9.

[73] Madrazo I, Drucker-Colín R, Díaz V, Martínez-Mata J, Torres C, Becerril JJ. Open microsurgical autograft of adrenal medulla to the right caudate nucleus in two patients with intractable Parkinson's disease. The New England Journal of Medicine 1987; 316(14) 831–4.

[74] Doudet DJ, Cornfeldt ML, Honey CR, Schweikert AW, Allen RC. PET imaging of implanted human retinal pigment epithelial cells in the MPTP-induced primate model of Parkinson's disease. Experimental Neurology 2004; 189(2) 361–8.

[75] Cepeda IL, Flores J, Cornfeldt ML, O'Kusky JR, Doudet DJ. Human retinal pigment epithelial cell implants ameliorate motor deficits in two rat models of Parkinson disease. Journal of Neuropathology and Experimental Neurology 2007; 66(7) 576–84.

[76] Liu S, Qu Y, Stewart TJ, Howard MJ, Chakrabortty S, Holekamp TF, et al. Embryonic stem cells differentiate into oligodendrocytes and myelinate in culture and after spinal cord transplantation. Proceedings of the National Academy of Sciences of the United States of America 2000; 97(11) 6126–31.

[77] Nistor GI, Totoiu MO, Haque N, Carpenter MK, Keirstead HS. Human embryonic stem cells differentiate into oligodendrocytes in high purity and myelinate after spinal cord transplantation. Glia 2005; 49(3) 385–96.

[78] Lee H, Shamy G Al, Elkabetz Y, Schofield CM, Harrsion NL, Panagiotakos G, et al. Directed differentiation and transplantation of human embryonic stem cell-derived motoneurons. Stem Cells 2007; 25(8) 1931–9.

[79] Kitagawa A, Nakayama T, Takenaga M, Matsumoto K, Tokura Y, Ohta Y, et al. Lecithinized brain-derived neurotrophic factor promotes the differentiation of embryonic stem cells in vitro and in vivo. Biochemical and Biophysical Research Communications 2005; 328(4) 1051–7.

[80] Lee SH, Lumelsky N, Studer L, Auerbach JM, McKay RD. Efficient generation of midbrain and hindbrain neurons from mouse embryonic stem cells. Nature Biotechnology 2000; 18(6) 675–9.

[81] Yan Y, Yang D, Zarnowska ED, Du Z, Werbel B, Valliere C, et al. Directed differentiation of dopaminergic neuronal subtypes from human embryonic stem cells. Stem Cells 2005; 23(6) 781–90.

[82] Sanchez-Pernaute R, Lee H, Patterson M, Reske-Nielsen C, Yoshizaki T, Sonntag KC, et al. Parthenogenetic dopamine neurons from primate embryonic stem cells restore function in experimental Parkinson's disease. Brain: A Journal of Neurology 2008; 131(Pt 8) 2127–39.

[83] Tabar V, Tomishima M, Panagiotakos G, Wakayama S, Menon J, Chan B, et al. Therapeutic cloning in individual parkinsonian mice. Nature Medicine 2008; 14(4) 379–81.

[84] Bozkurt A, Brook GA, Moellers S, Lassner F, Sellhaus B, Weis J, et al. In vitro assessment of axonal growth using dorsal root ganglia explants in a novel three-dimensional collagen matrix. Tissue Engineering 2007; 13(12) 2971–9.

[85] Bozkurt A, Lassner F, O'Dey D, Deumens R, Böcker A, Schwendt T, et al. The role of microstructured and interconnected pore channels in a collagen-based nerve guide on axonal regeneration in peripheral nerves. Biomaterials 2012; 33(5) 1363–75.

[86] Scott JB, Afshari M, Kotek R, Saul JM. The promotion of axon extension in vitro using polymer-templated fibrin scaffolds. Biomaterials 2011; 32(21) 4830–9.

[87] Bockelmann J, Klinkhammer K, von Holst A, Seiler N, Faissner A, Brook GA, et al. Functionalization of electrospun poly(ε-caprolactone) fibers with the extracellular matrix-derived peptide GRGDS improves guidance of schwann cell migration and axonal growth. Tissue Engineering Part A 2011; 17(3-4) 475–86.

[88] Musoke-Zawedde P, Shoichet MS. Anisotropic three-dimensional peptide channels guide neurite outgrowth within a biodegradable hydrogel matrix. Biomedical Materials 2006; 1(3) 162–9.

[89] Azemi E, Stauffer WR, Gostock MS, Lagenaur CF, Cui XT. Surface immobilization of neural adhesion molecule L1 for improving the biocompatibility of chronic neural probes: In vitro characterization. Acta Biomaterialia 2008; 4(5) 1208–17.

[90] Yue Z, Liu X, Molino PJ, Wallace GG. Bio-functionalisation of polydimethylsiloxane with hyaluronic acid and hyaluronic acid--collagen conjugate for neural interfacing. Biomaterials 2011; 32(21) 4714–24.

[91] Schnell E, Klinkhammer K, Balzer S, Brook G, Klee D, Dalton P, et al. Guidance of glial cell migration and axonal growth on electrospun nanofibers of poly-epsilon-caprolactone and a collagen/poly-epsilon-caprolactone blend. Biomaterials 2007; 28(19) 3012–25.

[92] Song M, Uhrich KE. Optimal micropattern dimensions enhance neurite outgrowth rates, lengths, and orientations. Annals of Biomedical Engineering 2007; 35(10) 1812–20.

[93] Chang WC, Sretavan DW. Novel high-resolution micropatterning for neuron culture using polylysine adsorption on a cell repellant, plasma-polymerized background. Langmuir: The ACS Journal of Surfaces and Colloids 2008; 24(22) 13048–57.

[94] Li GN, Hoffman-Kim D. Evaluation of neurite outgrowth anisotropy using a novel application of circular analysis. Journal of Neuroscience Methods 2008; 174(2) 202–14.

[95] Li GN, Liu J, Hoffman-Kim D. Multi-molecular gradients of permissive and inhibitory cues direct neurite outgrowth. Annals of Biomedical Engineering 2008; 36(6) 889–904.

[96] Adams DN, Kao EY-C, Hypolite CL, Distefano MD, Hu W-S, Letourneau PC. Growth cones turn and migrate up an immobilized gradient of the laminin IKVAV peptide. Journal of Neurobiology 2005; 62(1) 134–47.

[97] Mukhatyar VJ, Salmerón-Sánchez M, Rudra S, Mukhopadaya S, Barker TH, García AJ, et al. Role of fibronectin in topographical guidance of neurite extension on electrospun fibers. Biomaterials 2011; 32(16) 3958–68.

[98] Holmes TC, de Lacalle S, Su X, Liu G, Rich A, Zhang S. Extensive neurite outgrowth and active synapse formation on self-assembling peptide scaffolds. Proceedings of the National Academy of Sciences of the United States of America 2000; 97(12) 6728–33.

[99] Silva GA, Czeisler C, Niece KL, Beniash E, Harrington DA, Kessler JA, et al. Selective differentiation of neural progenitor cells by high-epitope density nanofibers. Science 2004; 303(5662) 1352–5.

[100] Gelain F, Bottai D, Vescovi A, Zhang S. Designer self-assembling peptide nanofiber scaffolds for adult mouse neural stem cell 3-dimensional cultures. PloS One 2006; 1 e119.

[101] Shi W, Nie D, Jin G, Chen W, Xia L, Wu X, et al. BDNF blended chitosan scaffolds for human umbilical cord MSC transplants in traumatic brain injury therapy. Biomaterials 2012; 33(11) 3119–26.

[102] Li X, Yang Z, Zhang A. The effect of neurotrophin-3/chitosan carriers on the proliferation and differentiation of neural stem cells. Biomaterials 2009; 30(28) 4978–85.

[103] Cho Y Il, Choi JS, Jeong SY, Yoo HS. Nerve growth factor (NGF)-conjugated electrospun nanostructures with topographical cues for neuronal differentiation of mesenchymal stem cells. Acta Biomaterialia 2010; 6(12) 4725–33.

[104] Madduri S, Papaloïzos M, Gander B. Trophically and topographically functionalized silk fibroin nerve conduits for guided peripheral nerve regeneration. Biomaterials 2010; 31(8) 2323–34.

[105] Xiong Y, Zeng Y-S, Zeng C-G, Du B-L, He L-M, Quan D-P, et al. Synaptic transmission of neural stem cells seeded in 3-dimensional PLGA scaffolds. Biomaterials 2009; 30(22) 3711–22.

[106] Koh HS, Yong T, Chan CK, Ramakrishna S. Enhancement of neurite outgrowth using nano-structured scaffolds coupled with laminin. Biomaterials 2008; 29(26) 3574–82.

[107] Yu LMY, Miller FD, Shoichet MS. The use of immobilized neurotrophins to support neuron survival and guide nerve fiber growth in compartmentalized chambers. Biomaterials 2010; 31(27) 6987–99.

[108] Moore K, MacSween M, Shoichet M. Immobilized concentration gradients of neurotrophic factors guide neurite outgrowth of primary neurons in macroporous scaffolds. Tissue Engineering 2006; 12(2) 267–78.

[109] Wang Y, Wei YT, Zu ZH, Ju RK, Guo MY, Wang XM, et al. Combination of hyaluronic acid hydrogel scaffold and PLGA microspheres for supporting survival of neural stem cells. Pharmaceutical Research 2011; 28(6) 1406–14.

[110] Von Philipsborn AC, Lang S, Loeschinger J, Bernard A, David C, Lehnert D, et al. Growth cone navigation in substrate-bound ephrin gradients. Development 2006; 133(13) 2487–95.

[111] Tu Q, Li L, Zhang Y, Wang J, Liu R, Li M, et al. The effect of acetylcholine-like biomimetic polymers on neuronal growth. Biomaterials 2011; 32(12) 3253–64.

[112] Bellamkonda R V. Peripheral nerve regeneration: an opinion on channels, scaffolds and anisotropy. Biomaterials 2006; 27(19) 3515–8.

[113] Xie J, Willerth SM, Li X, Macewan MR, Rader A, Sakiyama-Elbert SE, et al. The differentiation of embryonic stem cells seeded on electrospun nanofibers into neural lineages. Biomaterials 2009; 30(3) 354–62.

[114] Pérez-Garnés M, Martínez-Ramos C, Barcia JA, Escobar Ivirico JL, Gómez-Pinedo U, Vallés-Lluch A, et al. One-dimensional migration of olfactory ensheathing cells on synthetic materials: experimental and numerical characterization. Cell Biochemistry and Biophysics 2013; 65(1) 21–36.

[115] Stokols S, Tuszynski MH. Freeze-dried agarose scaffolds with uniaxial channels stimulate and guide linear axonal growth following spinal cord injury. Biomaterials 2006; 27(3) 443–51.

[116] Clements IP, Kim Y, English AW, Lu X, Chung A, Bellamkonda R V. Thin-film enhanced nerve guidance channels for peripheral nerve repair. Biomaterials 2009; 30(23-24) 3834–46.

[117] Norman JJ, Desai TA. Methods for fabrication of nanoscale topography for tissue engineering scaffolds. Annals of Biomedical Engineering 2006; 34(1) 89–101.

[118] Miller C, Jeftinija S, Mallapragada S. Synergistic effects of physical and chemical guidance cues on neurite alignment and outgrowth on biodegradable polymer substrates. Tissue Engineering 2002; 8(3) 367–78.

[119] Yang IH, Co CC, Ho CC. Controlling neurite outgrowth with patterned substrates. J Biomed Mater Res A 2011; 97(4) 451–6.

[120] Béduer A, Vieu C, Arnauduc F, Sol J-C, Loubinoux I, Vaysse L. Engineering of adult human neural stem cells differentiation through surface micropatterning. Biomaterials 2012; 33(2) 504–14.

[121] Yao L, Wang S, Cui W, Sherlock R, O'Connell C, Damodaran G, et al. Effect of functionalized micropatterned PLGA on guided neurite growth. Acta Biomaterialia 2009; 5(2) 580–8.

[122] Mahoney MJ, Chen RR, Tan J, Saltzman WM. The influence of microchannels on neurite growth and architecture. Biomaterials 2005; 26(7) 771–8.

[123] Seidlits SK, Lee JY, Schmidt CE. Nanostructured scaffolds for neural applications. Nanomedicine 2008; 3(2) 183–99.

[124] Goldner JS, Bruder JM, Li G, Gazzola D, Hoffman-Kim D. Neurite bridging across micropatterned grooves. Biomaterials 2006; 27(3) 460–72.

[125] Krsko P, McCann TE, Thach TT, Laabs TL, Geller HM, Libera MR. Length-scale mediated adhesion and directed growth of neural cells by surface-patterned poly(ethylene glycol) hydrogels. Biomaterials 2009; 30(5) 721–9.

[126] Sapelkin A V, Bayliss SC, Unal B, Charalambou A. Interaction of B50 rat hippocampal cells with stain-etched porous silicon. Biomaterials 2006; 27(6) 842–6.

[127] Johansson F, Carlberg P, Danielsen N, Montelius L, Kanje M. Axonal outgrowth on nano-imprinted patterns. Biomaterials 2006; 27(8) 1251–8.

[128] Yang F, Murugan R, Ramakrishna S, Wang X, Ma Y-X, Wang S. Fabrication of nano-structured porous PLLA scaffold intended for nerve tissue engineering. Biomaterials 2004; 25(10) 1891–900.

[129] Sangsanoh P, Waleetorncheepsawat S, Suwantong O, Wutticharoenmongkol P, Weeranantanapan O, Chuenjitbuntaworn B, et al. In vitro biocompatibility of schwann cells on surfaces of biocompatible polymeric electrospun fibrous and solution-cast film scaffolds. Biomacromolecules 2007; 8(5) 1587–94.

[130] Lim SH, Liu XY, Song H, Yarema KJ, Mao H-Q. The effect of nanofiber-guided cell alignment on the preferential differentiation of neural stem cells. Biomaterials 2010; 31(34) 9031–9.

[131] Xie J, Willerth SM, Li X, Macewan MR, Rader A, Sakiyama-Elbert SE, et al. The differentiation of embryonic stem cells seeded on electrospun nanofibers into neural lineages. Biomaterials 2009; 30(3) 354–62.

[132] Yang F, Murugan R, Wang S, Ramakrishna S. Electrospinning of nano/micro scale poly(L-lactic acid) aligned fibers and their potential in neural tissue engineering. Biomaterials 2005; 26(15) 2603–10.

[133] Corey JM, Lin DY, Mycek KB, Chen Q, Samuel S, Feldman EL, et al. Aligned electrospun nanofibers specify the direction of dorsal root ganglia neurite growth. Journal of Biomedical Materials Research Part A 2007; 83(3) 636–45.

[134] Thompson DM, Buettner HM. Neurite outgrowth is directed by schwann cell alignment in the absence of other guidance cues. Ann Biomed Eng 2006; 34(1) 161–8.

[135] Recknor JB, Sakaguchi DS, Mallapragada SK. Directed growth and selective differentiation of neural progenitor cells on micropatterned polymer substrates. Biomaterials 2006; 27(22) 4098–108.

[136] Alexander JK, Fuss B, Colello RJ. Electric field-induced astrocyte alignment directs neurite outgrowth. Neuron Glia Biology 2006; 2(2) 93–103.

[137] Sørensen A, Alekseeva T, Katechia K, Robertson M, Riehle MO, Barnett SC. Long-term neurite orientation on astrocyte monolayers aligned by microtopography. Biomaterials 2007; 28(36) 5498–508.

[138] Sun D, Lye-Barthel M, Masland RH, Jakobs TC. Structural remodeling of fibrous astrocytes after axonal injury. The Journal of Neuroscience: The Official Journal of the Society for Neuroscience 2010; 30(42) 14008–19.

[139] Runyan SA, Phelps PE. Mouse olfactory ensheathing glia enhance axon outgrowth on a myelin substrate in vitro. Exp Neurol 2009; 216(1) 95–104.

[140] Deumens R, Koopmans GC, Den Bakker CG, Maquet V, Blacher S, Honig WM, et al. Alignment of glial cells stimulates directional neurite growth of CNS neurons in vitro. Neuroscience 2004; 125(3) 591–604.

[141] Richardson JA, Rementer CW, Bruder JM, Hoffman-Kim D. Guidance of dorsal root ganglion neurites and Schwann cells by isolated Schwann cell topography on poly(dimethyl siloxane) conduits and films. Journal of Neural Engineering 2011; 8(4) 046015.

[142] Kofron CM, Liu YT, Lopez-Fagundo CY, Mitchel JA, Hoffman-Kim D. Neurite outgrowth at the biomimetic interface. Ann Biomed Eng 2010; 38(6) 2210–25.

[143] Sanghvi AB, Miller KP-H, Belcher AM, Schmidt CE. Biomaterials functionalization using a novel peptide that selectively binds to a conducting polymer. Nature Materials 2005; 4(6) 496–502.

[144] Wong JY, Langer R, Ingber DE. Electrically conducting polymers can noninvasively control the shape and growth of mammalian cells. Proceedings of the National Academy of Sciences of the United States of America 1994; 91(8) 3201–4.

[145] Song H-K, Toste B, Ahmann K, Hoffman-Kim D, Palmore GTR. Micropatterns of positive guidance cues anchored to polypyrrole doped with polyglutamic acid: a new platform for characterizing neurite extension in complex environments. Biomaterials 2006; 27(3) 473–84.

[146] Mattioli-Belmonte M, Giavaresi G, Biagini G, Virgili L, Giacomini M, Fini M, et al. Tailoring biomaterial compatibility: in vivo tissue response versus in vitro cell behavior. The International Journal of Artificial Organs 2003; 26(12) 1077–85.

[147] Richardson-Burns SM, Hendricks JL, Martin DC. Electrochemical polymerization of conducting polymers in living neural tissue. Journal of Neural Engineering 2007; 4(2) L6–L13.

[148] Wang X, Gu X, Yuan C, Chen S, Zhang P, Zhang T, et al. Evaluation of biocompatibility of polypyrrole in vitro and in vivo. Journal of Biomedical Materials Research Part A 2004; 68(3) 411–22.

[149] Asplund M, Thaning E, Lundberg J, Sandberg-Nordqvist AC, Kostyszyn B, Inganäs O, et al. Toxicity evaluation of PEDOT/biomolecular composites intended for neural communication electrodes. Biomedical Materials 2009; 4(4) 045009.

[150] Guimard NK, Gomez N, Schmidt CE. Conducting polymers in biomedical engineering. Progress in Polymer Science 2007; 32(8-9) 876–921.

[151] Sisken BF, Kanje M, Lundborg G, Herbst E, Kurtz W. Stimulation of rat sciatic nerve regeneration with pulsed electromagnetic fields. Brain Research 1989; 485(2) 309–16.

[152] Zhao M, Forrester J V, McCaig CD. A small, physiological electric field orients cell division. Proceedings of the National Academy of Sciences of the United States of America 1999; 96(9) 4942–6.

[153] Yamada M, Tanemura K, Okada S, Iwanami A, Nakamura M, Mizuno H, et al. Electrical stimulation modulates fate determination of differentiating embryonic stem cells. Stem Cells 2007; 25(3) 562–70.

[154] Alexander JK, Fuss B, Colello RJ. Electric field-induced astrocyte alignment directs neurite outgrowth. Neuron Glia Biology 2006; 2(2) 93–103.

[155] Schmidt CE, Shastri VR, Vacanti JP, Langer R. Stimulation of neurite outgrowth using an electrically conducting polymer. Proceedings of the National Academy of Sciences of the United States of America 1997; 94(17) 8948–53.

[156] Ghasemi-Mobarakeh L, Prabhakaran MP, Morshed M, Nasr-Esfahani MH, Ramakrishna S. Electrical stimulation of nerve cells using conductive nanofibrous scaffolds for nerve tissue engineering. Tissue Engineering Part A 2009; 15(11) 3605–19.

[157] Thompson BC, Richardson RT, Moulton SE, Evans AJ, O'Leary S, Clark GM, et al. Conducting polymers, dual neurotrophins and pulsed electrical stimulation--dramatic effects on neurite outgrowth. Journal of Controlled Release 2010; 141(2) 161–7.

[158] Xie J, Macewan MR, Willerth SM, Li X, Moran DW, Sakiyama-Elbert SE, et al. Conductive Core-Sheath Nanofibers and Their Potential Application in Neural Tissue Engineering. Advanced Functional Materials 2009; 19(14) 2312–8.

[159] Patel N, Poo MM. Orientation of neurite growth by extracellular electric fields. The Journal of Neuroscience: The Official Journal of the Society for Neuroscience 1982; 2(4) 483–96.

[160] Voge CM, Stegemann JP. Carbon nanotubes in neural interfacing applications. Journal of Neural Engineering 2011; 8(1) 011001.

[161] Young T-H, Lin U-H, Lin D-J, Chang H-H, Cheng L-P. Immobilization of L-lysine on microporous PVDF membranes for neuron culture. Journal of Biomaterials Science Polymer Edition 2009; 20(5-6) 703–20.

[162] Schmidt CE, Leach JB. Neural tissue engineering: strategies for repair and regeneration. Annual Review of Biomedical Engineering 2003; 5 293–347.

[163] Liu X, Yue Z, Higgins MJ, Wallace GG. Conducting polymers with immobilised fibrillar collagen for enhanced neural interfacing. Biomaterials 2011; 32(30) 7309–17.

[164] Shi G, Rouabhia M, Wang Z, Dao LH, Zhang Z. A novel electrically conductive and biodegradable composite made of polypyrrole nanoparticles and polylactide. Biomaterials 2004; 25(13) 2477–88.

[165] Gomez N, Schmidt CE. Nerve growth factor-immobilized polypyrrole: bioactive electrically conducting polymer for enhanced neurite extension. Journal of Biomedical Materials Research Part A 2007; 81(1) 135–49.

[166] Gomez N, Lee JY, Nickels JD, Schmidt CE. Micropatterned Polypyrrole: A Combination of Electrical and Topographical Characteristics for the Stimulation of Cells. Advanced Functional Materials 2007; 17(10) 1645–53.

[167] Elkin BS, Azeloglu EU, Costa KD, Morrison B. Mechanical heterogeneity of the rat hippocampus measured by atomic force microscope indentation. Journal of Neurotrauma 2007; 24(5) 812–22.

[168] Moore SW, Roca-Cusachs P, Sheetz MP. Stretchy proteins on stretchy substrates: the important elements of integrin-mediated rigidity sensing. Developmental Cell 2010; 19(2) 194–206.

[169] Seidlits SK, Khaing ZZ, Petersen RR, Nickels JD, Vanscoy JE, Shear JB, et al. The effects of hyaluronic acid hydrogels with tunable mechanical properties on neural progenitor cell differentiation. Biomaterials 2010; 31(14) 3930–40.

[170] Georges PC, Miller WJ, Meaney DF, Sawyer ES, Janmey PA. Matrices with compliance comparable to that of brain tissue select neuronal over glial growth in mixed cortical cultures. Biophysical Journal 2006; 90(8) 3012–8.

[171] Banerjee A, Arha M, Choudhary S, Ashton RS, Bhatia SR, Schaffer D V, et al. The influence of hydrogel modulus on the proliferation and differentiation of encapsulated neural stem cells. Biomaterials 2009; 30(27) 4695–9.

[172] Man AJ, Davis HE, Itoh A, Leach JK, Bannerman P. Neurite outgrowth in fibrin gels is regulated by substrate stiffness. Tissue Engineering Part A 2011; 17(23-24) 2931–42.

[173] Gunn JW, Turner SD, Mann BK. Adhesive and mechanical properties of hydrogels influence neurite extension. Journal of Biomedical Materials Research Part A 2005; 72(1) 91–7.

[174] Leach JB, Brown XQ, Jacot JG, Dimilla PA, Wong JY. Neurite outgrowth and branching of PC12 cells on very soft substrates sharply decreases below a threshold of substrate rigidity. Journal of Neural Engineering 2007; 4(2) 26–34.

[175] Stabenfeldt SE, LaPlaca MC. Variations in rigidity and ligand density influence neuronal response in methylcellulose-laminin hydrogels. Acta Biomaterialia 2011; 7(12) 4102–8.

[176] Cheng C-M, LeDuc PR, Lin Y-W. Localized bimodal response of neurite extensions and structural proteins in dorsal-root ganglion neurons with controlled polydimethylsiloxane substrate stiffness. Journal of Biomechanics 2011; 44(5) 856–62.

[177] Pfister BJ, Iwata A, Taylor AG, Wolf JA, Meaney DF, Smith DH. Development of transplantable nervous tissue constructs comprised of stretch-grown axons. Journal of Neuroscience Methods 2006; 153(1) 95–103.

[178] Pfister BJ, Bonislawski DP, Smith DH, Cohen AS. Stretch-grown axons retain the ability to transmit active electrical signals. FEBS Lett 2006; 580(14) 3525–31.

[179] Haq F, Keith C, Zhang G. Neurite development in PC12 cells on flexible micro-textured substrates under cyclic stretch. Biotechnol Prog 2006; 22(1) 133–40.

[180] Zelzer M, Alexander MR, Russell NA. Hippocampal cell response to substrates with surface chemistry gradients. Acta Biomaterialia 2011; 7(12) 4120–30.

[181] Lee SJ, Khang G, Lee YM, Lee HB. The effect of surface wettability on induction and growth of neurites from the PC-12 cell on a polymer surface. Journal of Colloid and Interface Science 2003; 259(2) 228–35.

[182] Lamour G, Eftekhari-Bafrooei A, Borguet E, Souès S, Hamraoui A. Neuronal adhesion and differentiation driven by nanoscale surface free-energy gradients. Biomaterials 2010; 31(14) 3762–71.

[183] Tate CC, Shear DA, Tate MC, Archer DR, Stein DG, LaPlaca MC. Laminin and fibronectin scaffolds enhance neural stem cell transplantation into the injured brain. Journal of Tissue Engineering and Regenerative Medicine 2009; 3(3) 208–17.

[184] Wei YT, Tian WM, Yu X, Cui FZ, Hou SP, Xu QY, et al. Hyaluronic acid hydrogels with IKVAV peptides for tissue repair and axonal regeneration in an injured rat brain. Biomedical Materials 2007; 2(3) S142–6.

[185] Nisbet DR, Rodda AE, Horne MK, Forsythe JS, Finkelstein DI. Neurite infiltration and cellular response to electrospun polycaprolactone scaffolds implanted into the brain. Biomaterials 2009; 30(27) 4573–80.

[186] Wong DY, Krebsbach PH, Hollister SJ. Brain cortex regeneration affected by scaffold architectures. Journal of Neurosurgery 2008; 109(4) 715–22.

[187] Lu D, Mahmood A, Qu C, Hong X, Kaplan D, Chopp M. Collagen scaffolds populated with human marrow stromal cells reduce lesion volume and improve functional outcome after traumatic brain injury. Neurosurgery 2007; 61(3) 596–602; discussion 602–3.

[188] Xiong Y, Qu C, Mahmood A, Liu Z, Ning R, Li Y, et al. Delayed transplantation of human marrow stromal cell-seeded scaffolds increases transcallosal neural fiber length, angiogenesis, and hippocampal neuronal survival and improves functional outcome after traumatic brain injury in rats. Brain Research 2009; 1263 183–91.

[189] Qu C, Xiong Y, Mahmood A, Kaplan DL, Goussev A, Ning R, et al. Treatment of traumatic brain injury in mice with bone marrow stromal cell-impregnated collagen scaffolds. Journal of Neurosurgery 2009; 111(4) 658–65.

[190] Huang K-F, Hsu W-C, Chiu W-T, Wang J-Y. Functional improvement and neurogenesis after collagen-GAG matrix implantation into surgical brain trauma. Biomaterials 2012; 33(7) 2067–75.

[191] Hou S, Xu Q, Tian W, Cui F, Cai Q, Ma J, et al. The repair of brain lesion by implantation of hyaluronic acid hydrogels modified with laminin. Journal of Neuroscience Methods 2005; 148(1) 60–70.

[192] Tian WM, Hou SP, Ma J, Zhang CL, Xu QY, Lee IS, et al. Hyaluronic acid-poly-D-lysine-based three-dimensional hydrogel for traumatic brain injury. Tissue Engineering 2005; 11(3-4) 513–25.

[193] Wang Y, Cooke MJ, Morshead CM, Shoichet MS. Hydrogel delivery of erythropoietin to the brain for endogenous stem cell stimulation after stroke injury. Biomaterials 2012; 33(9) 2681–92.

[194] Cui FZ, Tian WM, Hou SP, Xu QY, Lee I-S. Hyaluronic acid hydrogel immobilized with RGD peptides for brain tissue engineering. Journal of Materials Science Materials in Medicine 2006; 17(12) 1393–401.

[195] Mingyu C, Kai G, Jiamou L, Yandao G, Nanming Z, Xiufang Z. Surface modification and characterization of chitosan film blended with poly-L-lysine. Journal of Biomaterials Applications 2004; 19(1) 59–75.

[196] Crompton KE, Tomas D, Finkelstein DI, Marr M, Forsythe JS, Horne MK. Inflammatory response on injection of chitosan/GP to the brain. Journal of Materials Science Materials in Medicine 2006; 17(7) 633–9.

[197] Tate MC, Shear DA, Hoffman SW, Stein DG, LaPlaca MC. Biocompatibility of methylcellulose-based constructs designed for intracerebral gelation following experimental traumatic brain injury. Biomaterials 2001; 22(10) 1113–23.

[198] Bhang SH, Lee YE, Cho S-W, Shim J-W, Lee S-H, Choi CY, et al. Basic fibroblast growth factor promotes bone marrow stromal cell transplantation-mediated neural regeneration in traumatic brain injury. Biochemical and Biophysical Research Communications 2007; 359(1) 40–5.

[199] Guo J, Leung KKG, Su H, Yuan Q, Wang L, Chu T-H, et al. Self-assembling peptide nanofiber scaffold promotes the reconstruction of acutely injured brain. Nanomedicine: Nanotechnology, Biology, and Medicine 2009; 5(3) 345–51.

[200] Lesný P, De Croos J, Prádný M, Vacík J, Michálek J, Woerly S, et al. Polymer hydrogels usable for nervous tissue repair. Journal of Chemical Neuroanatomy 2002; 23(4) 243–7.

[201] Cui FZ, Tian WM, Fan YW, Hou SP, Xu QY, Lee I-S. Cerebrum Repair with PHPMA Hydrogel Immobilized with Neurite-Promoting Peptides in Traumatic Brain Injury of Adult Rat Model. Journal of Bioactive and Compatible Polymers 2003; 18(6) 413–32.

[202] Loh NK, Woerly S, Bunt SM, Wilton SD, Harvey AR. The regrowth of axons within tissue defects in the CNS is promoted by implanted hydrogel matrices that contain BDNF and CNTF producing fibroblasts. Experimental Neurology 2001; 170(1) 72–84.

[203] Wong DY, Hollister SJ, Krebsbach PH, Nosrat C. Poly(epsilon-caprolactone) and poly (L-lactic-co-glycolic acid) degradable polymer sponges attenuate astrocyte response and lesion growth in acute traumatic brain injury. Tissue Engineering 2007; 13(10) 2515–23.

[204] Lewitus DY, Smith KL, Shain W, Bolikal D, Kohn J. The fate of ultrafast degrading polymeric implants in the brain. Biomaterials 2011; 32(24) 5543–50.

[205] Azemi E, Lagenaur CF, Cui XT. The surface immobilization of the neural adhesion molecule L1 on neural probes and its effect on neuronal density and gliosis at the probe/tissue interface. Biomaterials 2011; 32(3) 681–92.

[206] Moon SU, Kim J, Bokara KK, Kim JY, Khang D, Webster TJ, et al. Carbon nanotubes impregnated with subventricular zone neural progenitor cells promotes recovery from stroke. International Journal of Nanomedicine 2012; 7 2751–65.

[207] King VR, Alovskaya A, Wei DYT, Brown RA, Priestley J V. The use of injectable forms of fibrin and fibronectin to support axonal ingrowth after spinal cord injury. Biomaterials 2010; 31(15) 4447–56.

[208] Liu T, Houle JD, Xu J, Chan BP, Chew SY. Nanofibrous collagen nerve conduits for spinal cord repair. Tissue Engineering Part A 2012; 18(9-10) 1057–66.

[209] Tsai EC, Dalton PD, Shoichet MS, Tator CH. Matrix inclusion within synthetic hydrogel guidance channels improves specific supraspinal and local axonal regeneration after complete spinal cord transection. Biomaterials 2006; 27(3) 519–33.

[210] Yao L, Damodaran G, Nikolskaya N, Gorman AM, Windebank A, Pandit A. The effect of laminin peptide gradient in enzymatically cross-linked collagen scaffolds on neurite growth. Journal of Biomedical Materials Research Part A 2010; 92(2) 484–92.

[211] Khaing ZZ, Milman BD, Vanscoy JE, Seidlits SK, Grill RJ, Schmidt CE. High molecular weight hyaluronic acid limits astrocyte activation and scar formation after spinal cord injury. Journal of Neural Engineering 2011; 8(4) 046033.

[212] Wei Y-T, He Y, Xu C-L, Wang Y, Liu B-F, Wang X-M, et al. Hyaluronic acid hydrogel modified with nogo-66 receptor antibody and poly-L-lysine to promote axon regrowth after spinal cord injury. Journal of Biomedical Materials Research Part B, Applied Biomaterials 2010; 95(1) 110–7.

[213] Shoichet MS, Tator CH, Poon P, Kang C, Baumann MD. Intrathecal drug delivery strategy is safe and efficacious for localized delivery to the spinal cord. Progress in Brain Research 2007; 161 385–92.

[214] Kang CE, Poon PC, Tator CH, Shoichet MS. A new paradigm for local and sustained release of therapeutic molecules to the injured spinal cord for neuroprotection and tissue repair. Tissue Engineering Part A 2009; 15(3) 595–604.

[215] Dodla MC, Bellamkonda R V. Anisotropic scaffolds facilitate enhanced neurite extension in vitro. Journal of Biomedical Materials Research Part A 2006; 78(2) 213–21.

[216] Stokols S, Tuszynski MH. The fabrication and characterization of linearly oriented nerve guidance scaffolds for spinal cord injury. Biomaterials 2004; 25(27) 5839–46.

[217] Chvatal SA, Kim Y-T, Bratt-Leal AM, Lee H, Bellamkonda R V. Spatial distribution and acute anti-inflammatory effects of Methylprednisolone after sustained local delivery to the contused spinal cord. Biomaterials 2008; 29(12) 1967–75.

[218] Prang P, Müller R, Eljaouhari A, Heckmann K, Kunz W, Weber T, et al. The promotion of oriented axonal regrowth in the injured spinal cord by alginate-based anisotropic capillary hydrogels. Biomaterials 2006; 27(19) 3560–9.

[219] Kataoka K, Suzuki Y, Kitada M, Ohnishi K, Suzuki K, Tanihara M, et al. Alginate, a bioresorbable material derived from brown seaweed, enhances elongation of amputated axons of spinal cord in infant rats. Journal of Biomedical Materials Research 2001; 54(3) 373–84.

[220] Fornaro M, Plescia J, Chheang S, Tallini G, Zhu Y-M, King M, et al. Fibronectin protects prostate cancer cells from tumor necrosis factor-alpha-induced apoptosis via the AKT/survivin pathway. The Journal of Biological Chemistry 2003; 278(50) 50402–11.

[221] Taylor L, Jones L, Tuszynski MH, Blesch A. Neurotrophin-3 gradients established by lentiviral gene delivery promote short-distance axonal bridging beyond cellular grafts in the injured spinal cord. The Journal of Neuroscience: The Official Journal of the Society for Neuroscience 2006; 26(38) 9713–21.

[222] Hyatt AJT, Wang D, Kwok JC, Fawcett JW, Martin KR. Controlled release of chondroitinase ABC from fibrin gel reduces the level of inhibitory glycosaminoglycan chains in lesioned spinal cord. Journal of Controlled Release 2010; 147(1) 24–9.

[223] Johnson PJ, Tatara A, Shiu A, Sakiyama-Elbert SE. Controlled release of neurotrophin-3 and platelet-derived growth factor from fibrin scaffolds containing neural progenitor cells enhances survival and differentiation into neurons in a subacute model of SCI. Cell Transplantation 2010; 19(1) 89–101.

[224] Tysseling-Mattiace VM, Sahni V, Niece KL, Birch D, Czeisler C, Fehlings MG, et al. Self-assembling nanofibers inhibit glial scar formation and promote axon elongation after spinal cord injury. The Journal of Neuroscience 2008; 28(14) 3814–23.

[225] Cigognini D, Satta A, Colleoni B, Silva D, Donegà M, Antonini S, et al. Evaluation of early and late effects into the acute spinal cord injury of an injectable functionalized self-assembling scaffold. PloS One 2011; 6(5) e19782.

[226] Guo J, Su H, Zeng Y, Liang Y-X, Wong WM, Ellis-Behnke RG, et al. Reknitting the injured spinal cord by self-assembling peptide nanofiber scaffold. Nanomedicine: Nanotechnology, Biology, and Medicine 2007; 3(4) 311–21.

[227] Patist CM, Mulder MB, Gautier SE, Maquet V, Jérôme R, Oudega M. Freeze-dried poly(D,L-lactic acid) macroporous guidance scaffolds impregnated with brain-derived neurotrophic factor in the transected adult rat thoracic spinal cord. Biomaterials 2004; 25(9) 1569–82.

[228] Hurtado A, Moon LDF, Maquet V, Blits B, Jérôme R, Oudega M. Poly (D,L-lactic acid) macroporous guidance scaffolds seeded with Schwann cells genetically modified to secrete a bi-functional neurotrophin implanted in the completely transected adult rat thoracic spinal cord. Biomaterials 2006; 27(3) 430–42.

[229] Hurtado A, Cregg JM, Wang HB, Wendell DF, Oudega M, Gilbert RJ, et al. Robust CNS regeneration after complete spinal cord transection using aligned poly-L-lactic acid microfibers. Biomaterials 2011; 32(26) 6068–79.

[230] Yang Y, De Laporte L, Zelivyanskaya ML, Whittlesey KJ, Anderson AJ, Cummings BJ, et al. Multiple channel bridges for spinal cord injury: cellular characterization of host response. Tissue Engineering Part A 2009; 15(11) 3283–95.

[231] Lowry N, Goderie SK, Lederman P, Charniga C, Gooch MR, Gracey KD, et al. The effect of long-term release of Shh from implanted biodegradable microspheres on recovery from spinal cord injury in mice. Biomaterials 2012; 33(10) 2892–901.

[232] Tuinstra HM, Aviles MO, Shin S, Holland SJ, Zelivyanskaya ML, Fast AG, et al. Multifunctional, multichannel bridges that deliver neurotrophin encoding lentivirus for regeneration following spinal cord injury. Biomaterials 2012; 33(5) 1618–26.

[233] Moore MJ, Friedman JA, Lewellyn EB, Mantila SM, Krych AJ, Ameenuddin S, et al. Multiple-channel scaffolds to promote spinal cord axon regeneration. Biomaterials 2006; 27(3) 419–29.

[234] Krych AJ, Rooney GE, Chen B, Schermerhorn TC, Ameenuddin S, Gross L, et al. Relationship between scaffold channel diameter and number of regenerating axons in the transected rat spinal cord. Acta Biomaterialia 2009; 5(7) 2551–9.

[235] Olson HE, Rooney GE, Gross L, Nesbitt JJ, Galvin KE, Knight A, et al. Neural stem cell-and Schwann cell-loaded biodegradable polymer scaffolds support axonal regeneration in the transected spinal cord. Tissue Engineering Part A 2009; 15(7) 1797–805.

[236] Wong DY, Leveque J-C, Brumblay H, Krebsbach PH, Hollister SJ, Lamarca F. Macroarchitectures in spinal cord scaffold implants influence regeneration. Journal of Neurotrauma 2008; 25(8) 1027–37.

[237] Chen BK, Knight AM, Madigan NN, Gross L, Dadsetan M, Nesbitt JJ, et al. Comparison of polymer scaffolds in rat spinal cord: a step toward quantitative assessment of combinatorial approaches to spinal cord repair. Biomaterials 2011; 32(32) 8077–86.

[238] Gelain F, Panseri S, Antonini S, Cunha C, Donega M, Lowery J, et al. Transplantation of nanostructured composite scaffolds results in the regeneration of chronically injured spinal cords. ACS Nano 2011; 5(1) 227–36.

[239] Novikov LN, Novikova LN, Mosahebi A, Wiberg M, Terenghi G, Kellerth J-O. A novel biodegradable implant for neuronal rescue and regeneration after spinal cord injury. Biomaterials 2002; 23(16) 3369–76.

[240] Novikova LN, Pettersson J, Brohlin M, Wiberg M, Novikov LN. Biodegradable poly-beta-hydroxybutyrate scaffold seeded with Schwann cells to promote spinal cord repair. Biomaterials 2008; 29(9) 1198–206.

[241] Woerly S, Pinet E, de Robertis L, Van Diep D, Bousmina M. Spinal cord repair with PHPMA hydrogel containing RGD peptides (NeuroGel). Biomaterials 2001; 22(10) 1095–111.

[242] Hejcl A, Lesný P, Prádný M, Michálek J, Jendelová P, Stulík J, et al. Biocompatible hydrogels in spinal cord injury repair. Physiological Research / Academia Scientiarum Bohemoslovaca 2008; 57 Suppl 3 S121–32.

[243] Senthilkumar KS, Saravanan KS, Chandra G, Sindhu KM, Jayakrishnan A, Mohanakumar KP. Unilateral implantation of dopamine-loaded biodegradable hydrogel in the striatum attenuates motor abnormalities in the 6-hydroxydopamine model of hemi-parkinsonism. Behavioural Brain Research 2007; 184(1) 11–8.

[244] Grondin R, Gash DM. Glial cell line-derived neurotrophic factor (GDNF): a drug candidate for the treatment of Parkinson's disease. Journal of Neurology 1998; 245(11 Suppl 3) 35–42.

[245] Garbayo E, Montero-Menei CN, Ansorena E, Lanciego JL, Aymerich MS, Blanco-Prieto MJ. Effective GDNF brain delivery using microspheres--a promising strategy for Parkinson's disease. Journal of Controlled Release 2009; 135(2) 119–26.

[246] Grandoso L, Ponce S, Manuel I, Arrúe A, Ruiz-Ortega JA, Ulibarri I, et al. Long-term survival of encapsulated GDNF secreting cells implanted within the striatum of parkinsonized rats. International Journal of Pharmaceutics 2007; 343(1-2) 69–78.

[247] Cherksey BD, Sapirstein VS, Geraci AL. Adrenal chromaffin cells on microcarriers exhibit enhanced long-term functional effects when implanted into the mammalian brain. Neuroscience 1996; 75(2) 657–64.

[248] Borlongan C V, Saporta S, Sanberg PR. Intrastriatal transplantation of rat adrenal chromaffin cells seeded on microcarrier beads promote long-term functional recovery in hemiparkinsonian rats. Experimental Neurology 1998; 151(2) 203–14.

[249] Borlongan C V, Thanos CG, Skinner SJM, Geaney M, Emerich DF. Transplants of encapsulated rat choroid plexus cells exert neuroprotection in a rodent model of Huntington's disease. Cell Transplantation 2008; 16(10) 987–92.

[250] Delcroix GJ-R, Garbayo E, Sindji L, Thomas O, Vanpouille-Box C, Schiller PC, et al. The therapeutic potential of human multipotent mesenchymal stromal cells combined with pharmacologically active microcarriers transplanted in hemi-parkinsonian rats. Biomaterials 2011; 32(6) 1560–73.

Chapter 8

The Role of Acellular Flowable Matrix in Tissue Regeneration

Dragica Maja Smrke and Danijela Semenič

Additional information is available at the end of the chapter

http://dx.doi.org/10.5772/59406

1. Introduction

Wound healing is a complex dynamic biological process which has four partly overlapping phases: hemostasis, inflammation, proliferation and remodeling. These phases involve a large number of cell types, extracellular components, growth factors and cytokines. [1]

The extracellular matrix (ECM) of mammalian tissues has been used as a scaffold to facilitate the repair and reconstruction of numerous tissues. Such scaffolds are prepared in many forms including sheets, powders, and hydrogels. [2]

The ECM is a complex network structure that surrounds and supports cells. It is filled with ECM molecules like proteins and proteoglycans, which are secreted by the cells. Cell receptors bind both soluble and tethered signaling cues from the ECM environment, while simultaneously, cells send out signals to actively construct and degrade their microenvironment for remodeling. Thus, the ECM acts not only as a mechanical scaffold for the cells, but also a bioactive and dynamic environment that mediates cellular functions. [3,4,5]

It is highly desirable to synthesize scaffolds to mimic the structure and biofunctions of the natural ECM. [3,6,7,8]

2. ECM-like scaffolds

A widely-cited challenge in the field of tissue engineering is to provide a blood vessel network to facilitate oxygen, nutrient, biochemical and waste exchange for thick tissues beyond the range of diffusion. In the absence of a blood vessel network, necrosis will quickly occur beyond the surface of any implanted, metabolic tissue. [9,10]

Extracellular matrix (ECM) proteins, such as fibronectin, laminin and collagen IV, play important roles in many cellular behaviors, including cell adhesion and spreading. Understanding their adsorption behavior on surfaces with different natures is helpful for studying the cellular responses to environments. [11]

Various polymers, including natural, synthetic and natural/synthetic hybrid polymers, have been used to make hydrogels via chemical or physical crosslinking. Recently, bioactive synthetic hydrogels have emerged as promising scaffolds because they can provide molecularly tailored biofunctions and adjustable mechanical properties, as well as an extracellular matrix-like microenvironment for cell growth and tissue formation. [3]

Glycosaminoglycans (GAGs) are ubiquitously present at the cell surface and in extracellular matrix, and crucial for matrix assembly, cell-cell and cell-matrix interactions. The supramolecular presentation of GAG chains, along with other matrix components, is likely to be functionally important but remains challenging to control and to characterize, both in vivo and in vitro. [12] The ECM component hyaluronic acid (HA) possesses a non-sulfated glycosaminoglycan (GAG) structure and is widely distributed throughout the ECM of all connective tissues. HA plays an essential role in many biological processes such as tissue hydration, nutrient diffusion, proteoglycan organization and cell differentiation.

Various proteins have been used to make natural-hydrogel tissue-engineering scaffolds. Among them, collagen, the most abundant protein in mammals, is a representative natural polymer to fabricate natural hydrogels. Collagen can be degraded naturally by metallomatrix proteinases (MMPs) – specifically, collagenase – allowing for local degradation controlled by cells present in the engineered tissue. [3]

2.1. Injectable hydrogels

Injectable hydrogels derived from the extracellular matrix (ECM) of decellularized tissues have recently emerged as scaffolds for tissue-engineering applications. [13]

ECM hydrogels provide advantages such as injectability, the ability to fill an irregularly shaped space, and the inherent bioactivity of native matrix. However, material properties of ECM hydrogels and the effect of these properties upon cell behavior are neither well understood nor controlled. [2] The objective of the study, made by Wolf et al. was to prepare and determine the structure, mechanics, and the cell response in vitro and in vivo of ECM hydrogels prepared from decellularized porcine dermis and urinary bladder tissues. Dermal ECM hydrogels were characterized by a more dense fiber architecture and greater mechanical integrity than urinary bladder ECM hydrogels, and showed a dose dependent increase in mechanical properties with ECM concentration. In vitro, dermal ECM hydrogels supported greater C2C12 myoblast fusion, and less fibroblast infiltration and less fibroblast mediated hydrogel contraction than urinary bladder ECM hydrogels. Both hydrogels were rapidly infiltrated by host cells, primarily macrophages, when implanted in a rat abdominal wall defect. Both ECM hydrogels degraded by 35 days in vivo, but UBM hydrogels degraded more quickly, and with greater amounts of myogenesis than dermal ECM. These results show that ECM hydrogel properties can be varied and partially controlled by the scaffold tissue source, and that these properties can markedly affect cell behavior. [2]

Seif-Naraghi et al. introduce the potential for using a decellularized ECM-derived hydrogel for the improved delivery of heparin-binding growth factors. Immobilization of growth factors on a scaffold has been shown to increase their stability and activity. This can be done via chemical crosslinking, covalent bonding, or by incorporating natural or synthetic growth factor-binding domains similar to those found in vivo in sulfated glycosaminoglycans (GAGs). Many decellularized ECM-derived hydrogels retain native sulfated GAGs, and these materials may therefore provide an excellent delivery platform for heparin-binding growth factors. In his study, the sulfated GAG content of an ECM hydrogel derived from decellularized pericardial ECM was confirmed by Fourier transform infrared spectroscopy and its ability to bind basic fibroblast growth factor (bFGF) was established. Delivery in the pericardial matrix hydrogel increased retention of bFGF both in vitro and in vivo in ischemic myocardium compared to delivery in collagen. In a rodent infarct model, intramyocardial injection of bFGF in pericardial matrix enhanced neovascularization by approximately 112% compared to delivery in collagen. Importantly, the newly formed vasculature was anastomosed with existing vasculature. Thus, the sulfated GAG content of the decellularized ECM hydrogel provides a platform for incorporation of heparin-binding growth factors for prolonged retention and delivery. [13]

2.2. Acellular flowable matrix

In recent years development of new technologies and therapies aimed to treat hard-to-heal wounds. The formation of a sinus tract complicates these difficult-to-treat diabetic wounds even further. In the lower extremity, sinus tract wounds are a significant risk for osteomyelitis. Typical treatment includes extensive surgical debridement with radical dissection, cavity packing, and occlusive dressing. Cellular and acellular bioengineered grafts have demonstrated some promise as alternatives for the treatment of such wounds. [14,15,16,17]

However, these grafts are provided as sheets in planar form. As such, they are harder to use and less effective in sinus tract wounds because of the irregularly shaped tunnels or extensions in these wounds. Ideally, the matrix material is capable of conforming to the wound area and providing the necessary bulk and scaffold to support the woundhealing process. [17]

An injectable form of an acellular human dermal regenerative tissue matrix Graftjacket® Xpress Flowable Soft Tissue Scaffold was designed to treat difficult tunneling wounds in a minimally invasive manner. This injectable matrix is a micronized form of a human acellular dermal regenerative tissue matrix. The injectable human matrix is composed of a micronized or powdered version of the graftjacket membrane. The matrix membrane is a human acellular scaffold derived from human dermal tissue. It is biocompatible and biologically permissive. This dermal matrix retains its native extracellular matrix proteins such as collagens I, III, IV, and VII; elastin; and proteoglycans in cell-friendly forms.

Although the sheet form of the scaffold does demonstrate intact vascular channels that allows for rapid angiogenesis, those channels are destroyed in the micronization process. The flowable matrix allows for nutritional diffusion and cell attachment to facilitate cell migration as well as proliferation. These properties result in the rapid revascularization and cellular repopulation of the matrix scaffold. [17]

Graftjacket® Xpress FSTS supports the body's repair of damaged or inadequate integumental tissue, such as deep dermal wounds or diabetic ulcers. Graftjacket® FSTS is supplied as a dried, acellular dermal particulate in a syringe, before aplication it is reconstituted through syringe adapter with sterile saline solution and applied directly into the debrided, clean wound through a flexible applicator tip. The product is recomended for burns, chronic wounds as diabetic foot ulcers, pressure ulcers, surgical wounds ans venous ulcers.

Another interesting commercially available product is Integra® Flowable Wound Matrix. It is an advanced wound care device comprised of a granulated Integra® Wound Matrix, which consists of collagen and glycosaminoglycan. Collagen is of bovine origin. The Integra® Flowable Wound Matrix is hydrated with saline prior to application. It is designed for use in deep soft tissue or tunneling wounds. Gel-like consistency allows intimate contact between the grafting material and wound bed, composition and method of administration allows for complete coverage in deep crevice wounds. It provides a resorbable scaffold on which cells can attach, migrate, proliferate and differentiate.

Integra® Flowable Wound Matrix is indicated for the management of wounds including: partial-and full-thickness wounds, pressure ulcers, venous ulcers, diabetic ulcers, chronic vascular ulcers, tunneled/undermined wounds, surgical wounds (donor sites/grafts, post-laser surgery, podiatric, wound dehiscence), trauma wounds (abrasions, lacerations, second-degree burns, skin tears) and draining wounds.

3. Review of recent clinical experience with acellular matrix

Schmidt et al. in his retrospectively study concluded, that acellular porcine dermal collagen matrix is a feasible and reliable biological patch material for reconstruction of the thoracic wall. Excellent wound healing and long-term stability were achieved even in large defects or complete sternal replacements. Biological collagen matrixes have emerged as an alternative to the routinely used synthetic materials by major thoracic wall resections. Histological examination showed integration, neovascularization, and long-term persistence of the collagen matrix on late reoperation of one patient. [18]

The results of the multicenter prospective study from Kavros et al. demonstrate that the use of fetal bovine acellular dermal matrix (PriMatrix) integrated with standard-of-care therapy is a successful treatment regimen to heal diabetic foot ulcers (DFUs). In the study were included patiets with chronic DFU that ranged in area from 1 to 20 cm and failed to heal more than 30% during a 2-week screening period when treated with moist wound therapy. For qualifying subjects, PriMatrix was secured into a clean, sharply debrided wound; dressings were applied to maintain a moist wound environment, and the DFU was pressure off-loaded. Wound area measurements were taken weekly for up to 12 weeks, and PriMatrix was reapplied at the discretion of the treating physician. 55 patiets were enrolled at 9 US centers with 46 subjects progressing to study completion. Ulcers had been in existence for an average of 286 days, and initial mean ulcer area was 4.34 cm. Of the subjects completing the study, 76% healed by 12 weeks with a mean time to healing of 53.1 ± 21.9 days. The mean number of applications for these healed wounds was 2.0 ± 1.4, with 59.1% healing with a single application of PriMatrix and 22.9% healing with 2 applications. For subjects not healed by 12 weeks, the average wound area reduction was 71.4%. [19]

Goyal et al. evaluate in the study the degree of patient acceptance with acellular dermal matrix (ADM) allograft in the treatment of buccal gingival recession and compare it with subepithelial connective tissue graft. Obtaining predictable and aesthetic root coverage has become an important part of periodontal therapy. The search for the appropriate root coverage techniques has resulted in many different approaches. In the study thirty patients with Miller's class II recessions were treated and randomly assigned to the test group (ADM) and control group (subepithelial connective tissue graft). All patients underwent full periodontal evaluation and pre-surgical preparation, including oral hygiene instructions and scaling and root planing. The exposed roots were thoroughly planed and covered by a graft without any further root treatment. Results were evaluated based on the parameters measuring patient satisfaction and clinical outcome after 6 months of the surgical procedure. Postoperatively, significant root coverage, reduction in probing depth, gain in clinical attachment level, and increase in widths of keratinized tissue and attached gingiva were observed on intra-group comparison. There was no significant difference in any of the parameters between test and control groups. The subepithelial connective tissue graft and ADM graft were able to successfully treat gingival recession defects; however, the ADM showed better patient acceptance than the connective tissue graft. [20]

Wang et al. made an interesting prospective randomized multi-center study with comparison of two differently processed acellular dermal matrix products for root coverage procedures. They have compared 2 acellular dermal matrix (ADM) materials produced by different processing techniques, freeze-dried (FDADM) and solvent-dehydrated ADM (SDADM), in their ability to correct Miller's Class I and II recession defects. Eighty subjects from four study centers, each with a single maxillary anterior Miller's class I or II recession defect were enrolled. Subjects were randomly assigned and treated with coronally advanced flap (CAF)+FDADM (N=42) or CAF+SDADM (N=38). Gingival thickness, recession depth, recession width, probing pocket depth, clinical attachment level, Gingival index, Plaque index, patient discomfort and wound healing index were recorded before surgery, immediately post surgery, and over 2, 4, 12, 24 and 52 weeks postoperatively. When evaluating the clinical parameters after one year, both groups showed significant ($P<0.05$) improvement for most of the parameters evaluated when compared to baseline (Day 0). For example, percentage of root coverage was 77.20% ± 29.10% for CAF+FDADM and 71.01% ± 32.87% for CAF+SDADM. On the other hand, no significant differences were observed between the two materials for any clinical parameter tested or for patient satisfaction except for PD on the mesial side of the defects ($p=0.03$). As a conclusion both ADM materials, freeze-dried or solvent-dehydrated, can be used successfully to correct Miller's class I or II recession defects. There were no statistically significant differences between groups for any of the clinical parameters tested. [21]

In a prospective, randomized study made in 2004 by Brigido and colleagues, Graftjacket tissue matrix, an acellular regenerative tissue matrix, has been used for wound closure. He studied the efficacy of this tissue product in wound repair compared with conventional treatment. Therefore, researchers used diabetic foot ulcers to evaluate the efficacy of GraftJacket tissue matrix in wound repair. Only a single administration of the tissue matrix was required. After 1 month of treatment, preliminary results demonstrate that this novel tissue matrix promotes faster healing at a statistically significant rate over conventional treatment. Because wounds in this series of patients are deep and circulation around the wound is poor, the preliminary results suggest that this tissue matrix could be applicable to other types of orthopedic wounds. [22]

Another multicenter, retrospective study from Winters el al. presents the use of a human acellular dermal regenerative tissue matrix (Graftjacket regenerative tissue Matrix) as an alternative treatment for 100 chronic, full-thickness wounds of the lower extremity in 75 diabetic patients. Wound locations included the foot (86.0%), ankle (8.0%), and lower extremity (6.0%). Mean wound age was 20.4 weeks (1.3-191.4 weeks). University of Texas (UT) wound classifications included 15 (15.0%) 1A, 1 (1.0%) 1B, 1 (1.0%) 1C, 2 (2.0%) 1D, 18 (18.0%) 2A, 8 (8.0%) 2B, 5 (5.0%) 2C, 3 (3.0%) 2D, 3 (3.0%) 3A, 7 (7.0%) 3B, 3 (3.0%) 3C, and 34 (34.0%) 3D. The mean time to matrix incorporation, 100% granulation, and complete healing was 1.5 weeks (0.43-4.4 weeks), 5.1 weeks (0.43-16.7 weeks), and 13.8 weeks (1.7-57.8 weeks), respectively. The overall matrix success rate, as defined by full epithelialization, was 90.0%. One failed wound subsequently healed approximately 7 weeks after matrix reapplication. The healing rate was 91.0%, as 91 of the 100 wounds healed. No statistically significant differences were observed between UT classifications and time to matrix incorporation, 100% granulation, and complete healing. Absence of matrix-related complications and high rates of closure in a wide array of diabetic wounds suggest that this matrix is a viable treatment for complex lower extremity wounds. Lack of any statistically significant differences between UT grades and wound outcome end points lends further support to the universal applicability of this matrix, with successful results in both superficial diabetic wounds and in wounds penetrating to the bone or joint. [23]

Integra bilayer wound matrix (IBWM) is a bilayer skin replacement system composed of a dermal regeneration layer and a temporary epidermal layer. It is used to treat various types of deep, large wounds via an inpatient procedure in an operating room. Prospective pilot study with Integra bilayer wound matrix (IBWM) in the treatment of diabetic foot ulcers was made by Yao and colleagues. They sought to determine ease of use and effectiveness of IBWM in an outpatient clinical setting when treating diabetic foot ulcers. In addition, no epidermal autografting was performed in conjunction with the IBWM after silicone release, as is common in the inpatient setting. Eleven patients with diabetic foot ulcers were enrolled. One patient was discontinued from the study owing to noncompliance leading to a serious adverse event. Therefore, ten patients who received the study intervention were included in the per-protocol population reported herein. The mean patient age was 60.6 years, with an average 11-year history of diabetes mellitus. Each ulcer was located on the plantar aspect of the foot. No infection was reported during the study. Patients treated with IBWM showed progressive wound healing over time: the greatest mean wound reduction was approximately 95% in week 12. Seven of ten patients (70%) achieved complete wound closure by week 12. No recurrent ulcers were reported during follow-up. These results are consistent with the hypothesis that IBWM is easy to use, safe, and effective when used on diabetic foot ulcers in an outpatient clinical setting without the secondary procedure of autografting for closure. [24]

Purpose of retrospective comparative study from Papa et al in 2014 was to evaluate the results of reconstruction of diabetic feet by split thickness skin graft (STSG) and by dermal substitute Integra® covered by STSG in terms of vascularity of the reconstructed wound-bed by measurements of tissue oxygenation (TcPO2). 23 patients were included into the study (12 were reconstructed by STSG only and 11 with Integra® and STSG three weeks later). Wound beds reconstructed by Integra® showed on average 10 mmHg higher TcPO2. The study estimated in an objective way, by TcPO2 value measurements, the oxygenation of the wound bed in diabetic feet after reconstruction by STSG only and after adding dermal substitute Integra® to the wound bed before final STSG coverage. During first month after reconstruction no statistically significant differences were found. After 3 months TcPO2 studies revealed statistically significant higher oxygen tissue pressure in diabetic feet covered by Integra® plus STSG. These findings endorse in an objective way the clinical findings already reported while using the dermal substitute. It remains to explain the role of this increase of oxygen tissue pressure in redefine the indications for the use of dermal substitutes in reconstruction of poor vascularized regions. [25]

3.1. Review of most recent clinical experience with acellular flowable tissue matrix

Retrospective study from Brigido et al. about an acellular flowable dermal replacement scaffold used on lower extremity sinus tract wounds showed successful treatment potencial. Injectable human dermal matrix has been developed for the treatment of complex diabetic sinus tract wounds. In this retrospective series, 12 patients with deep tunneling wounds were treated with GRAFTJACKET Xpress Scaffold and followed for 12 weeks. Complete wound healing was achieved in 10 of 12 patients within the 12-week evaluation. The average time to complete healing was 8.5 weeks, whereas the average time to depth healing was 7.8 weeks. The data from the study suggest that this injectable human dermal matrix has unique properties that allow it to facilitate healing of complex tunneling diabetic foot ulcers. The material was easy to prepare and inject into the wound, thereby preventing the necessity of extensive surgical exposure. The matrix supports neo-subcutaneous tissue formation and allows the body to rapidly repair these wounds. [17]

We treated two patients with lower limb chronic wounds of different etiology: unhealed ulcer off the stump following transmetatarsal amputation of the foot in a patient with diabetes and a patient with foot ulcer after traumatic foot amputation. Both patients were previously treated for two years with modern wound dressings, that were changed according to the appearance of the bottom of the wound and the amount of exudate. Despite optimal choice of dressing ulcers persisted. After appropriate preparation of the wound bed and removal of hyperceratotic edges we applied Integra ™ Flowable Wound Matrix, which was covered with a non-adhesive silicone dressing and polyurethane foam. After seven days, we measured the the perimeter and the surface of the wound. The extent of the wound in the first patient was reduced by 40% and the surface by 61%. In the second patient, the treatment was less effective, with reduction of the perimeter by 7% and surface area by 11%. We hope to offer the new treatment option to a larger number of patients and gather more representable data.

4. Conclusions

This review provides an outlook on some of the advanced wound healing options and summarizes established treatment strategies of using acellular matrix and acellular flowable wound matrix. Due to high rates of morbidity and mortality, chronic ulcers pose a global health problem requiring substantial resources. Conventional methods of treatment are insufficient when dealing with complex and non-healing ulcers. Advanced regenerative methods should be considered in patients with delayed healing of ulcers, that did not reduce in size after conventional standard 4-week treatment. The regenerative treatment should be used additonally to the conventional methods, maintaining the gold standards of appropriate debridement, infection control and moist wound healing.

Author details

Dragica Maja Smrke* and Danijela Semenič

*Address all correspondence to: dsmrke@gmail.com

Department of Surgical Infections, University Clinical Centre Ljubljana, Slovenia

References

[1] Baltzis D, Eleftheriadou I, Veves A. Pathogenesis and Treatment of Impaired Wound Healing in Diabetes Mellitus: New Insights. *Adv Ther*. 2014; 31(8) 817-36

[2] Wolf MT, Daly KA, Brennan-Pierce EP, Johnson SA, Carruthers CA, D'Amore A, Nagarkar SP, Velankar SS, Badylak SF. A hydrogel derived from decellularized dermal extracellular matrix. Biomaterials 2012; 33(29) 7028-38

[3] Zhu J, Marchant RE. Design properties of hydrogel tissue-engineering scaffolds. Expert Rev Med Devices, 2011; 8(5) 607-626

[4] Scott JE. Extracellular matrix, supramolecular organization and shape. J Anat. 1995; 187, 259–269

[5] Rhodes JM, Simons M. The extracellular matrix and blood vessel formation; not just a scaffold. J Cell Mol Med 2007; 11(2) 176-205

[6] Ma PX. Biomimetic materials for tissue engineering. Adv Drug Deliv Rev 2008; 60(2) 184–198

[7] Chen R, Hunt JA. Biomimetic materials processing for tissue-engineering processes. J Mater Chem 2007; 17(38) 3974–3979

[8] Tibbitt MW, Anseth KS. Hydrogels as extracellular matrix mimics for 3D cell culture. Biotechnol Bioeng 2009; 103(4)

[9] Jain RK, Au P, Tam J, Duda DG, Fukumura D. Engineering vascularized tissue. Nat Biotechnol 2005; 23 821–823

[10] Allen P, Melero-Martin J, Bischoff J. Type I collagen, fibrin and Puramatrix matrices provide permissive environments for human endothelial and mesenchymal progenitor cells to form neovascular networks. J Tissue Eng Regen Med. 2011; 5(4) e74–e86

[11] Lin JH, Chang HY, Kao WL, Lin KY, Liao HY, You YW, Kuo YT, Kuo DY, Chu KJ, Chu YH, Shyue JJ. Effect of Surface Potential on Extracellular Matrix Protein Adsorption. Langmuir. 2014; 11

[12] Migliorini E, Thakar D, Sadir R, Pleiner T, Baleux F, Lortat Jacob H, Coche-Guerente L, Richter RP. Well-defined biomimetic surfaces to characterize glycosaminoglycan-mediated interactions on the molecular, supramolecular and cellular levels. Biomaterials. 2014; 35(32), 8903-15

[13] Seif-Naraghi SB, Horn D, Schup-Magoffin PJ, Christman KL. Injectable extracellular matrix derived hydrogel provides a platform for enhanced retention and delivery of a heparin-binding growth factor. Acta Biomater. 2012; 8(10) 3695-703

[14] Veves A, Falanga V, Armstrong DG, Sabolinski ML. Graftskin, a human skin equivalent, is effective in the management of noninfected neuropathic diabetic foot ulcers: a prospective randomized multicenter clinical trial. Diabetes Care. 2001; 24 290-295

[15] Marston WA, Hanft J, Norwood P, Pollak R. The efficacy and safety of Dermagraft in improving the healing of chronic diabetic foot ulcers: results of a prospective randomized trial. Diabetes Care 2003; 26 1701-1705

[16] Brigido SA. The use of an acellular dermal regenerative tissue matrix in the treatment of lower extremity wounds: a prospective 16-week pilot study. Int Wound J. 2006; 3 181-187

[17] Brigido SA, Schwartz E, McCarroll R, Hardin-Young J. Use of an acellular flowable dermal replacement scaffold on lower extremity sinus tract wounds: a retrospective series. Foot Ankle Spec. 2009; 2(2) 67-72

[18] Schmidt J, Redwan B, Koesek V, Heitplatz B, Bedetti B, Aebert H,Wiebe K. Thoracic Wall Reconstruction with Acellular Porcine Dermal Collagen Matrix. Thorac Cardiovasc Surg. 2014; 28

[19] Kavros SJ, Dutra T, Gonzalez-Cruz R, Liden B, Marcus B, McGuire J, Nazario-Guirau L. The Use of PriMatrix, a Fetal Bovine Acellular Dermal Matrix, in Healing Chronic Diabetic Foot Ulcers: A Prospective Multicenter Study. Adv Skin Wound Care. 2014; 27(8) 356-62

[20] Goyal N, Gupta R, Pandit N, Dahiya P. Analysis of patient acceptance following treatment of Miller's class II gingival recession with acellular dermal matrix and connective tissue graft. J Indian Soc Periodontol. 2014; 18(3) 352-6

[21] Wang HL, Romanos GE, Geurs NC, Sullivan A, Suárez-López Del Amo F, Eber RM. Comparison of Two Differently Processed Acellular Dermal Matrix Products for Root Coverage Procedures: A Prospective Randomized Multi-Center Study. J Periodontol. 2014; 26, 1-25

[22] Brigido SA, Boc SF, Lopez RC. Effective management of major lower extremity wounds using an acellular regenerative tissue matrix: a pilot study. Orthopedics. 2004; 27(1) 145-9

[23] Winters CL, Brigido SA, Liden BA, Simmons M, Hartman JF, Wright ML. A multicenter study involving the use of a human acellular dermal regenerative tissue matrix for the treatment of diabetic lower extremity wounds. Adv Skin Wound Care. 2008; 21(8) 375-81

[24] Yao M, Attalla K, Ren Y, French MA, Driver VR. Ease of use, safety, and efficacy of integra bilayer wound matrix in the treatment of diabetic foot ulcers in an outpatient clinical setting: a prospective pilot study. J Am Podiatr Med Assoc. 2013; 103(4) 274-80

[25] Papa G, Spazzapan L, Pangos M, Delpin A, Arnez ZM. Compared to coverage by STSG grafts only reconstruction by the dermal substitute Integra® plus STSG increases TcPO2 values in diabetic feet at 3 and 6 months after reconstruction. G Chir. 2014; 35(5-6) 141-5

Section 3

Tissues

Chapter 9

An Organ Regeneration Platform for Industrial Production of Hollow Neo-Organs

Joydeep Basu

Additional information is available at the end of the chapter

http://dx.doi.org/10.5772/59465

1. Introduction

Organ regeneration technologies aim to restore the original structure and functionality of a diseased organ. In general, healing responses within mammals are characterized by fibrosis and scar tissue formation, not regeneration. Nevertheless, developing mammalian fetuses during the first trimester present wound healing without fibrosis and scar tissue formation (Adzick and Lorenz, 1994). Additionally, compensatory hyperplasia of mammalian kidney or liver secondary to partial nephrectomy or hepatectomy, remodeling of epidermis or bone consequent to injury and regeneration of limb digit tips in humans and mice post-amputation are all examples of regenerative outcomes in adult mammals indicative of an innate regenerative potential within adult mammals (reviewed by Roy and Gatien, 2008).

However, model organisms such as *Hydra*, planaria, zebrafish, *Xenopus* and urodeles (salamanders) present the clearest examples of regenerative outcomes secondary to injury. In these systems, cell-based strategies harnessing pluripotent and tissue specific stem cells as well as dedifferentiation have been leveraged to mediate the regeneration of whole limbs and organs (reviewed by Tanaka and Reddien, 2011). Systematic experimentation with limb regeneration in urodeles has permitted the decipherment of key mechanistic pathways of regeneration at the molecular level. Activation of salient signaling cascades including p53, TGF-β, Delta, ppRB and Wnt/β-catenin have all been associated with limb regeneration (reviewed by Roy and Gatien, 2008). These signaling pathways catalyze a sequence of instructive interactions between mesodermal and ectodermal cell populations that are ultimately responsible for lineage specification (reviewed by Wessels, 1977). In addition, the methodical depletion of macrophages within the first 24 hours subsequent to limb amputation in urodeles has been demonstrated to lead to permanent failure of limb regeneration, extensive fibrosis and disregulation of transcriptional patterns associated with synthesis of extra-cellular matrix (ECM) components (Godwin et al., 2013). This specific sequence of cellular events associated with regeneration of the urodele limb as defined below recapitulates aspects of embryonic organogenesis and may serve as a model system for establishing the existence of similar pathways in mammals:

Stages in regeneration of urodele limb secondary to amputation:

1. The open wound is enclosed by wound epithelium to form a permissive epithelial structure referred to as the apical ectodermal cap (AEC).
2. Up-regulation of matrix metallo-proteinase (MMP) expression catalyzes structural reorganization of the ECM.
3. Dedifferentiation of cell populations takes place proximal to the plane of amputation.
4. Proliferation and migration of dedifferentiated cells is observed under the AEC.
5. Induction of a *blastema*, a mass of mesenchymal cells that will eventually re-differentiate to create the new muscles, bones, nerves and tendons required to regenerate a functional limb.

An understanding of these stages has already been applied to accelerate regenerative outcomes in mammals. For example, application of MMP1 to digit remnants of adult mice with amputation at the mid-second phalanx significantly improved regeneration of soft tissue and observed rates of wound closure. More multi-potent progenitor cells, capillary vasculature and neuromuscular related tissues were also noted (Mu et al., 2013). Furthermore, recent data on regenerative outcomes in mammals from tissue engineering of bladder, esophagus and intestine provides additional evidence of the existence of a regenerative pathway in adult mammals mimicking aspects of that observed in urodeles, including formation of a neo-blastema. This regenerative pathway is characterized by a dependence on adequate vascularization and innervation at the site of regeneration. Importantly, these observations provide insight into a potential mechanism of action for tissue engineered products characterized broadly as instructive signaling between mesenchymal cells of the regenerative construct and host epithelial cell populations. This insight may be harnessed to facilitate development of novel neo-organ products. Here, we review recent progress in regeneration of tubular organs with a particular focus on the gastrointestinal tract that highlights how Tengion's Organ Regeneration Platform™ may harness these fundamental regenerative pathways.

2. Tengion's organ regeneration platform

Tubular organ regeneration involves a specific, temporal sequence of cellular infiltration, vasculogenesis, neurogenesis and the defined differentiation of mucosal, stromal and parenchymal laminar tissue architectures (reviewed by Basu and Ludlow, 2010). Strategies for organ and tissue regeneration must therefore achieve the dual objectives of triggering a true regenerative response while ameliorating any tendency towards fibrotic repair. The methodology first pioneered for regeneration of the bladder may serve as a foundational platform for the regeneration of any tubular organ including penis, vagina, lung, small intestine, stomach and vessels.

Illustrated by Tengion's Neo-Bladder Augment™ (NBA), this approach initially applied committed cell populations of target organ sourced (bladder)-derived urothelial cells (UC) and smooth muscle cells (SMC) complexed with a biodegradable, synthetic scaffold to trigger *de novo* organogenesis (Oberpenning et al., 1999). The NBA is composed of a biodegradable scaffold seeded with autologous, bladder derived UC and SMC. During implantation, the NBA is wrapped with omentum to facilitate vascularization of the developing neo-organ (Figure 1). This is a critical requirement for successful organ regeneration (Basu & Ludlow, 2010; Basu & Ludlow, 2011; Basu & Ludlow, 2012). Analysis of the matured construct at the cellular level highlights committed SMC interacting with polymer fibers of the biodegradable scaffold (Figure 2). Over time (~3 months), a laminarly organized neo-organ develops. Variations of this approach have recently been described, and the key methodological elements arising from these principal recent reports demonstrating *in vivo* regeneration of neo-bladder or neo-bladder related tissue has been reviewed in detail (Basu & Ludlow, 2010; Basu & Ludlow, 2011; Basu & Ludlow, 2012).

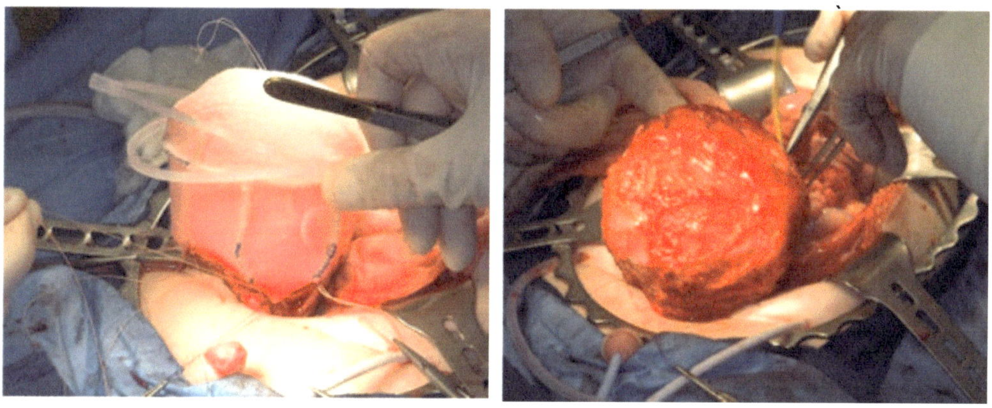

Figure 1. Tengion Neo-Bladder Augment (NBA) during implantation (left) and subsequent to wrapping with omentum (right).

Examination of the temporal sequence of neo-bladder regeneration in canine clinical studies illustrates the dichotomy in outcomes between implantation of acellular and cellularized scaffolds (Jayo et al., 2008a,b). This distinction is a principal theme of most studies describing the regeneration of tubular organs. Cell seeded scaffolds mediated a regenerative response within one month post-implantation, characterized by induction of heavily vascularized, smooth muscle-like parenchyma. In contrast, acellular scaffolds triggered a principally fibrotic, reparative outcome characterized by randomly organized collagen fibers with minimal vascularization. Baseline urodynamics were reconstituted within four months of implantation with cell seeded scaffolds, whereas the urodynamic profile of animals implanted with acellular scaffolds remained abnormal throughout the trial period (Jayo et al., 2008a). In a related dog

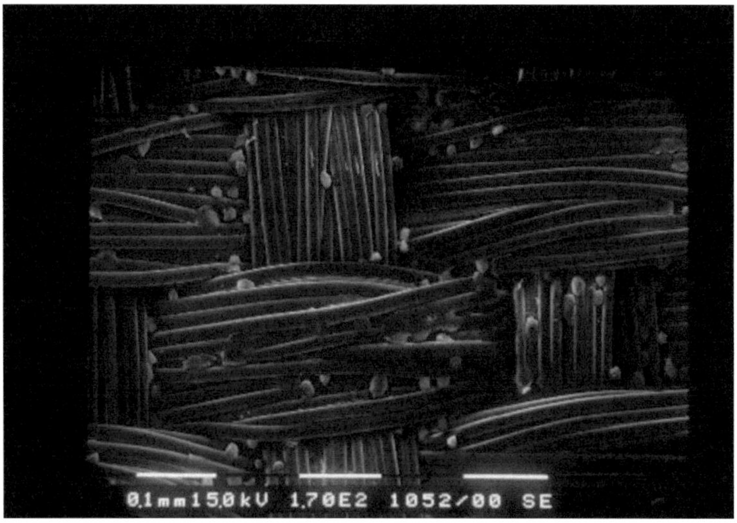

Figure 2. SMC interacting with woven mesh of PGA (poly-glycolic acid), a biodegradable polymer. 170X

study, restoration of tri-laminar bladder wall architecture occurred within three months post-implantation and normal compliance characteristics of a urinary bladder wall developed by 12 months (Jayo et al., 2008b). Regenerated bladders were functionally and structurally stable for up to two years post-implantation (Jayo et al., 2008b). Importantly, although the construct volume was constant at implantation within variably sized dogs, the ratio of the regenerated bladder's volume to host body weight adapted to the recipient animal's size, demonstrating that the neo-organ responds to homeostatic mechanisms regulating organ volume (Jayo et al., 2008b).

Although these canine studies utilized both bladder-derived urothelial and smooth muscle cells, urothelial cells have been shown to *not* be essential for bladder regeneration, thereby greatly facilitating process development and commercialization (Basu & Ludlow, 2010). Neo-urinary constructs seeded only with smooth muscle cells and not urothelial cells nevertheless catalyze complete regeneration of *de novo* urinary-like tissue with patent urothelium, presumably through migration of epithelial cells in *trans* via urine flow or from neighboring tissue (Basu et al., 2012a). However, use of bladder-derived smooth muscle cells is problematic in patients presenting with bladder related malignancies. Therefore, a number of alternate sources of smooth muscle cells have been investigated. Such alternate cell sources may have broad application in the regeneration of additional, laminarly organized tubular organs. Tengion has identified adipose and peripheral blood as alternate sources of smooth muscle cells usable for induction of neo-urinary tissue (Basu et al., 2011a; Basu et al., 2012a). As with other early signaling events during organogenesis or regeneration, instructive interactions between this mesenchymal cell population associated with the regenerative construct and epithelial cell populations within the host are understood to be critical for observed regenerative outcomes.

http://dx.doi.org/10.5772/59465

3. The need for urinary neo-organs

At its most fundamental level, the bladder is responsible for mediating storage and subsequent efflux of urine within a dynamically expandable and contractable container. Though relatively simple in terms of overall histology and structural organization (Figure 3), diseases impacting the bladder have the potential to significantly affect individual quality of life, resulting in continual incontinence or inability to effectively void urine as needed. Several congenital anomalies may result in abnormal bladder development requiring surgical intervention, including posterior urethral valves, bilateral ectopic ureters, bladder extrophy, cloacal extrophy, and spina bifida (myelomeningocele). The resultant clinical outcomes include incontinence and increased risk of renal failure from high pressures in the genito-urinary system.

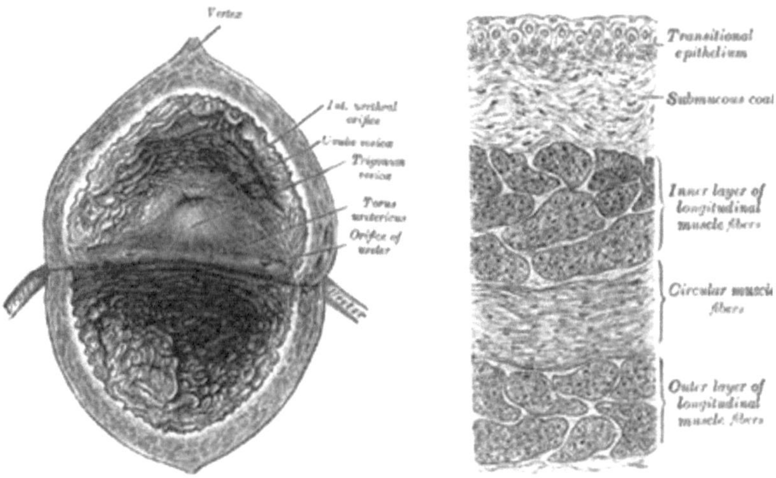

Figure 3. Macroscopic (left) and histological (right) organization of the bladder, illustrating tri-laminar cellular architecture of this hollow, muscular organ. Reproduced from Wikipedia Commons (Open Access) under terms of Open Access agreement.

The current standard of therapy for pediatric patients is bladder augmentation through enterocystoplasty, a procedure which involves the surgical removal of a section of large bowel that is then connected to the existing bladder to increase compliance, decrease pressure, and improve overall urine capacity. These surgeries are relatively complex and costly. Even in patients with good technical outcomes, the procedure is associated with numerous immediate risks and potential chronic complications. A similar surgical procedure is performed in adults requiring bladder replacement, typically secondary to the onset of bladder related malignancies. To this end, cancer of the bladder may be manifested as a broad spectrum of disease presenting across distinct bladder compartments.

The most common types of bladder cancers originate from the internal epithelial lining of the bladder, the urothelium. Less common cancers of the bladder may include squamous cell carcinoma, adenocarcinoma, sarcoma, and small cell carcinoma. Estimates predict that almost 71,000 new cases of bladder cancer will be diagnosed annually in the USA, with some 14,000 patients succumbing to the disease. Although many patients will have superficial urothelial tumors involving little or no metastasis within the smooth muscle layer, approximately 25% of bladder cancers will invade the detrusor musculature, with the majority of these presenting initially as invasive cancers with metastatic potential.

Invasive cancers often require multi-modality therapy which may involve extirpative surgery, i.e. radical cystectomy and construction of a urinary diversion. Despite the risk of adverse effects, there are approximately 10,000 of these procedures performed per year in the United States, including in 10% of children with congenital bladder abnormalities and 90% of adults with acquired disorders such as bladder cancer. In some cases of severe bladder cancer or other pelvic or abdominal cancers, removal of the entire bladder is indicated. In these circumstances, current standard of care also involves reconstruction of a bladder-like replacement using bowel tissue. Application of bowel tissue for reconstruction of urinary neo-organs is clearly problematic for a number of fundamental reasons, including the fact that bowel is a principally absorptive organ, whereas bladder is designed to store and excrete urine. Exposure of bowel tissue to urine or bladder tissue to bowel-derived micro-organisms has the potential to trigger multiple secondary complications. These may include any of the following:

- **Bowel complications**. Early complications are usually related to the bowel surgery required to harvest tissue for reconstructive use and typically consist of leaks, fistulas and obstructions. Because vitamin B12 is absorbed in the bowel tissue, loss of this tissue can result in anemia and neurologic abnormalities. Additionally, mal-absorption of salts and lipids can lead to diarrhea. Patients with neurogenic bladder are prone to bowel movement problems even before surgery and the removal of bowel tissue may either exacerbate existing conditions or create new motility problems. These difficulties further contribute to substantial physical and psychological morbidity within these patients.

- **Absorption issues**. Use of bowel tissue often leads to electrolyte and metabolic imbalances, which can cause bone loss. Furthermore, certain drugs taken by the patient may be reabsorbed by the implanted bowel tissue, potentially leading to unintended toxic levels of these substances within the bloodstream. The exposure of intestinal surface to urine also results in the inappropriate absorption of ammonium, chloride and hydrogen ions as well as potassium loss, leading to chronic metabolic imbalances or abnormalities.

- **Infection**. Persistent and recurrent infections are typical in patients with bowel tissue reconstruction. Such infections may reach the kidney and become life-threatening. Additionally, bacteria normally found in bowel tissue can serve as a source of infection and septic complications when repositioned into the urinary tract.

- **Stone formation**. One of the consequences of persistent infection is the development of stones. Stones are hard masses which can cause pain, bleeding, obstruction of urine or infections.

- **Mucus**. Bowel tissue, when repositioned in the urinary tract, continues to secrete mucus. Mucus increases the risk of stone formation and the viscosity of urine, and in the case of bladder augments, may require bladder irrigation and more frequent catheterization.
- **Cancer**. Malignancy, although rare, is a well-recognized complication following bladder augmentation using bowel tissue (enterocystoplasty), as well as from other reconstructive surgeries that incorporate bowel segments into the genitourinary tract.

In addition to cancers and developmental abnormalities, patients may present with neurogenic bladder or dysfunctional bladder due to some form of neurologic disease or condition. Treatment may often require an augmentation of the bladder in order to relieve high pressures and incontinence. Current therapies for neurogenic bladder include medical management through a combination of medication and clean intermittent catheterization and, in advanced cases, surgery. Surgical procedures, such as bladder augmentation, are often considered when other medical and less-invasive treatments fail to adequately lower bladder pressure or reduce the frequency of incontinent episodes.

Ultimately, it is self-evident that the ideal unit of anastamosis for urinary-like tissue is other urinary-like tissue. However, the lack of such material has generally precluded the widespread leveraging of this option. There is therefore clearly a compelling medical need for an improved approach that would eliminate or at least substantially reduce the complications potentially associated with the current standard of care. To this end, identification of bladder-like materials that may be applied towards bladder reconstruction in place of bowel tissue has been attempted. Pilot experiments in 1917 to augment bladder in dogs leveraged fascia (Neuhof, 1917); since then, numerous other biomaterials candidates have been evaluated, including skin, bladder sub-mucosa, small intestinal sub-mucosa, omentum, dura and peritoneum. Synthetic materials candidates have included polyvinyl sponge, Teflon, gelatin, collagen, vicryl and silicone. Failure to achieve successful outcomes with such biomaterials types may be attributed to physical or mechanical failure, lack of biocompatibility and the induction of fibrosis and scarring leading to contraction of the implant and reduction in effective volume over time. Unfortunately, of these candidates the best biomaterial mimicking the physical and mechanical properties of bladder tissue has been bowel. There is therefore a clear and present need for additional, novel technology platforms.

3.1. TE/RM methodologies for bladder replacement and augmentation

Tissue engineering and regenerative medicine approaches offer an alternative, potentially superior methodology to the use of bowel tissue for urinary diversion or replacement. In this methodology, the patient's own cells would be sourced from a bladder biopsy (or other, alternate cell source, see below) and applied to an appropriate, degradable biomaterial scaffold to create a neo-organ or organ-like construct that, upon implantation within the patient and anastamosis to native components of the urinary system, would lead to regeneration of functional, urinary-like neo-tissue capable of storing urine and mediating voiding of urine as needed in response to appropriate neuronal signaling. Such a cell/biomaterial construct would catalyze the regeneration of urinary-like neo-tissue recapitulating native, laminarly organized bladder wall histo-architecture composed of a luminal urothelial layer and multiple smooth muscle layers, appropriately vascularized and innervated (Figure 3). Regeneration of urinary-like neo-tissue is accompanied by progressive degradation of the biomaterial, such that a seamless transition is achieved between the degrading biomaterial and the regenerating urinary-like neo-tissue.

In preliminary experiments to demonstrate the formation of tissue engineered urothelial-like structures in the rabbit, bladder-derived urothelial cells were used to seed meshes of non-woven polyglycolic acid (PGA), which were subsequently implanted within the peritoneal cavity of athymic mice. Upon recovery, structures composed of degrading biopolymer lined with urothelial cells were observed (Atala et al., 1992). In follow-up studies, combinations of bladder-derived smooth muscle cells and urothelial cells were used to seed tubular-like structures composed of non-woven PGA mesh. Implantation of these constructs subcutaneously within athymic rabbits led to regeneration of urinary-like tubular organoids composed of urothelial cells lining a central lumen and surrounded by layers of smooth muscle cells, as observed within native bladder tissue. Evidence of neo-vascularization was also noted (Atala, et al., 1993). These studies provided preliminary proof of concept to support the potential for *in vivo* regeneration of urinary-like neo-organs through implantation of cell seeded, synthetic bio-polymeric scaffolds.

Although current strategies for creation of bladder-like neo-organs leverage principally non-bladder cell sources, initial work on neo-bladder tissue engineering was dependent on patient-derived bladder biopsies as a source of urothelial and smooth muscle cells. For this approach to be commercially viable, the expansion dynamics of cellular growth for both biopsy-derived urothelial and smooth muscle cell populations must be established. Although smooth muscle cells could be reliably expanded from small bladder biopsies without difficulty, the demonstration that a single biopsy-derived source of bladder urothelial cells could also be expanded to the numbers required for effective seeding of urinary neo-organs was critical for establishing the preliminary bioprocess potential of this methodology (Cilento, et al., 1994). The alternative would involve multiple biopsy sampling to generate sufficient cell numbers for urinary neo-organ seeding, greatly decreasing the attractiveness of this technology for practical application in the clinic.

From a biomaterials perspective, the application of synthetic biopolymers such as PGA for the seeding of urothelial and bladder-derived smooth muscle cells permitted development of modified polymers with continuously tunable physical and mechanical characteristics. To this end, the temporal sequence of polymer hydrolysis may be manipulated by altering the nature and sequence of individual monomer units. In addition, coating by other polymers such as PLGA may be applied to further fine-tune the physical properties of the biomaterial scaffold. Finally, the open, fibrous networked structure of the biomaterial (Figure 4) facilitates angiogenesis and neo-vascularization of the developing neo-organ. Taken together, this binary cell/synthetic biopolymer construct was the foundational technology platform needed for initiating large animal clinical trials and proof-of-concept trials in man.

http://dx.doi.org/10.5772/59465

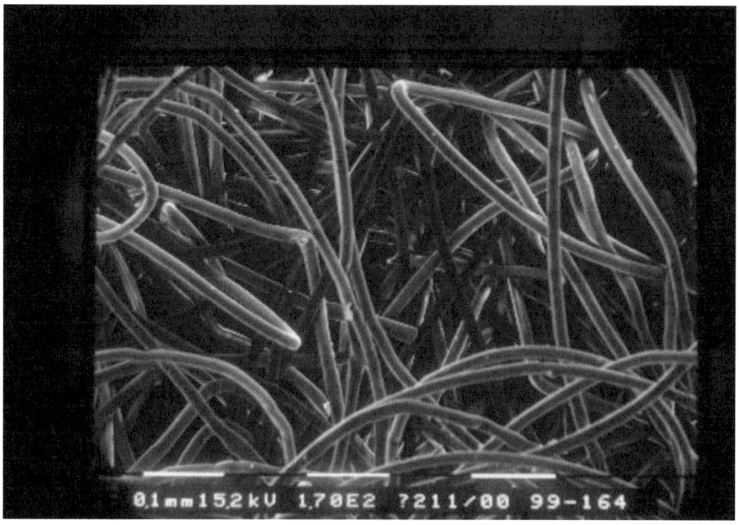

Figure 4. Biodegradable PGA felt, with open "spaghetti"-like structure. 170X

3.2. Demonstration of Neo-Bladder formation in large animals

Subsequent to initial proof-of-concept studies showing *de novo* regeneration of urinary-like structures in a small animal model (rabbit), the next step in successful process development is evaluation of the technology in a large animal clinical model such as swine or dog. For the Neo-Bladder Replacement, a key proof-of concept study was provided using a canine total cystectomy (bladder removal) model. 14 beagles were sub-divided into three groups: A (*n*=2), underwent cystectomy without further intervention. B (*n*=6) underwent cystectomy followed by implantation with acellular, bladder-shaped biopolymer scaffold and group C (*n*=6) was subjected to cystectomy and implantation with cell-seeded biopolymeric scaffold. To create the constructs, PGA/PLGA biomaterial scaffolds (Figure 1) were shaped by hand into bladder-shaped structures. Urothelial and smooth muscle cells were sourced from bladder biopsies and expanded separately to numbers sufficiently meaningful to warrant seeding onto the scaffolds (10^6 cells per cm^2 of polymer surface). Urothelial cells were used to seed the interior, luminal surface of the construct, and smooth muscle cells were applied to the exterior, non-luminal surface.

Upon maturation, these constructs were implanted within group C animals. Implanted neo-bladder constructs were wrapped extensively with omentum during surgery, as omentum is a well-established source of pro-angiogenic growth factors (Litbarg, et al., 2007). All animals were monitored over a period of up to 11 months post-implantation and extensive urodynamic and radiographic measurements taken to document urologic functionality. At the completion of the study period, animals were sacrificed and bladder-like structures examined histologically for evidence of tri-laminar bladder wall architecture from regenerated neo-organs. Structural and functional differences in regenerative outcomes were observed between all three groups of canines. Group A dogs regained only minimal urinary reservoir volumes without approaching pre-cystectomy parameters. Group B dogs presented principally fibrotic neo-bladders with a regenerated luminal urothelial cell layer but minimal regeneration of bladder wall musculature. Urological dynamics were also significantly affected. This ability of acellular constructs to regenerate a normal luminal urothelial cell layer is noteworthy, providing a preliminary indication that prior seeding of constructs with urothelial cells may not in fact be a requirement to achieve full regeneration of native-like urinary neo-tissue from the implanted neo-organ construct. We will re-visit this observation and its implications later in this chapter. However, canines implanted with cell seeded, bio-polymeric neo-bladder constructs were able to regenerate histologically and functionally appropriate neo-bladders with native-like tri-laminar wall histo-architecture and native-like urodynamics (Oberpenning, et al., 1999).

It may also be useful at this point to clearly define functional outcomes that are commonly referred to as "regenerative" versus "reparative" in the context of TE/RM. Organ regeneration involves the replacement and restoration of cellular components and fully developed tissue mass as well as the reconstruction of the organizational, architectural and functional characteristics associated with the particular organ. In this regard, outcomes observed with group C canines above are principally regenerative. In contrast, a reparative outcome is associated with incomplete tissue replacement and may be distinguished by the extensive deposition of collagen and concomitant fibrosis. Outcomes observed with group B canines represent principally reparative pathways. This study and many others demonstrate clearly that regenerative outcomes are mediated by cell seeded constructs only. Acellular constructs are unable to catalyze neo-organ regeneration. This is an important foundational principle for continued development of neo-organ regeneration technology.

Analysis of the dynamics of neo-bladder regeneration in subtotal cystectomized canines serves to further illustrate the dichotomy in outcomes between implantation of acellular and cellularized scaffolds. In another such study, bladder shaped scaffolds composed of woven PGA felt or PLGA (poly-lactic-*co*-glycolic acid) seeded with autologously sourced bladder-derived urothelial cells and smooth muscle cells facilitated a regenerative response within one month post-implantation, as characterized by induction of an extensively vascularized, smooth muscle-like parenchyma. In contrast, acellular PGA/PLGA scaffolds triggered a principally fibrotic, reparative outcome featuring disorganized collagen fibers with minimal vascularization. Baseline urodynamics were reconstituted within four months post-implantation with cell-seeded scaffold, whereas urodynamic profiles of animals implanted with acellular scaffolds remained abnormal throughout the nine-month study (Jayo, et al., 2008).

In a related cystectomized canine study, native-like tri-laminar bladder wall tissue architecture was observed at three months post-implantation with a bladder shaped non-woven PGA felt scaffold seeded with $1.5X10^8$ each of autologously sourced bladder-derived urothelial cells and smooth muscle cells, and normal compliance characteristics of a urinary bladder had developed by 12 months. Regenerated bladders in animals receiving these cell-seeded scaffolds have shown functional and structural stability for up to two years post-implantation. Importantly, although the volume of the cell-seeded scaffold was held constant in this particular study, implantation of the construct within dogs of different sizes that had gained varying amounts

http://dx.doi.org/10.5772/59465

of weight over the course of the study yielded organs that, as measured by the ratio of bladder capacity to body weight, adapted to the individual recipient animal's size, demonstrating that the regenerated neo-organ was capable of responding to homeostatic mechanisms regulating organ volume (Jayo, et al., 2008).

3.3. Neo-bladder replacement in human pediatric patients-1st clinical trials of a neo-organ

Studies in canines as outlined above established proof-of-concept for the application of cell seeded biodegradable polymeric scaffolds for regeneration of functional, native-like neo-bladders in a large animal cystectomy model (Oberpenning, et al., 1999). Additional data suggested that smooth muscle cells sourced from diseased bladder could potentially be applied successfully to regenerate neo-urinary tissue (Lai, et al., 2002). These data laid the groundwork for initiation of a proof of concept clinical trial in man. In this seminal study, seven pediatric patients presenting with myelomeningocele (a form of spina bifida) were recruited to receive the first ever human neo-organ implants. As previously described in canines, both urothelial and smooth muscle cells were isolated and expanded from autologously sourced bladder biopsies. Up to 5 cell passages over 7-8 weeks was required to generate enough cells to seed the neo-bladder scaffold. Using a sterile pipette, the scaffold exterior was seeded with bladder-derived smooth muscle cells at a seeding density of $5X10^7$ cells/cm^3. After a 48hr recovery period, the luminal surface of the scaffold was seeded with urothelial cells at a density of $5X10^7$ cells/cm^3. The construct was matured in a tissue culture incubator at 37 degrees C for 3-4 days, prior to implantation. Subsequent to implantation, the engineered neo-bladder was cycled (i.e. subjected to serial volume expansion and contraction) as part of regular post-operative care for up to 3 weeks post-implantation; the mechanical forces induced across the neo-bladder during cycling have been found to augment regenerative outcomes. Engineered neo-bladders were found to functionally rescue urologic dynamics and were associated with tri-laminar bladder wall architectures upon histological examination of bladder biopsies recovered at 31 months post-implantation. A number of different scaffold iterations were evaluated, with changes being made over the course of the study to accommodate new data being made available from this and other related studies. Ultimately, an omentum wrapped, collagen/PGA scaffold was found to present best overall regenerative outcomes (Atala, et al., 2006).

3.4. Definition of the cell source used for seeding Neo-Bladders: Adipose

Adipose tissue contains a heterogenous mix of cell types, including adipocytes, endothelial cells, pericytes, SMC, and MSC. Adherent cells that were derived from adipose tissue using different conditions have been referred to as MSC without applying a systematic approach to defining the isolated cell composition; however, distinctly different cell types can share phenotypic characteristics. As more cell-based therapies move from preclinical to clinical evaluation, establishing a cell composition profile that provides a degree of distinction from other cell types will likely become increasingly important for protecting intellectual property rights, gaining regulatory approval, and scaling up for manufacturing. The relevance of cell type characteristics defined from analysis of native tissue or initial cell isolates to expanded cell populations is unclear because gene and protein expression patterns can be altered by isolation and *in vitro* expansion. Such differences have been documented for MSC and we have observed the same to be true for adipose-sourced smooth muscle-like cells (Ad-SMC).

The more important questions from a product development perspective are what characteristics are associated with the cell population that will be administered and how indicative are those characteristics with regard to cell identity and product potency. Our comparison of Ad-SMC and MSC leveraged established definitions of SMC and MSC and focused on the analysis of smooth muscle phenotype, growth kinetics, *in vitro* tri-lineage differentiation bioactivity and functional responses to small molecules that affect SMC-specific signaling pathways. Unlike methodologies for MSC, isolation and expansion of Ad-SMC is specifically promoted by growth of adipose SVF derived cells at low cell densities in the absence of positive/negative selection through magnetic bead based separation, inductive cytokines or growth factors, high serum (>10%) concentrations or preselected lots of serum. The specific influences of media formulations on isolation and expansion of cells with MSC-like bioactivity has been extensively documented (Basu, et al., 2011). Expansion of Ad-SMC is therefore the default outcome from culture of SVF-derived cell populations in principally basal media formulations at high cell densities not supplemented by inductive cytokines, preselected lots of serum or high concentrations of FBS. This property greatly facilitates process development and manufacturing.

From where do Ad-SMCs originate and what is their relationship to MSC? Adipose is a heavily vascularized tissue and a number of studies have implicated the perivascular niche as a potential source of both MSC as well as smooth muscle and endothelial cells. Pericytes with MSC differentiation potential have been isolated directly from blood vessels as well as from multiple organ systems throughout the body. However, although ACTA2/SMAA+cells have been localized to all capillaries, arterioles and venules of the adipose-derived vascular bed, expression of STRO-1, a key MSC-specific marker, is tightly associated with endothelium and additionally found only within a subset of blood vessels. Furthermore, expression of the stem cell-specific markers Oct4 and telomerase was observed only rarely, suggesting that truly pluripotent progenitors are uncommon within adipose. In their entirety, these observations point to MSC, endothelium, and smooth muscle occupying distinct spaces within the broader peri-vascular niche. Nevertheless, there remains the potential for considerable ebb-and-flow across developmental lineages. For example, endothelial cells appear capable of lineage switching towards a smooth muscle cell like phenotype in response to TGF-β or the depletion of pro-angiogenic factors and loss of endothelial cell-cell contact. In addition, adherent cell types with endothelial and smooth muscle phenotypes as well as limited mesenchymal differentiation potential have been identified to circulate in adult peripheral blood. Such circulating smooth muscle cells may contribute to the population of adipose-derived smooth muscle cells, although we have been unable to purify them directly from human adult peripheral blood in meaningful numbers (our unpublished observations).

Given that MSC in long term culture also follow a smooth muscle cell-like differentiation pathway, we believe that taken together, the published data as well as our observations are consistent with the peri-vascular niche of adipose SVF as a source for a broad continuum of smooth muscle cells, smooth muscle progenitors, MSC and partially lineage committed MSC-like cell types with variable and overlapping degrees of proliferative and differentiation potential. Modulation of SVF derived cell phenotypes and functionality away from MSC and towards SMC is possible through selection of media formulations during cell isolation and expansion. Our studies provide methodological validation for the utility of non-bladder sources of smooth muscle-like cells for applications in urologic regenerative medicine, thereby bypassing the potential for isolation and expansion of transformed SMC associated with bladder biopsies derived from patients presenting with bladder-related malignancies.

3.5. Other approaches to tissue engineering Neo-Bladders

The sequential process development of the smooth muscle cell seeded/biodegradable synthetic biopolymer scaffold neo-bladder regenerative platform as outlined above represents by far and away the most clinically advanced such neo-organ product candidate developed to date. No other methodology has even come close to clinical trials in humans or to commercial manufacture. However, it is still worthwhile at this point to critically examine some alternative approaches that have been presented in the literature and discuss why these have proven to be ultimately unsuitable for commercialization.

From a biomaterials perspective, we have already discussed why the application of synthetic, biodegradable polymers is preferable for industry. Briefly, such biomaterials may be sourced or manufactured with defined chemical composition and offer reliable and reproducible physical and mechanical properties that are continuously tunable by modification of basic polymer chemistries. In contrast, naturally sourced biomaterials such as bladder-derived sub-mucosa present variable chemical compositions that are donor specific, and may not be readily sourced, as is in fact the case with human bladder-derived sub-mucosa.

This has led to the use of xenogeneically sourced biomaterials as substitute. Notwithstanding the fact that porcine derived intestinal sub-mucosa is an established product for multiple surgical applications, scaffolds made of decellularized bladder sub-mucosa were actually evaluated in the first human neo-bladder clinical trials head to head against scaffolds composed of principally synthetic polymers (Atala, et al., 2006). The most favorable regenerative outcomes were generated from implanted neo-bladders composed principally of PGA, whereas the least favorable regenerative outcomes were observed within patients implanted with neo-bladders constructs composed of decellularized bladder sub-mucosa. The course of the clinical trial was appropriately modified to incorporate these findings. Taken together, these observations reinforce the application of synthetic polymers over natural sourced biomaterials for commercial viability.

From a cell sourcing perspective, a number of investigators are actively exploring the utility of smooth muscle cells derived by the directed differentiation of MSC isolated from bone marrow or adipose. A key characteristic of such stem and progenitor cell populations is a requirement for exposure to combinations of exogenous growth factors, extracellular matrix components, or other defined factors to drive differentiation along defined developmental lineages. For example, in one recent report, adipose-derived MSC was differentiated into smooth muscle-like cells using inductive media containing 100U/ml heparin for up to 6 weeks prior to seeding polymeric bladder dome-like scaffold structures that demonstrated some evidence of functionality in a rat cystectomy model (Jack, et al., 2009). In additional, related studies, TGF-β or small molecule agonists targeting the TGF-β signaling pathway (e.g., sphingosyl-phosphorylcholine, bradykinin and angiotensin II) have also been used to induce a smooth muscle-like phenotype from adipose or bone marrow derived MSC (Gong, et al., 2008; Kim, et al., 2008a; Jeon, et al., 2006; Kim, et al., 2008b). In a less targeted approach, epigenomic reprogramming with the DNA demethylating agent 5-azaC, has been used to direct bone marrow-derived MSC towards a cardiomyocyte-like phenotype (Xu,, et al., 2004). Dedifferentiated adipocytes may also be driven along a smooth muscle lineage using TGF-β and have been reported to contribute towards bladder tissue regeneration in a mouse bladder injury model (Sakuma, et al., 2009). Finally, methods for TGF-β induced differentiation of smooth muscle cells from bone-marrow derived cells have been described (Kanematsu, et al., 2005; Becker, et al., 2008). These studies invariably involve demonstrations of *in vitro* directed differentiation of SMC-like cells derived from MSC, using various combinations of growth factors and cytokines over an extended time period. Any *in vivo* data is typically based in small animals, is often subcutaneous and therefore of little relevance to product development and manufacturing.

Finally, the use of the peritoneal cavity as a living bioreactor for maturation of tubular organs has been proposed. It is well established that implantation of foreign material into the peritoneal cavity may frequently trigger a fibrotic response, with encapsulation of the foreign matter by multiple layers of fibroblasts. We have observed similar, non-regenerative outcomes associated with implantation of certain synthetic biopolymers such as poly-caprolactate within the renal parenchyma (Basu et al., 2012). However, this normally unwelcome outcome might be leveraged for induction of tubular neo-tissue formation.

In the first such demonstration of this approach, silastic tubing was implanted into the peritoneal cavity of rats/rabbits. At 2 weeks post-implantation, the tube was observed to be coated with several layers of myofibroblasts, collagen and a layer of mesothelium. Removal of the silastic tube followed by eversion of the cellular construct resulted in formation of a vessel-like structure, with laminar wall architecture mimicking that of native vasculature. Anastamosis of this neo-vessel with the vasculature of the original host animal demonstrated evidence of function and patency for up to 4 months post-implantation (Campbell, et al., 1999).

Similarly, neo-organs resembling bladder, uterus and vas deferens could be engineered by grafting biomaterials templates of appropriate shape within the peritoneal cavity of rabbits. After 2-3 weeks of implantation within the environment of this “living bio-reactor”, neo-organ like frameworks composed of myo-fibroblastic layers could be harvested and grafted onto the appropriate organs of the original host to demonstrate functionality and eventual regeneration of urinary-like tissue (Campbell, et al., 2008). Although interesting, it is difficult to envision how such a strategy can ever be feasible in practice. Clearly, undergoing major surgery and all associated post-operative care and monitoring simply to create the neo-organ implant will not be an attractive option, given the alternatives currently under development. Therefore, we do not anticipate that tissue engineering of bladder using the peritoneal cavity as a living bioreactor will become a commercially viable strategy.

http://dx.doi.org/10.5772/59465

3.6. Key learnings from development of Neo-Bladder: Factors facilitating commercial viability of organ regeneration platform

The Neo-Bladder experience at Tengion highlights certain key factors that may impact the commercial success of an organ regeneration platform. These same principles are pertinent to discussion of non-bladder tubular organs as well as to regenerative technologies targeting solid organs such as kidney, liver and pancreas. We outline these principles below:

1. A biodegradable, synthetic scaffold based on well-characterized biomaterials (e.g., PGA, PLGA) that can be manufactured reproducibly with defined characteristics is desirable.

2. A population of committed cell types (e.g., smooth muscle cells) that is easily isolatable and expandable is required for scaffold seeding to trigger regeneration *in vivo*.

3. A purified population of defined stem cells in neither needed nor desirable. Cost of goods increases significantly with requirements to monitor stem cell potential and to control directed differentiation.

4. Engineering of an entire organ *in vitro* or within the peritoneal cavity is neither needed nor desirable. Instead, an *in vitro* cell seeded scaffold is adequate to trigger the innate regenerative response in the mammalian body, resulting in *de novo* organogenesis *in vivo*, including in humans.

4. Neo-urinary conduit: Introduction

The possibility of creating a TE/RM product facilitating urinary diversion without leveraging native GI tissue was the rationale underlying the development of the Neo-Urinary Conduit (NUC). In its simplest configuration, the NUC is a cell/scaffold construct that, upon implantation within the body and attachment to native ureters, mediates efflux of urine from the kidneys directly to the external surface of the body. As with the Neo-Bladder Replacement and related TE/RM products, the NUC construct serves as a template to catalyze the regeneration of native-like urinary tissue concomitant to degradation of the biomaterial scaffold following implantation.

4.1. GLP preclinical analysis of de novo neo-urinary conduit formation in a porcine cystectomy model

32 Gottingen swine with total cystectomy and incontinent ureterostomy (8 animals per data point composed of 4 males and 4 females each) were used in a GLP preclinical analysis to determine the safety and functionality of tissue-engineered NUC constructs seeded with autologous smooth muscle cells derived from the bladder, blood or adipose tissue. Of the 32 animals, the first group (4 males, 4 females) was implanted with NUC seeded with bladder-derived smooth muscle cells. A second group was implanted with NUC scaffold seeded with adipose-derived SMCs, a third group was implanted with a NUC scaffold seeded with blood-derived SMCs, and the 4th group was implanted with unseeded NUC scaffold only. Device

effect and performance was monitored through ultrasound imaging, pyelogram, as well as urine and blood analysis at different time-points of the study. At the completion of the recovery period (Day 84+/-5), all animals were euthanized and a necropsy performed for harvesting the kidneys, conduits, and associated organs and tissues for histological preparation and pathological examination.

4.2. Alternate cell sourcing of SMC for seeding of NUC

Tissue engineering principles have been successfully applied to provide implantable cell-seeded matrices for use in the reconstruction, repair, augmentation or replacement of laminarily organized luminal organs and tissue structures, such as a bladder or a bladder segment or component, typically composed of urothelial and smooth muscle layers. Smooth muscle cells may be derived from the patient's own tissue, including the bladder, urethra, ureter and other urogenital tissue. However, there are challenges associated with dependence upon the development and maintenance of cell culture systems from the primary organ site as the basic unit for developing new and healthy engineered tissues, as for example during treatment of cancerous bladder tissue. Clearly, such cancerous cells are inappropriate for populating an implantable neo-bladder scaffold or matrix. We have therefore attempted to identify and characterize alternative sources of smooth muscle cells capable of reconstituting urologic characteristics de novo upon implantation on appropriate synthetic, biopolymer scaffold constructs, and to demonstrate functionality *in vivo.*

Numerous studies have indicated that adherent, fibroblast-like cells with typical smooth muscle cell characteristics may be recovered from the mononuclear fraction of peripheral blood, cord blood or bone marrow. These smooth muscle-like cells may be pericytes, which are recruited to developing arterioles, capillaries and venuoles. Additionally, it is well established that adipose tissue is a readily available source of adherent cell types and may therefore also represent an alternative source of smooth muscle cells useful for urologic application. Indeed, adipose isolated during abdominoplasty precedures is rich with capillaries, providing a potential source of isolatable smooth muscle-like pericytes. Although smooth muscle cells have also been isolated from other tissue sources such as skeletal muscle and omentum, we chose to focus on recovery of smooth muscle cells from peripheral blood and adipose, as these represent the source tissue with maximum potential clinical utility. A porcine cystectomy model was selected to evaluate the performance of peripheral blood and adipose-derived smooth muscle cells relative to bladder-derived smooth muscle cells upon application in a cell/scaffold composite.

Direct plating of the peripheral blood-derived mononuclear fraction from swine resulted in outgrowth of colonies with typical smooth muscle cell morphology. 100% of animals screened (n=24) generated smooth muscle cell colonies, with 2.44 X10^3-2.37 X10^6 smooth muscle cells recovered at passage zero from 50mls of peripheral blood. Recovery of smooth muscle cells was unaffected by changes in media formulation, cell density or surface coatings (data not shown). A similar approach was used to investigate the potential application of subcutaneous or lipoaspirate-derived adipose as a source of smooth muscle cells. The stromal-vascular fraction (SVF) of adipose represents a heterogenous population of cells including endothelial

http://dx.doi.org/10.5772/59465

cells, smooth muscle cells as well as progenitor cells with limited mesenchymal potential. We were able to generate colonies (expandable into monolayers) of smooth muscle cells from porcine adipose with 100% efficiency (*n*=24), with a cell recovery rate of 1.37 $X10^5$-4.36 $X10^5$ cells/g adipose tissue. In comparison, smooth muscle cells could be isolated from bladder tissue with a recovery rate of 1.29 $X10^6$-9.3 $X10^6$ cells/g bladder tissue. Expansion of smooth muscle cell colonies from peripheral blood or adipose resulted in the formation of a cell monolayer with a typical whirled, "hill-and-valley" organization characteristic of cultured bladder-derived smooth muscle cells (Figure 5). Enrichment of smooth muscle cells was facilitated by use of high cell densities and high glucose media, which has been shown to specifically select against the growth and expansion of mesenchymal stem cells (Basu, et al., 2011).

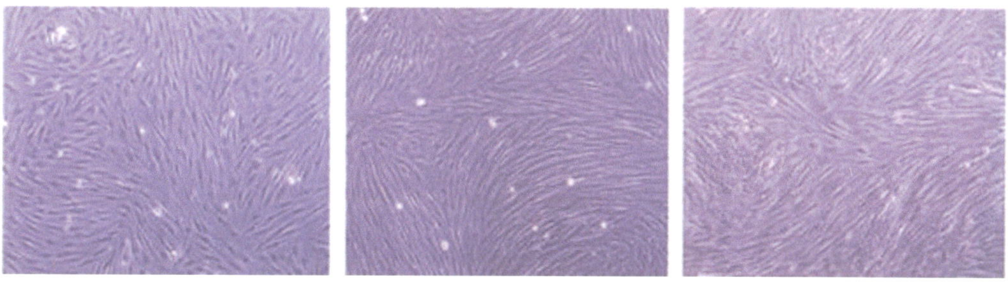

Figure 5. Healthy SMC in culture present a well-established appearance of flattened, spindle-like, fibroblastic morphology, with characteristic "hill-and-valley" organization.

Analysis of the functional properties of peripheral blood or adipose-derived smooth muscle cells *in vitro* demonstrates that they are indistinguishable from bladder-derived smooth muscle cells. Increased expression of proteins associated with smooth muscle contractility is a characteristic feature of smooth muscle cell differentiation and maturation. Myocardin is a key transcription factor required for smooth muscle cell differentiation and acts to mediate the expression of smooth muscle markers essential for contractility including SM22, α-smooth muscle actin (SMαA), smooth muscle myosin heavy chain (SMMHC) and calponin (CNN). Expression of the smooth muscle markers SMMHC and CNN is generally regarded as diagnostic of mature smooth muscle cells.

Application of adipose and peripheral blood-derived smooth muscle cells for urologic regenerative medicine is contingent on being able to secure adequate cell numbers within an acceptable time frame. Towards this end, we have observed that smooth muscle cell colonies (from a 50ml sample of porcine peripheral blood or 7-25g porcine adipose) are identifiable within 7 days post seeding, and may be passaged within 14 days. In fact, tens of millions of smooth muscle cells may be recovered from bladder, peripheral blood or adipose within 2-4 weeks (*n*=24). Bladder and adipose-derived smooth muscle cells were expanded for 2 passages prior to harvesting of cells for seeding a synthetic, neo-urinary conduit scaffold. Peripheral blood-derived smooth muscle cells were expanded for 3-4 passages to generate equivalent cell numbers. On average, 30-40 $X10^6$ smooth muscle cells were used to seed a neo-urinary conduit scaffold.

We have previously shown that bladder-derived smooth muscle cells may be used to seed a synthetic, biopolymer scaffold which upon implantation into an *in vivo* clinical model of bladder cystectomy resulted in the regeneration of a fully functional *de novo* bladder augment (Jayo, et al., 2008a,b). However, because use of bladder-derived smooth muscle cells may not be clinically ideal, we proceeded to evaluate the *in vivo* clinical efficacy of peripheral blood and adipose-derived smooth muscle cells in a 3 month porcine clinical model of urinary incontinence. Bladder, adipose and peripheral blood-derived smooth muscle cells were used to seed PGA/PLGA-based scaffolds to create a regenerative, neo-urinary conduit permitting efflux of urine from the ureters directly to the external body surface.

We observed that constructs composed of smooth muscle cells obtained from blood or bladder sources regenerated a patent conduit composed of an urothelial cell lining and smooth muscle layer that did not result in alterations to the upper urinary tract. No evidence was found for elevated creatinine, metabolic abnormalities or altered hematological parameters. Histological characteristics of the regenerated urological tissue forming the neo-urinary conduit were generally similar regardless of the origin of the smooth muscle cell population. In contrast, scaffold-only implanted animals developed patent urothelial-lined conduits composed primarily of fibrous connective tissue and limited smooth muscle development. This group also had a high frequency of hydroureter and hydronephrosis. In both groups, early post-operative management of the conduit lumen and stoma was required to maintain patency for the study duration (Basu, et al., 2012).

This study demonstrates that a synthetic, biopolymeric scaffold composite seeded with autologous smooth muscle cells derived from multiple potential cell sources (blood, fat or bladder) is capable of being used to re-create a patent neo-urinary conduit. The regenerated conduit was not associated with sequela commonly found with urinary diversions generated from intestinal tissue or with non-cell seeded synthetic scaffolds. This ability to create urologic structures *de novo* from synthetic scaffolds seeded by peripheral blood or adipose-derived smooth muscle cells will greatly facilitate the translation of urologic tissue engineering technologies into clinical practice.

4.3. Clinical trials of NUC

Based on successful outcomes in GLP porcine cystectomy models as outlined above, Tengion has initiated Phase I clinical trials of NUC constructs in human patients requiring urinary by-pass. This Phase I study "Incontinent Urinary Diversion Using an Autologous Neo-Urinary Conduit" is currently recruiting patients (as of time of writing), with the objective of implanting up to 10 patients by 2015. The objective of the study is to evaluate if NUC constructs, made in the laboratory by a combination of the patient's own adipose derived smooth muscle cells in combination with defined, degradable biomaterial scaffolds, may be used to form a functional conduit to safely facilitate passage of urine from kidneys to outside the body subsequent to radical cystectomy in patients presenting with bladder cancer.

Primary outcome indices over a 12 month post-implantation timeframe include structural integrity and conduit patency. CT scans will be used to demonstrate that urine may flow safely through the NUC construct. Additional measures of primary outcomes up to 12 months post-implantation include an evaluation of any product or procedure related adverse events. Similarly, secondary outcome indices will include analysis of NUC structural integrity and patency over a 12-60 month post-implantation timeframe. CT scan and renal ultrasound will be applied to demonstrate that urine is able to flow safely through the NUC construct up to 60 months post-implantation. Procedural and product related adverse events will also be monitored to 60 months post-implantation. Finally, the overall safety of the NUC construct will be assessed through evaluation of non-product/procedural related adverse events and patient vital signs. Eligibility for this clinical trial has been adjusted to be as inclusive as possible while ruling out any patients that present with secondary conditions likely to negatively impact regenerative outcomes:

4.4. Regeneration of muco-cutaneous region at skin-conduit junction

Successful implantation of NUC requires the regeneration of a native-like muco-cutaneous region at the skin/conduit junction following implantation. A muco-cutaneous junction is a region where the *luminal mucosa* transitions to skin. Such muco-cutaneous junctions occur throughout the body at orifices including mouth, nostrils, anus, eyelids and components of the genito-urinary system. Typically, such junctions involve transitioning of epithelium to epidermis, *lamina propia* to dermis and smooth muscle to skeletal muscle. An *in vitro* tissue engineered human lip-like element reproducing the transitional morphology of human lip including epidermal skin, vermillion and oral mucosa has recently been reported (Peramo et al., 2012). This observation notwithstanding, it should be appreciated that the regeneration of a muco-cutaneous junction at the skin/conduit interface of a synthetic neo-organ such as the NUC has never before existed in nature. This indicates that context-appropriate signaling between construct and host is nevertheless possible leading to regeneration of cell types relevant to the neo-organ.

In the case of the urinary system, urine exits the body via the *urethral meatus*, a distinct structure incorporating features that defend the opening against local and/or ascending infections, emptying into the vaginal vestibule in females and the *fossa navicularis* in males. Specifically, this muco-cutaneous junction is a non-keratinized, stratified squamous epithelium composed of glycogen rich cells that provide substrate for a protective endogenous lacto-bacterial flora. Also, as the epithelium nears the skin, it is associated with acid phosphatase activity and lysozyme-like immuno-reactivity indicative of the presence of macrophages that secrete bacterial compounds. In the pre-clinical study in cystectomized pigs discussed previously, implanted NUC was observed to catalyze formation of a native-like transition between the urinary mucosa and skin epithelium with structural features resembling that of muco-cutaneous junctions observed in native urethras (Basu et al., 2012a). These observations demonstrate that the *de novo* regeneration of transitional junctions including muco-cutaneous junctions may be catalyzed by regenerative constructs such as Tengion's Organ Regeneration Platform™, opening the door to regeneration of components of the gastro-intestinal tract, where many such transitional zones are located.

5. The gastro-intestinal tract: Introduction

Individual components of the GI represent locally specialized, iterative variations of essentially the same laminarly organized tubular histologic architecture as the bladder, and therefore should be amenable to regenerative therapies based on the same platform technologies successfully applied to the bladder. However, from a commercial perspective, the small intestine represents by far the most pressing current clinical need, with small bowel transplantation representing an unsatisfactory current standard of care for pediatric small bowel syndrome (SBS) (http://digestive.niddk.nih.gov/ddiseases/pubs/shortbowel/). Additionally, esophageal replacement or augmentation for esophageal cancer may also be commercially viable.

Signaling interactions between endodermal epithelium and mesenchymal cells derived from splanchic mesoderm mediate axial differentiation of the primitive gut tube during embryogenesis into distinctive functional domains. These individual components of the GI including the oral cavity, esophagus, stomach, small intestine and colon represent locally specialized iterative variations of essentially the same laminarly organized tubular histologic architecture as the bladder. The molecular genetics underlying these differentiative signaling cascades is complex, but is broadly understood to involve members of the *hedgehog* family of ligands, including *sonic hedgehog*, *desert hedgehog* and *indian hedgehog*. *Hedgehog* proteins bind the multipass transmembrane receptor *patched-1*, thereby blocking *patched-1*-mediated inhibition of the seven-pass membrane protein *smoothened*. *Smoothened*-mediated signal transduction regulates the transcriptional response to *hedgehog* activation within the target tissue. In mouse models and human patients, mutations in *hedgehog* signaling proteins and downstream targets lead to a variety of severe malformations of the GI tract. Genetic analysis of GI formation has been extensively reviewed in detail elsewhere (Van den Brink, 2007). Conservation of signaling between mesenchymal and epithelial cell populations during organogenesis of bladder and GI suggests that the organ regeneration platform appropriate for bladder regeneration may also function in an independent positional context to catalyze regeneration of the GI, including but not limited to small intestine and esophagus.

The intestinal epithelium is the most highly regenerative tissue within adult mammals and may therefore be expected to be most amenable towards tissue engineering or regenerative medicine methodologies. Perhaps the best described strategy for regeneration of intestine and other components of the gastrointestinal tract involves the use of *in vivo* derived organoid units, formed from incompletely disassembled clusters of epithelial and mesenchymal cells generated through partial digestion of intestinal epithelium and therefore likely incorporating resident intestinal stem cells. In one such study, autologous organoids were derived from the intestine of 6 week old mini-pigs and used to seed PLLA scaffold tubes that were subsequently matured within the peritoneal cavity of the original donor. Seven weeks post-implantation, this tissue-engineered small intestine recapitulated the gross overall laminar organization of native small intestine (SI) (Sala, et al., 2009). Significantly, acellular scaffolds did not result in the regeneration of tissue engineered gastrointestinal structures. These data notwithstanding, anastamosis of these tissue engineered small intestines to host SI within a large animal model

http://dx.doi.org/10.5772/59465

remains to be demonstrated. Additionally, up to 10cm of host SI was harvested to derive donor organoids that are not readily expandable *in vitro*. Whether organoid units capable of seeding a scaffold structure may be isolated from diseased human intestine, and how much diseased donor material may be needed, remain factors yet to be elucidated. The requirement to leverage the peritoneal cavity as a bioreactor for tissue engineering may also impede widespread application of this approach. The bladder-derived organ regeneration platform of bio-polymeric scaffold seeded with smooth muscle cells may be applicable for regeneration of SI. To this end, stomach derived smooth muscle cells were used to seed a collagen-based scaffold prior to implantation within surgically isolated ileal loops of dogs for eight weeks, prior to re-anastamosis to the native intestine. Acellular collagen scaffold was used as a control. By 12 weeks post-surgery, macroscopic analysis of the cell seeded scaffold implantation site demonstrated regeneration of neo-mucosa resembling native mucosa. However, in animals containing an acellular scaffold, the implant site remained ulcerated up to 12 weeks post-implantation. Additional histological data showed significantly enhanced vascularization, epithelialization and organization of the circular muscle layer at the cell seeded scaffold defect site relative to acellular control (Nakase, et al., 2006).

Increasing the number of smooth muscle cells seeded onto the scaffold increased the area of regenerated SI tissue, although no concomitant increase in the thickness of the smooth muscle layer was observed (Nakase, et al., 2007). Nevertheless, these data suggest that a simple regenerative platform composed of biodegradable scaffold nucleated with smooth muscle cells may be adequate to facilitate SI regeneration. Although this approach must be reproduced using a directly anastamosed tubular scaffold and alternate sources of smooth muscle cells, if successful, this methodology represents the most straightforward, clinically relevant and commercially viable strategy for regeneration of the SI. This organ regeneration platform technology may also be leveraged for regeneration of the esophagus. In one such example, patch defects were created in the abdominal esophagus of 27 female rats, subsequently implanted with gastric acellular matrix (GAM). Of the 24 survivors, none showed evidence of regeneration of the lamina muscularis mucosa even 18 months post-implantation (Urita, et al., 2007). In contrast, a study of a canine model of esophageal resection and replacement demonstrated that PGA tubes seeded with a mixture of keratinocytes and fibroblasts triggered regeneration of smooth muscle laminar organization similar to native esophagus within 3 weeks post-implantation, whereas acellular PGA tubes formed esophageal strictures were associated with near complete obstruction within two to three weeks (Nakase, et al., 2008). In another dog study, cervical esophageal defects were patched with either small intestinal submucosa (SIS) alone, or SIS seeded with autologous oral mucosal epithelial cells. After four weeks, dogs implanted with cell seeded SIS showed almost complete re-epithelialization and minimal evidence of inflammation and, by eight weeks post-surgery, regeneration of the underlying smooth muscle layer. Acellular SIS grafted animals presented only partial re-epithelialization and a more extensive inflammatory response by four weeks, and no muscular regeneration by eight weeks. Attempts to introduce an acellular SIS tubular construct into the cervical esophagus of piglets were also unsuccessful, demonstrating scarification and a minimal regenerative response (Doede, et al., 2009).Progress has also been made in efforts to tissue engineer the stomach. Stomach derived organoid units, (analogous to the SI organoids used to tissue engineer the SI) upon seeding of a biopolymeric scaffold, triggered reconstitution of the gastric and muscularis mucosa in stomach tissue engineered within the peritoneal cavities of swine (Sala, et al., 2009). In another study, circular defects were created in the stomach of seven dogs and a composite biodegradable scaffold ("New-sheet"), soaked with either autologous peripheral blood or bone marrow aspirate, was sutured over the defect. By 16 weeks post implantation, the defect site had formed regenerated stomach with evidence of re-epithelialization, formation of villi, vascularization and fibrosis within the sub-mucosal layer. However, minimal regeneration of the smooth muscle layer was observed, as shown by expression of smooth muscle α-actin, though not calponin, a marker of mature smooth muscle cells (Araki, et al., 2009).Though strictly not a tubular organ, the anal sphincter is a component of the gastrointestinal tract and is critical in regulating patency of the large intestine. Recent efforts to engineer the anal sphincter leverage the same general platform used to catalyze bladder regeneration. To this end, smooth muscle cells isolated from human internal anal sphincter were seeded onto fibrin gels poured around a central mold. Cell mediated contraction of the gel around the mold resulted in the formation of a 3D cylindrical tube of sphincteric smooth muscle tissue. Although *in vivo* anastamosis remains to be demonstrated, this bioengineered anal sphincter demonstrated contractile properties and response to defined neurotransmitters consistent with the functionality of native anal sphincter (Somara, et al., 2009; Hashish, et al., 2010). Use of alternatively sourced smooth muscle cells may facilitate the transition of engineered sphincter towards commercial production.

Preclinical rodent studies were initiated in our laboratory to evaluate the ability of constructs composed of smooth muscle cell seeded PLGA coated PGA patches to support regeneration of esophagus. In these studies, small (3-5 mm^2) defects were introduced within the esophagus of rodents, such that the entire esophageal wall was removed within the defect. Injuries were subsequently repaired with cell seeded patch constructs and allowed to regenerate *in vivo* for up to 10 weeks post-implantation. Evidence of early regenerative outcomes was observed at 8 days post-implantation in the form of developing bundles of smooth muscle. Complete regeneration of longitudinal and circularly oriented musculatures and luminal epithelia were observed at 10 weeks post-implantation of construct. Importantly, as with previous studies, vascularization of the regenerative construct by wrapping with omentum during implantation was crucial for a successful regenerative outcome. These small animal data establish the utility of SMC coated PGA felt-based biomaterials for tissue engineering and regeneration of the esophagus (Basu et al., 2012b; Basu et al., 2013).

For the small intestine (SI), patch based constructs composed of adipose derived smooth muscle cell seeded PGA felts were observed to support full regenerative outcomes upon implantation within rodent models of SI injury. Regeneration of intestinal epithelia with microvilli as well as regenerated bundles of smooth muscle cells were observed within 3 months post-implantation. Studies were also initiated to evaluate the regenerative potential of cell seeded biomaterials constructs within the context of a tube. In addition to supporting the proliferation and migration of SI derived cells, such tubular constructs must have appropriate physical and mechanical properties that complement those of native tissue. In particular, within the GI tract, tubular constructs must be capable of supporting peristaltic movement of food or fecal matter. Commercially sourced PGA braided tubes were observed to have physical properties reminiscent of SI, and were therefore applied in the context of tubular cell seeded constructs for evaluation of regenerative outcomes within rodent SI. As with the patch, complete regeneration of intestinal epithelia and partial regeneration of bundles of smooth muscle cells was observed at 3 months post-implantation. Implanted animals were capable of eating normally, gained weight and passed fecal material in a manner comparable to healthy controls. A detailed analysis of regenerative outcomes in rodents implanted with SMC seeded tubular constructed has been previously presented (Basu et al., 2011b). As with all other tubular organs, presence of a cellular component on the regenerative construct is essential to a successful regenerative outcome. This data notwithstanding, observed difficulties in regeneration of SI musculature from simple constructs derived from biomaterials used in bladder regeneration suggest that there are elements unique to SI currently missing from the current generation of construct prototypes. One such potential factor is the role of the enteric nervous system in facilitating regenerative outcomes within the GI tract.

6. Summary

1. Tengion's foundational Organ Regeneration Platform™ composed of adipose-sourced smooth muscle cells complexed with a synthetic, biodegradable scaffold, is broadly applicable to regeneration of multiple tubular organs.
2. Constructs function by manipulating fundamental signaling pathways between mesenchymal and epithelial cell populations to trigger regenerative outcomes mimicking organogenesis during development and regeneration in organisms such as urodeles, including induction of a neo-blastema.
3. Tubular organ regeneration is influenced by adequate vascularization and cellular constituents present at the local site.
4. Mucosal regeneration is influenced by blood supply with organ specific events controlling biological events.

This ability to apply the same fundamental signaling elements that mediate the regenerative process through application of the same combination of cells and biomaterials towards regeneration of multiple tubular neo-organs will greatly facilitate development of a manufacturing platform with commonalities throughout bio-processing.

Author details

Joydeep Basu*

Address all correspondence to: joydeep.basu@tengion.com

Process Research and Translation, Tengion, Inc, Ste G, Winston-Salem, USA

References

[1] Adzick, N.S. and Lorenz, H.P. (1994). Cells, matrix, growth factors and the surgeon. The biology of scarless fetal wound repair. *Ann Surg* 220, 10-18

[2] Asnaghi M.A. *et al.* (2009) A double-chamber rotating bioreactor for the development of tissue engineered hollow organs: from concept to clinical trial. *Biomaterials* 30, 5260-5269

[3] Atala *et al.*, 2006. Tissue engineered autologous bladders for patients needing cystoplasty. *Lancet* 306: 1241-46

[4] Basu, J. and Ludlow, J.W. (2010). Platform technologies for tubular organ regeneration. *Trends Biotech* 28, 526-33

[5] Basu, J., and Ludlow, J.W. (2011). Tissue engineering of tubular and solid organs: an industry perspective. In *Advances in Regenerative Medicine*, ed. S. Wislet-Gendebein, Intech Open

[6] Basu, J., Genheimer, C.W., Guthrie, K.I., Sangha, N., Quinlan, S.F., Bruce, A.T., Reavis, B., Halberstadt, C., Ilagan, R.M., Ludlow, J.W. (2011a) Expansion of the human adipose-derived stromal vascular cell fraction yields a population of smooth muscle-like cells with markedly distinct phenotypic and functional properties relative to mesenchymal stem cells. *Tissue Eng Part C* 17, 843-60

[7] Basu, J., Mihalko, K.L., Payne, R., Rivera, E., Knight, T., Genheimer, C.W., Guthrie, K.I., Sangha, N., Jayo, M.J., Jain, D., Bertram, T.A., Ludlow, J.W. (2011b). Regeneration of rodent small intestine tissue following implantation of scaffolds seeded with a novel source of smooth muscle cells. *Regen Med* 6, 721-31

[8] Basu, J., and Ludlow, J.W., (2012). *Developments in tissue engineered and regenerative medicine products, a practical approach.* Woodhead Publishing

[9] Basu, J., Jayo, M.J., Ilagan, R.M., Guthrie, K.I., Sangha, N., Genheimer, C.W., Quinlan, S.F., Payne, R., Knight, T., Rivera, E., Jain, D., Bertram, T.A. (2012a). Regeneration of native like neo-urinary tissue from non-bladder cell sources. *Tissue Eng Part A* 18, 1025-34

[10] Basu, J., Mihalko, K.L., Payne, R., Rivera, E., Knight, T., Genheimer, C.W., Guthrie, K.I., Sangha, N., Jayo, M.J., Jain, D., Bertram, T.A., Ludlow, J.W. (2012b). Extension of bladder based organ regeneration platform for tissue engineering of esophagus. *Med Hypotheses* 78, 231-4

[11] Basu, J., Mihalko, K.L., Rivera, E.A., Guthrie, K.I., Genheimer, C.W., Sangha, N., Ludlow, J.W. (2013) Tissue engineering of esophagus and small intestine in rodent injury models. *Methods Mol Biol* 1001, 311-24

[12] Basu, J., and Bertram, T (2014). Regenerative medicine of the gastrointestinal tract. *Toxicol Pathol* 42, 82-90

[13] Basu J, Genheimer CW, Rivera EA, Payne R, Mihalko K, Guthrie K, Bruce AT et al., 2011. Functional evaluation of primary renal cell/biomaterial Neo-Kidney Augment prototypes for renal tissue engineering. *Cell Transplant*. 20:1771-1790

[14] Becker *et al.*, 2008. TGFbeta1 and epithelial-mesenchymal interactions promote smooth muscle gene expression in bone-marrow stromal cells: possible application in therapies for urological defects. *Int J Artif Organs* 31, 951

[15] Campbell *et al.*, 1999. Novel vascular graft grown within recipient's own peritoneal cavity. *Circ Res* 85: 1173-8

[16] Campbell *et al.*, 2008. The peritoneal cavity as a bioreactor for tissue engineering visceral organs: bladder, uterus and vas deferens. *J Tissue Eng Regen Med* 2: 50-60

[17] Cilento *et al.*, 1994. Phenotypic and cytogenetic characterization of human bladder urothelia expanded in vitro. *J Urol* 152: 665-670

[18] Doede, T., Bondartschuk, M., Joerck, C., Schulze, E., Goerning, M. (2009). Unsuccessful alloplastic esophageal replacement with porcine small intestinal submucosa. *Artif Organs* 33, 328-333

[19] Genheimer *et al.*, 2010. Increased urothelial cell detection in the primary bladder smooth muscle cell cultures with dual MACs/qRT-PCR approach. *Appl Immunohistochem Mol Morphol* 19: 184

[20] Godwin, J.W., Pinto, A.R., Rosenthal, N.A. (2013). Macrophages are required for adult salamander limb regeneration. *PNAS* 110, 9415-9420

[21] Gong *et al.*, 2008. Influence of culture medium on smooth muscle cell differentiation from human bone marrow derived mesenchymal stem cells. *Tissue Eng Part A* 15, 319

[22] Hashish M. *et al.* (2010). Surgical implantation of a bioengineered internal anal sphincter. *J Ped Surg* 45, 52-58

[23] He *et al.*, 2007. Concise review: multipotent mesenchymal stromal cells in blood. *Stem Cells* 25: 69

[24] Jack *et al.*, 2009. Urinary bladder smooth muscle engineered from adipose stem cells and a three dimensional synthetic composite. *Biomaterials* 30: 3259

[25] Jayo, M.J., Jain, D., Wagner, B.J., Bertram, T.A. (2008a). Early cellular and stromal responses in regeneration versus repair of a mammalian bladder using autologous cell and biodegradable scaffold technologies. *J Urol* 180, 392-7

[26] Jayo, M.J., Jain, D., Ludlow, J.W., Payne, R., Wagner, B.J., McLorie, G., Bertram, T.A. (2008b). Long term durability, tissue regeneration and neo-organ growth during skeletal maturation with a neo-bladder augmentation construct. *Regen Med* 3, 671-82

[27] Jeon *et al.*, 2006. Sphingosylphosphorylcholine induces differentiation of human mesenchymal stem cells into smooth muscle like cells through a TGF-beta dependant mechanism. *J Cell Sci* 119

[28] Kanematsu *et al.*, 2005. Induction of smooth muscle cell like phenotype in marrow derived cells among regenerating urinary bladder smooth muscle cells. *Am J Pathol* 166, 565

[29] Kim *et al.*, 2008a. Bradykinin-induced expression of alpha-smooth muscle actin in human mesenchymal stem cell. *Cell Signal* 20, 1882

[30] Kim *et al.*, 2008b. Angiotensin II-induced differentiation of adipose tissue derived mesenchymal stem cells to smooth muscle cells. *Int J Biochem Cell Biol* 40, 2482

[31] Knop, E., Knop, N., Zhivov, A., Kraak, R., Korb, D.R., Blackie, C., Greiner, J.V., Guthoff, R. (2011). The lid wiper and muco-cutaneous junction anatomy of the human eyelid margins: an *in vivo* confocal and histological study. *J Anat* 218, 449-461

[32] Lai *et al.*, 2002. Phenotypic and functional characterization of in vivo tissue engineered smooth muscle from normal and pathological bladders. *J Urol* 168: 1853-1858

[33] Litbarg *et al.*, 2007. Activated omentum becomes rich in factors that promote healing and tissue regeneration. *Cell Tissue Res* 328: 487-97

[34] Macchiarini P. *et al.* (2008) Clinical transplantation of a tissue-engineered airway. *Lancet* 372, 2023-2033

[35] Metharom *et al.*, 2008. Myeloid lineage of high proliferative potential human smooth muscle outgrowth cells circulating in blood and vasculogenic smooth muscle-like cells in vivo. *Atherosclerosis* 198: 29-38

[36] Mu, X., Bellayr, I., Pan, H., Choi, Y., Li, Y. (2013). Regeneration of soft tissues is promoted by MMP1 treatment after digit amputation in mice. *PloS One* 8, e5915

[37] Nakase, Y., Hagiwara, A., Nakamura, T., Yamagishi, H. (2006). Tissue engineering of small intestinal tissue using collagen sponge scaffolds seeded with smooth muscle cells. *Tissue Eng* 12, 403-412

[38] Nakase, Y., Nakamura, T., Kin, S., Ikada, Y., Hagiwara, A. (2007). Endocrine cell and nerve regeneration in autologous in situ tissue-engineered small intestine. *J Surg Res* 137, 61-68

[39] Nakase, Y., Nakamura, T., Kin, S., Hagiwara, A. (2008). Intra-thoracic esophageal replacement by in situ tissue-engineered esophagus. *J Thorac Cardiovasc Surg* 136, 850-859

[40] Neuhof H. Fascial transplantation into visceral defects: an experimental and clinical study. *Surg Gynecol Obst* 25: 383

[41] Oberpenning, F., Meng, J., Yoo, J.J., Atala, A. (1999) De novo reconstitution of a functional mammalian urinary bladder by tissue engineering. *Nat Biotechnol* 17, 149-155

[42] Peramo, A., Marcelo, C.L., Feinberg, S.E. (2012). Tissue engineering of lips and mucocutaneous junctions: *in vitro* development of tissue engineered constructs of oral mucosa and skin for lip reconstruction. *Tissue Eng Part C* 18, 273-282

[43] Rivera, E., Bivalacqua, T, Schoenberg, M., Steinberg, G., Smith, N., Robertson, A., Basu, J., Bruce, A., Guthrie, K., Payne, R., Spencer, T., Jain, D., Bertram, T. (2013). Clinical translation of a tissue engineered urinary diversion using a biodegradable scaffold seeded with adipose smooth muscle cells. *Stem Cell Trilogy*, London, UK

[44] Roy, S and Gatien, S. (2008). Regeneration in axolotls: a model to aim for! *Exp Gerontol* 43, 968-973

[45] Sato, T., Vries, R.G., Barker, N., Stange, D.E., Clevers, H. (2009). Single Lgr5 stem cells build crypt-villus structures in vitro without a mesenchymal niche. *Nature* 459, 262-265

[46] Sala, F.G., Kunisaki, S.M., Ochoa, E.R., Vacanti, J., Grikscheit, T.C. (2009). Tissue-engineered small intestine and stomach form from autologous tissue in a preclinical large animal model. *J Surg Res* 156, 205-212

[47] Sakuma *et al*, 2009. Mature, adipocyte derived, dedifferentiated fat cells can differentiate into smooth muscle like cells and contribute to bladder tissue regeneration. *J Urol* 182,16

[48] Somara S. *et al* (2009). Bioengineered internal anal sphincter derived from isolated human internal and sphincter smooth muscle cells. *Gastroenterology* 137, 53-61

[49] Stocum, D.L. (2011). The role of peripheral nerves in urodele limb regeneration. *Eur J Neurosci* 34, 908-16

[50] Tanaka, E.M. and Reddien, P.W. (2011). The cellular basis for animal regeneration. *Dev Cell* 21, 172-185

[51] Urita, Y., Komoro, H., Chen, G., Shinya, M., Kaneko, S., Kaneko, M., Ushida, T. (2007). Regeneration of the esophagus using gastric acellular matrix: an experimental study in a rat model. *Pediatr Surg Int* 23, 21-26

[52] Van den Brink, G. (2007). Hedgehog signaling in development and homeostasis of the gastrointestinal tract. *Physiol Rev* 87, 1343-1375

[53] Warner, B.W. (2004). Tissue engineered small intestine. A viable clinical option? *Ann Surg* 240: 755-756

[54] Wei, R.Q., Tan, B., Tan, M.Y., Yang, Z.M. (2009). Grafts of porcine small intestinal sub-mucosa with cultured autologous oral mucosal epithelial cells for esophageal repair in a canine model. *Exp Biol Med (Maywood)* 234, 453-461

[55] Wessels 1977, Tissue Interactions and Development, Benjamin/Cummins

[56] Xu *et al.*, 2004. Mesenchymal stem cells from adult human bone marrow differentiate into a cardiomyocyte phenotype in vitro. *Exp Biol Med* 229, 623

Chapter 10

Bioengineering of Vascular Conduits

A. Pontini, M.M. Sfriso, M.I. Buompensiere,
V. Vindigni and F. Bassetto

Additional information is available at the end of the chapter

http://dx.doi.org/10.5772/59148

1. Introduction

Tissue engineering research was applied in the last few years to vascular conduit field, aiming to obtain a suitable and ready to use substitute for vessel replacement. Based on the possibilities to obtain *in vitro* a biocompatible structure it is established that is theoretically and experimentally possible to provide vessels that can be employed to replace both diseased than damaged native blood vessels overcoming the massive worldwide clinical need and the poor supply of natural graft and, the same time, offering a better long term performance than the artificial conduits. The challenges reported in literature about the approach of tissue engineering for replacing blood vessels are continuously increasing. It has been reported natural vessel like structure, with similar elastic wall properties that are necessary for the cyclic blood flow loading with similar native vascular diameter to allow a perfect match with the host vessel. Fundamental are also the result obtained in term of antithrombotic lumen [1, 2].

Particularly important achievement were reached for cardiovascular application but also the potential range of application could easily been expanded to all microsurgical and vascular applications [3]. Tissue engineering has been projected as an alternative treatment to these problems by replacing the damaged tissue or organ function with constructs which are bio fabricated based on the required tissue or organ features [4]. In particular, cardiovascular tissue engineering is more valuable and relevant compared to other fields of tissue engineering mainly because it increases life expectancy, preserve the extremities, and provide solutions to a large number of disease [5]. Tissue engineering could at least been see as an interdisciplinary field that applies the principles of engineering and life sciences towards the development of functional substitutes for damaged tissues.

It is strictly related to the fundamental concept of utilizing the body's natural biological response to tissue damage in conjunction with engineering principles [6, 7]. Besides, tissue engineering is planned to produce biomimetic constructs, which resemble normal tissues.

Moreover, the main objective of tissue engineering is the restoration of function through the delivery of living elements which become integrated in the patient [8].Tissue engineering strategies have three basic components: firstly, the cells or source which must express the appropriate genes and maintain the appropriate phenotype in order to preserve the specific function of the tissue, secondly, the bioreactive agents or signals that induce cells to function, and thirdly, the scaffolds that house the cells and act as substitute for the damaged tissue [9, 10]. The source may either be embryonic stem cells (ESC) or adult stem cells(ASC) in origin, the scaffolds may be categorized as synthetic, biological, or composite, and the signals may include growth factors/cytokines, adhesion factors, and bioreactors [11]. Develop a bioengineered tissue is due to a precise clinical needs and different ways were searched to obtain the answer (Figure 1). So, joint to the ideal construct is also the research of the best tissue engineering approaches in terms of biocompatibility, feasibility and costs.

So from 1999 till now several studies, both in vitro than in vivo, were presented, particularly inherent on vessel scaffold, which a large number of information about cell sources, technology application, outcomes and future perspectives [12].

Until nowadays the most encouraging result are limited in basic and experimental research but the rapidly increasing of optimal result permit to access to numerously clinical trials. Biomaterials technologies for vascular replacement must obtain an ideal graft that could overcome the needing of autologous vessel, that often it's not applicable, but, at the same time, providing similar properties. In fact the ideal bioengineered vascular substitute must be not thrombogenic, overall in small caliber vessel, and also when a long graft is needed.

Thrombosis mechanism into vascular substitutes, especially in artificial one, is the main cause of obliteration and subsequent failure of most microvascular prostheses. Autologous native vessel, are the most currently used material for small-diameter arterial replacement. Immune acceptance is a major advantage offered by this technique but the time of dissecting, harvesting and preparing autologous graft limited the microvascular emergency surgery and, in elective surgery, it could be possible that no one suitable vessel could be harvested.

For that reason, the tissue engineering was applied to improved prosthetic performance at the blood-biomaterial interface. Different approach were described to optimize vascular bioengineered conduits as completely bio-resorbeable vascular prostheses with the capacity for induce regeneration and growth of a new vascular segment, biologic scaffold enhanced by stem cell seeding, decelularizzed native vessel with or without cell enhancement. In vitro ed in vivo study of all these different approaches shown the possibility to overcome the limitations of the artificial prostheses that are nonviable and based on allogenic materials lacking the capacity of growth, repair, and remodeling. The use of bio engineered vascular conduit are fundamental in small caliber vessel where the artificial replacement is affected by a very low patency rate meanwhile the possibility to obtain bioengineered large vessel replacement is actually less important due to the satisfactory result and still less expensive use of artificial or homograft conduit.

Synthetic prostheses offer to microsurgery a possible solution for microvascular need of a ready to use and simple to manage small diameter vessel. Availability in multiple different diameters and lengths, uncomplicated storage associated with easy handling are some of main advantages of such grafts; nevertheless, inherent thrombogenicity and compliance mismatch could represent their drawbacks. The research aim is to obtain and ideal prostheses, particularly in term of biocompatibility and satisfactory patency rate in long period of time.

In fact, similar outcomes as large vessel replacement were not achieved in microvascular surgery.

Best performance was obtained when the blood flow is high and the resistance is low, because those conditions allowed to overcome possible thrombogenic events that occur in large part in small-diameter prostheses. Multiple strategies were studied to overcome these limitations applying tissue engineering techniques. Ideally artificial conduit ready to use should be composed of viable tissue, able to contract in response to hemodynamic forces and chemical stimulation, and secrete physiological blood vessel substances. Anastomoses using artificial prostheses should also allow complete healing without immunologic reaction, remodeling according to surroundings environment, and even have the ability to grow when placed in children.

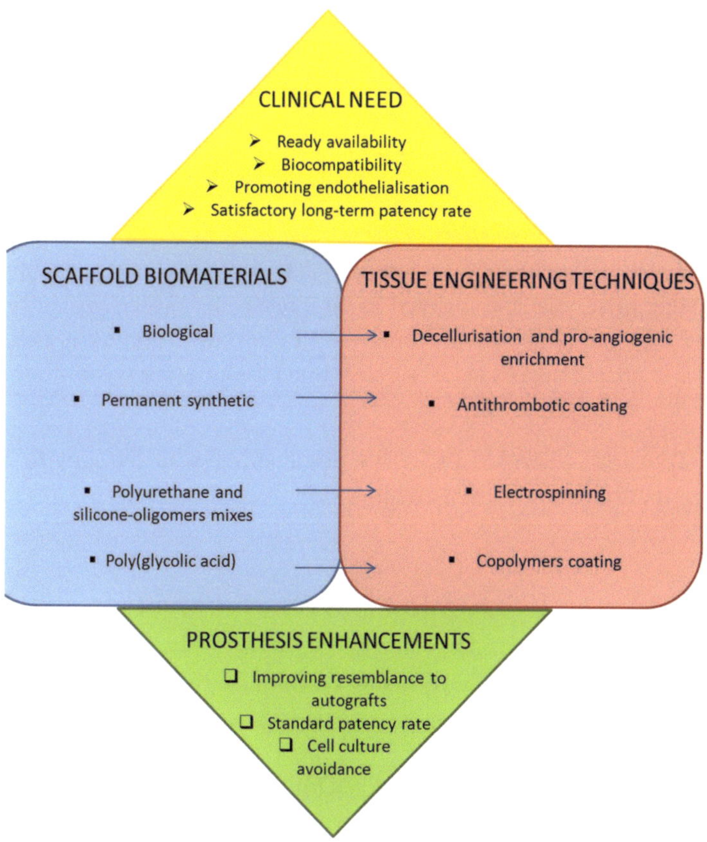

Figure 1. Clinical need and bioengineering of vascular conduit. Adapted from: Pontini et al. "Alternative Conduits for Microvascular Anastomoses" Surg Innov. 2013 Aug 20;21(3):277-282

2. Blood flow and functional implication for vascular replacement

Blood vessels are the conduit that allows the transportation in the human body to the organs and tissues of blood, oxygen, nutrients and catabolities removal. Small caliber arteries (< 6 mm), account for most of deaths in the United States every year [1] Atherosclerosis, due to numerously damage mechanism of *media and intima tunica* in vessel lumen, cause a progressive occlusion that could lead to a severe blood flow impairment and organ failure. Most common vascular disease based on critic vessel damage could be the cardiac infarction subsequent to a coronary artery occlusion, claudication or chronic ulcer because peripheral arterial disease, or stroke due to occlusion of carotid or cerebral arteries. Arterial replacement is widely considered a common treatment for vascular disease, accounting for more than 1.4 million arterial bypass operations performed annually in the United States [2] The "gold standard" treatment is based on autologous vessels employment such saphenous veins and mammary arteries but one-third of patients are lacking for suitable veins, often for general vascular disease or past harvesting in vascular procedures. Nevertheless autologous vein graft have shown high patency rate failure in the long term period principally due to intimal hyperplasia [13]. Need for vascular grafts are also important in reconstructive surgery, vascular trauma, organ transplantation, so a large number of vascular conduit are needed in clinical daily practice. Besides are must to be considered that a significant morbidity and high economical costs are associated with autologous vessel preparation.

Multiple factor are at least involved in a widely recognized need for an efficient, readily available and simple to manage small-diameter vascular graft. The first step on not autologous vessel replacement was constituted by artificial vessel based on different permanent material as polyurethane, polyethylene terephthalate and polytetrafluoroethylene (ePTFE). All these prosthetic materials have proved to be inferior to autologous conduits, especially for small caliber. Low patency rate outcome with important thrombosis risk, infection and low performance at anastomosis site have determined the progressive discharged of artificial conduits. [14]

The biological approach provided by tissue engineering was thought to allow a better performance, compatibility and host matching. [15]

3. Bioengineered vessel: definition and development

Vessel replacement is a clinical need in many vascular field. Blood vessel diseases, such as atherosclerosis and arteritis as reported from Ross in 1993 and Wilcox in 1996. The Chronic Venous Insufficiency (CVI) as described from Moriyama in 2011 and thrombosis remain globally the major vascular problems. Therapies for such diseases often require replacement of those vessels with vascular grafts. Autologous arteries or veins are the best substitutes for small-diameter (internal diameter < 6 mm) vessels as shown also from Tu and colleagues in 1997. However, in some cases of acute vascular disease, amputation or previous harvest, the implant of autologous vessels could be limited of the biodisponibility of patient. The most important described application were the arterial occlusive disease such coronorapathy and ischemic periferical condition treated with artery by-pass [16]. Furthermore were described use also in thoracic surgery and for dialysis access [17, 18].

Literature review of the historical approach in vascular bioengineering shown multiple and different approach worldwide to obtain the best biocompatible and long lasting replacement. The research start from modification of synthetic material and was develop to biologic material assessment. There have been so many attempts to develop a small-diameter vascular graft made of synthetic or natural polymers. The synthetic polymeric materials include polyethylene terephthalate and expanded polytetrafluoroethylene (ePTFE) as described from Teebken in 2002. Although these polymeric vascular grafts have been successfully employed to replace blood vessels above 6 mm in ID, these polymeric grafts cannot be used for treatment of small-diameter vascular diseases due to thrombus formation as demonstrated from Veith in 1986 and then from Chard in 1987.

Coating of the intimal side with antithrombogenic materials, such as heparin was the approach for example of Devine in 2001, polyethylene oxide was the attempt of Kidane in 1999, or, previously, with endothelial cells as described from James in 1998, has been applied to solve this problem as we also reported below. Unfortunately these approaches still remaining doubtfull in vivo and in long-term and are considered unsuccessful.

For that reason tissue engineered blood vessels (TEBV) arising as a promising approach to address the shortcoming of such problems. Many design criteria have been proposed for the development of blood vessels scaffold as it's possible to read in the works of Conte et al. in 1998; Mitchell et al in 2003 and Teebken et al. also in 2003. Scaffold must be biocompatible, i.e. non thrombogenic, non-immunogenic, and resistant to infection, all of which are associated with a confluent, quiescent, non-activated endothelium. Furthermore, it must induce an acceptable healing response that does not result in inflammation, hyperplasia, or fibrous capsule formation, and, ideally, leads to the integration of the graft into the body such that it eventually becomes indistinguishable from a native vessel.

It must possess appropriate mechanical properties, which include physiological compliance, the ability to withstand long-term hemodynamic stress without failure, and no susceptibility to permanent creep that can lead to aneurysm formation.

Scaffold must have an appropriate permeability to water, solutes, and cells and must exhibit physiological properties, such as vasoconstriction/relaxation responses.

Finally, easy handling and suturability are crucial for such vessels to be viable from a surgical standpoint.

These design criteria are quite challenging given the demanding mechanical environment of the cardiovascular system.

Although different approaches attempt to meet these criteria in different ways, it is widely held that 3 components are necessary for these criteria to be met:

- a biocompatible component with high tensile strength to provide mechanical support (collagen fibers or their analogue);

- a biocompatible elastic component to provide recoil and prevent aneurysm formation (elastin fibers or their analogue);
- a non-activated, confluent endothelium to prevent thrombosis.

In 1986, Weinberg and Bell generated what was widely regarded as the first tissue-engineered blood vessel substitute, consisting of cultures of bovine endothelial cells (ECs), smooth muscle cells (SMCs) and fibroblasts embedded in a collagen gel. However, the graft lacked sufficient strength and was unsuitable for implantation. This construct was evaluated in vivo as an arterial implant only after reinforcement with Dacron® as shown from Matsuda in 1995. Various methods of improving the mechanical properties of collagen gels (e.g., crosslinking agents such as glutaraldehyde) have been investigated, but none has proven to yield a structurally stable tissue-engineered vascular grafts (TEVG) as reported from Charulatha in 2003. As an alternative to collagen for natural ECM-based scaffolds, fibrin holds particular promise because of its ability to induce collagen and elastin synthesis and improved mechanical properties as shown from Swartz in 2005. Furthermore, encouraging result, even if in larger diameter vessel, was achieved by combining fibrin gels with biodegradable polymeric scaffolds followed by seeding of autologous arterial-derived cells, from the group of Tschoeke in 2009 as also endothelialized vessels have been successfully implanted in the carotid arteries of sheep from Koch and his team in 2010.

Decellularized tissue, often in the form of a Xenogenic, can serve as a naturally available scaffold. Examples of such scaffolds were realized by Lantz in 1993, who used the small intestinal submucosa (SIS) as a vascular implant. The SIS was decellularized and then implanted in aorta, carotid and femoral arteries of dogs. The grafts resulted completely endothelialized at 28 days post-implantation. At 90 days, the grafts were histologically similar to normal arteries and veins and contained a smooth muscle media and a dense fibrous connective tissue adventitia. Follow-up periods of up to 5 years found no evidence of infection, intimal hyperplasia, or aneurysmal dilation. One infection-challenge study suggested that SIS may be infection resistant, possibly because of early capillary penetration of the SIS (2 to 4 days after implantation) and delivery of body defenses to the local site.

Kaushal, in 2001, has employed a decellularized porcine iliac arteries, seeded them with endothelial progenitor cells (EPCs), and implanted the constructs into ovine carotid arteries. These TEVG constructs remained patent out to 130 days and were remodeled into neovessel, whereas the unseeded control group occluded within 15 days. These results indicate that decellularized vascular scaffolds are susceptible to early failure unless first undergoing endothelialisation or additional modification.

In fact, Simon in 2003, shown as elements of the ECM are exposed to physical and chemical stresses during the process of decellularization, which can adversely affect the biomechanical properties of the ECM. This deterioration might ultimately lead to degenerative structural graft failure. Additional drawbacks of decellularized materials included the inability to modify the ECM content and architecture, the variability among donor sources, and the risk of viral transmission from animal tissue.

http://dx.doi.org/10.5772/59148

In 2011 Quint has developed a unique method of developing decellularized tissue for small-diameter arterial grafts using biodegradable polymers. They developed a different approach to arterial tissue engineering that can substantially reduce the waiting time for a graft. Tissue-engineered vessels (TEVs) were grown from banked porcine smooth muscle cells that were allogenic to the intended recipient, using a biomimetic perfusion system. The engineered vessels were then decellularized, leaving behind the mechanically robust extracellular matrix of the graft wall. The acellular grafts were then seeded with cells that were derived from the intended recipient, EPC or EC, on the graft lumen. TEV were then implanted as end-to-side grafts in the porcine carotid artery, which is a rigorous test-bed due to its tendency for graft occlusion. The EPC-and EC-seeded TEV all remained patent for 3D in this study, whereas the contralateral control vein grafts were patent in only 3/8 implants. Going along with the improved patency, the cell-seeded TEV demonstrated less neointimal hyperplasia and fewer proliferating cells than did the vein grafts. Proteins in the mammalian target of rapamycin signaling pathway tended to be decreased in TEV compared with vein grafts, implicating this pathway in the TEV's resistance to occlusion from intimal hyperplasia.

These results indicate that a readily available, decellularized tissue-engineered vessel can be seeded with autologous endothelial progenitor cells to provide a biological vascular graft that resists both clotting and intimal hyperplasia.

Decellularized xenografts have been identified as potential scaffolds for small-diameter vascular substitutes. Xiong, for example, in 2013 shown a work that has aimed to develop and investigate a biomechanically functional and biocompatible acellular conduit using decellularized porcine saphenous arteries (DPSAs), through a modified decellularization process using Triton X-100 solution and serum-containing medium. Histological and biochemical analysis indicated a high degree of cellular removal and preservation of the extracellular matrix. Bursting pressure tests showed that the DPSAs could withstand a pressure of 1854 ± 164 mm Hg. Assessment of in vitro cell adhesion and biocompatibility showed that porcine pulmonary artery endothelial cells were able to adhere and proliferate on DPSAs in static and rotational culture. After interposition into rabbit carotid arteries in vivo, DPSAs showed patency rates of 60% at 1 month and 50% at 3 months. No aneurysm and intimal hyperplasia were observed in any DPSAs. All patent grafts showed regeneration of vascular elements, and thrombotic occlusion was found to be the main cause of graft failure, probably due to remaining xenoantigens.

The purpose of this work was to evaluate the effects of using a decellularization protocol in samples of rabbit and human arteries and veins, involving mechanical processes and enzymatic reactions in order to obtain a scaffold suitable for the implantation in an organism recipient. Subsequently, a further purpose of this thesis is to obtain a new type of scaffold derived from skeletal muscle decellularization.

Resuming all the mentioned aspect we can confirm that autologous vessels and vascular allografts are the most reliable vessel source, but often their supply are insufficient for their widespread application [19]. For that reason we saw the use of synthetic prosthetic vessel, before tissue engineering onset. The principal vessel typology needed in a large number of vascular substitution is represented by small caliber vessel (< 2 mm) and synthetic material failed because a low patency rate due to thrombosis [20].

As listed before, the thrombogenic surface of many engineered materials becomes particularly relevant in microvascular grafts.

Patency properties of vascular grafts could be considered the key point to obtain a conduit with a relevant chance of stable replacement of damaged vessel. In fact, thrombosis is the main mechanism of obliteration and subsequent failure of most microvascular anastomoses using artificial conduits. Various methods have been recorded to avoid it, such as coatings with antithrombotic drugs, as heparin, hirudin, aspirin, or tissue factor pathway inhibitor [21]. There have been attempts to emulate the endothelial cellular surface which, coated with heparan sulphate proteoglycan, produces a negative surface charge which helps to prevent platelet adherence. Some prostheses are therefore coated internally with heparin sulphate, which is quickly degraded, and some materials with an electronegative surface have been created, with uncertain results [22]. So far, many researchers have described seeding endothelial cells in conduits.

Laube et al. reported a patency rate of 90% in 27 months for ePTFE prostheses used in coronary bypass, after additional incubation with endothelial cells which allowed them to adhere to the material [23]. The major limitation of this method is the need for cell cultures and withdrawal of tissue from the patient, and in any case it remains a two-step procedure. A tubular structure of ePTFE is also left in place, with the risk of later infection. Constructs composed entirely of cells (Tissue Engineered Blood Vessels: TEBV) have been studied to overcome these complications [24]. Although the method promises amazing results, it is time-consuming and very expensive (Fig. 2).

To avoid the cost of cell cultures, many researchers have tried to improve endothelial coverage of prostheses by coating them with endothelial-friendly compounds with good haemocompatibility. E-PTFE prostheses have been coated with perlecan [6] and endothelial-specific adhesion proteins such as fibrin–and hirudin [25, 26].

Fibronectin coating seems to be a successful method, apart from loss of lining at high flow rates. This is why a functional ligand for fibronectin was used, with covalent binding of short peptide sequences (Arg-Gly-Asp, RGD) to improve cell adhesion. Instead of coating prostheses with the above substances, another possibility would be to use absorbable, already biocompatible biomaterials, to make entire prostheses.

In spite of all these experiments, endothelialisation in various types of vascular prostheses has been shown in animals but never satisfactorily in humans.

The type of material is not the only essential point to allow the endothelialisation but it's fundamentally correlate with the physiopathology of endothelialisation, which takes place in three main ways: trans-anastomotic endothelialisation; transmural, and due to 'fall-out' of circulating pluripotent cells. Therefore, trying to enhance endothelialisation means acting on each of these three modalites of cell growth.

Trans-anastomotic endothelialisation (TAE) appears to be very difficult in humans. Early studies on synthetic prostheses report that they cannot be longer than 0.5 cm, even after prolonged implantation. In spite of a long period of observation, internal endothelialisation has not been observed in humans, except in sites of anastomosis [27].

Several factors have been observed to influence this, such as species, senescence, anatomic dimensions of the vessel, and prosthetic materials, but even in animals TAE is limited [28].

Study of endothelial cells, both human and canine, compared in vitro, suggest that human cells have a greater potential for migration but a lower capacity for adhesion, which may explain the lack of re-endothelialisation in vivo, when blood flow may obstruct cell adhesion [29].

Instead, the transmural pathway seems to enhance rapid endothelialisation, according to recent studies on materials with sufficient porosity. Pore size takes on importance in these studies, since the prosthesis must be sufficiently large to allow cell growth, but not too large to cause loss of intercellular adhesion [30].

Materials with differently sized pores inside and outside the conduit have even been experimented, in order to obtain an biocompatible surface internally and a colonizable one externally.

Pore size also alters the haemocompatibility of biomaterials, as well as their compliance and degradation time. An optimal pore size for vascular engineering has been hypothesized, ranging from 30 to 50 microns. It appears that smaller pores would not allow growth of endothelial cells, and larger ones would cause excessive leakage of blood.

Pores in the walls of prosthetic materials can also avoid intimal hyperplasia.

It has been hypothesized that a thrombus initially deposited on the walls of the prothesis later organizes itself into muscle-like tissue, which then gives rise to intimal hyperplasia.

The precocious growth of endothelial tissue would avoid thrombosis and thus the consequent cascade of events leading to intimal hyperplasia. Increased pore size causes increased failure compliance of the material.

Several studies have shown that vascular implants with fibers organized in a circular fashion do not cause dilation. Even surgical technique could determine an influence in local hyperplasia at the anastomotic site [31].

Anyway, this is local, since cells undergo mechanical stress and are thus conditioned in their spatial orientation. "Fall-out healing" leads to the formation of endothelial islands, with no connection with the formation of trans-anastomotic or transmural tissue. This is a late phenomenon that appears to be the mechanism for repairing small vascular lesions however, recent studies show how this mechanism may be enhanced, by attracting EPC cells to participate [32] One of the last procedures presented to improve the patency rate is constituted from nano modification of the vascular lumen with heparin addition to obtain a very low trombogenic vessel The employment and the wide possibility offered in the vascular field from the use of nanotechnology was also investigated [33].

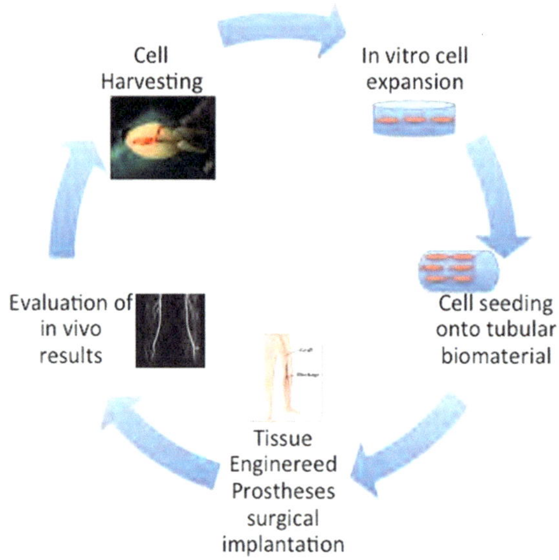

Figure 2. The in vivo –in vitro production of a tissue engineered prostheses. Adapted from: Pontini et al. "Alternative Conduits for Microvascular Anastomoses" Surg Innov. 2013 Aug 20;21(3):277-282

4. Methods and result in vascular tissue engineering

Many research groups have approached the problem of developing the ideal prosthesis in a variety of ways. Their main activities ranges around the triad scaffold-cells-growth factor. Scaffolds are ideal biomaterials for conduits, and cells can be seeded and cultivated on them, after preconditioning with various growth factors. To date, despite numerous scaffolds that have been manufactured trough varied forms of tissue engineering techniques, the construction of an entirely biomimetic blood vessels is still underway. To achieve a successful clinical application of tissue-engineered blood vessels, the bio fabrication of vascular grafts necessitates a vigorous yet time-efficient biotechnological process [34].

Several tissue engineering strategies have emerged to address biological flaws at the blood-material interface of the synthetic scaffolds, hence, paving the way to vascular cell seeding and design of bioactive polymers for *in situ* regeneration.

Moreover, advances in biomaterial design have been directed towards the generation of suitable materials that does not only mimic the native vascular tissue's mechanical properties but also promote cell growth, inhibit thrombogenicity, and facilitate extracellularr matrix production. In addition, an important characteristic of artificial scaffolds in advanced biomaterial vessels substitutes is not just the tolerance of the cells but the capacity to mimic the natural ECM in order to regulate extent and strength of cell adhesion, growth activity, cel ldifferentiation, and maturation to the desired phenotype [35, 36].

The extracellular matrix proteins such as collagen, elastin, fibronectin, vitronectin, and laminins which mediate cell-material adhesion have been thoroughly assessed in an earlier review [37].Materials for vascular replacements should be biomimetic in such a way that they should be resistant not only to thrombosis, but also to inflammation, and neointimal proliferation, and for all intents and purposes, they should resemble the native vessels [37]. For these reasons, it is necessary to investigate the physical, chemical, and biological properties and modifications of materials to further understand the molecular mechanism of the cell material interaction [37].

The lack of endothelial cells on the luminal surface of the artificial grafts contributes to synthetic graft thrombogenicity and promotes intimal proliferation within the graft. Endothelial cell (EC) seeding on the synthetic grafts has been attempted to mitigate these problems. The first group to perform endothelial cell (EC) isolation and their subsequent transplantation into vascular graft were Herring and his group [38]. Current researches indicate the significance of such process in vascular tissue engineering. The polymer surfaces which have been formerly investigated for endothelial attachment, proliferation, and function had been listed in an earlier review [39]. On the other hand, the synthetic polymers for reconstructing blood vessels for clinical practice which are based on polyethylene terephthalate (PET) or polytetrafluoroethylene (PTFE) had been previously reviewed [37].

Furthermore, blood vessel stem cells have been studied in combination with recent and alternative types of scaffolds/polymers. Parallel to this, in scaffold-based blood vessel engineering, bioreactors and pulsatile flow systems, designed by many scientists, have been found to progress the mechanical property of the engineered blood vessels by augmenting the deposition and remodeling of extracellular matrix as well as the maturation and differentiation of self-assembled micro tissues [40].

Bioreactors, which were originally designed for industrial use, have high degree of reproducibility, control, and automation for specific experimental bioprocesses and these have been the reasons for their transfer to large scale applications including vascular tissue engineering. The bioreactors allow scientists to manipulate the environment and the parameters such as pH, temperature, pressure, nutrient supply, and waste removal in order mimic the *in vivo* physiological condition and allow biological or biochemical processes to occur and subsequently develop the desired tissue [41].

Taken together, the formation of a microvasculature within a tissue-engineered organ or tissue will depend on multiple factors: the biochemical environment, EC type, the micro-architecture presented by the scaffold material, and mechanical signals [42]. Due to the goal of developing biomimetic blood vessel scaffolds, many groups have designed such biomaterials.

The polymers used in scaffold fabrication for tissue engineered blood vessels started from polyglycolic (PGA) to varied types such as polyglycolic acid-poly-L-lactic acid (PGAPLLA), Collagen/Elastin, chitosan, Poly (glycerol sebacate) (PGS), and very recently polyglycolide knitted fiber, and an L-lactide and ε-caprolactone copolymer sponge crosslinked to amniotic fluid. Furthermore, amniotic membranes have been used as scaffolds which signify that scaffold based tissue-engineered blood vessels can be fabricated from autologous cells at a reduced manufacturing period.

Resuming the most important that have been used as scaffolds, we could generally subdivided them into four main categories: permanent material included polyurethane material, resorbeable material and biological ones as allografts, xenografts, and derived products.

4.1. Permanent materials

Synthetic polymers such as e-PTFE and Dacron have not provided a satisfactory results in small diameter vessels (<6 mm). In literature are present a wide number of works on such implants as also the problems linked to the permanent synthetic polymers which are often difficult to overcome. Many works report variable results in terms of long-term patency. These materials have been implanted in humans, but do not develop an endothelialised surface, thus causing platelet adhesion and the development of a fibrin layer which may lead to thrombosis. Later failure may also be due to thrombosis after stenotic occlusion of the vessel consequent upon the development of endothelial hyperplasia. Several methods have been applied in the past to reduce thrombosis risk by application of tissue engineering procedure as application of anti-thrombotic drugs in their surface or surface ligands [43]. Bordenave et al illustrates how this procedure, in time, has moved from one-stage to two-stage techniques although, in spite of discouraging results, not much space was devoted to clinical practice, mainly because of its three most serious limitations: the impossibility of executing these techniques in emergencies; the need for prior withdrawal of cells; and the need for a GLP laboratory to treat human cells [15].

4.2. Polyurethane materials

These polymers are biocompatible and highly versatile, since their tensile strength and radial compliance vary according to segment composition, stiff segments being responsible for tensile strength and soft segments for elasticity [31] Originally produced as permanent biomaterials, they do deteriorate in vivo, due to oxidation and enzymatic and cell-mediated degradation, with the result that their biostability is under revision.

The differing composition of PU segments may lead to products with various degrees of biostability. PU have been combined with highly crystalline segments such as polycarbonates and silicon oligomers to increase their stability [34].

The most relevant development of small diameter vascular prostheses composed by PU revealed a total of 22 articles on polyurethanes, 14 in vitro, 4 on production of material and its medical properties, and only 4 in vivo.

The cellular compatibility of several PU (associated with other substances) has also been studied according to method of preparation, e.g., the use of porous structures. Electrospinning has been applied to other materials in the field of vascular engineering, and produces small diameter fibers with good tensile strength on the final material.

As regards chemical composition, PU has been combined with silk fibroin, showing better histocompatiblity of pure PU after implant in rat muscular tissue [44]. Many experiments have also been made on the mixed-composition PU PDMS (polydimethylsiloxane), a silicon-based polymer. In this case, PDMS not only increased biostability but also increased haemocompatibility and immunocompatibility. In vivo studies show encouraging long-term viability: in one, a PEUU/PDMS polymer was created with the spray phase inversion technique in a tubular form with two-phase porosity [45].

It showed good re-endothelialisation 24 months after implant. Another in vivo study in this series used poly(ester urethane)urea (PEUU) combined with a thrombogenic polymer not similar to a phospholipid, poly(2-methacryloyloxyethyl phosphorylcholine-co-methacryloyloxyethyl butylurethane) (PMBU), to create a fibrillar scaffold by electrospinning, with good tensile strength and compliance. The association with PMBU made the PU less prone to platelet deposition and hypertrophy of muscle cells. The in vivo patency of 1.3-mm conduits implanted in rat aorta after 8 weeks varied from 40% for pure PU to 67% for PU PMBU [46].

4.3. Bioresorbable materials

These materials may be synthetic or biopolymers already constituting the extracellular matrix. The most common absorbable biomaterials are polyesters. This category contains poly(α-hydroxylester poly(L-lactic acid) (PLLA), poly(-glycolic acid) (PGA), polylactone polyorthoesters (POE) and polycarbonates. When these materials are implanted in vivo, their polymeric structure is subject to a hydrolysis process and metabolization of the resulting products, such as lactic and glycolic acids.

Their safety and biocompatibility are now established.

However, the mechanical resistance of these products does not reach the desired levels – an anticipated outcome, as PGA was originally in the form of a non-woven fabric, and thus does not have measurable tensile strength [28]. Most studies have therefore concentrated on preconditioning methods to increase resistance, e.g., use of pulsatile flow bireactors, alternative techniques of cell culture, and administration of various growth factors [47.]Another substantial problem with PGA is its stiffness, which does not confer the elastic properties of native arterial tissues. In this case too, the use of copolymers has improved results. A fibrillar scaffold based on polyglycolic or polylactic acid coated with a 50:50 L-lactate or L-caprolactone (PCLA/PGA or PCLA/PLA) copolymer has been specifically tested for vascular repair, resulting in compliance closer to that of the original vessel with better surgical handling [48].

One study showed how PGA-based matrices have greater cellularity and production of proteins of extracellular matrices based on PHAV and P4HB. The authors explained this phenomenon as due to the higher porosity of PGA (> 90%), yielding a contact surface greater than that of cells.

To support the remodeling process in vivo, a biomaterial that functions only as a temporary absorbable guide, similar to an in vivo "Artery-Bioregeneration Assist Tube" (ABAT), which can promote the sequential and complete regeneration of vascular structures at the implantation site, entirely made of Hyaluronic Acid was used in different in vivo experimental model [49].

Other example of bioresorbeable material used for vascular purposed could be represented from collagen, which the author have experienced in small caliber vessel replacement in vivo experimental studies (Fig. 3)

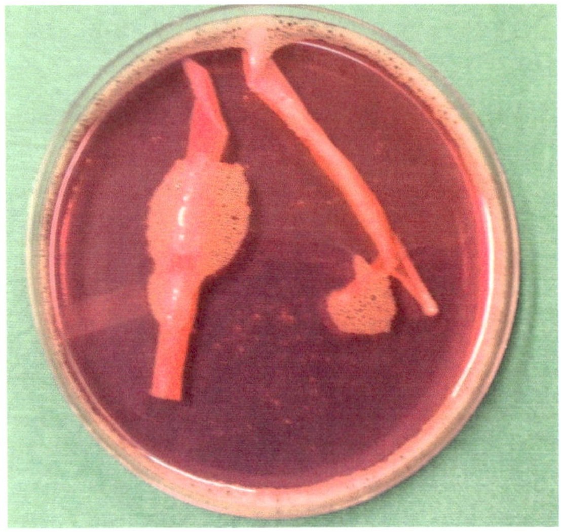

Figure 3. Example of pre-in vivo experimental implant of a biocompatible artificial collagen based artery scaffold. Protocol study of the author.

4.4. Biological materials

These biomaterials are widely available and they are of course excellent substrates for cell adhesion. In addition, the processing method can retain all their advantageous mechanical properties (tensile strength, elasticity) [50]. As the main disadvantages are possible residual antigenicity and infection after implant, techniques for their decellularization and sterilization have been refined [51]. Most recent article published on biological graft mainly deal with materials already naturally present as tubular structures in the body (arteries, veins, urethers) and submitted to decellularization. They are often studied as allo-or xenografts, and enriched with cells, bFGF, heparin and VEGF to improve patency in the long term. Of special interest for the physiopathology of tissue healing after implant is one study reporting trends after implants of decellularised porcine arteries in rat, concluding that the initial inflammation due to integration in tissues does not interfere with long-term modelling. One in vitro study examines the creation of a biotube produced by reaction to a foreign body1-3. These studies show the good mechanical properties of this biomaterial, but also the poor long-term patency of conduits [52, 53].

Decellularization have so represented one of the most reliable procedure to obtain an ideal scaffold for vascular replacement, in particular for its peculiar property to retain native ECM that is the fundamental aspect for cell seeding and cell host colonization (Fig.4).

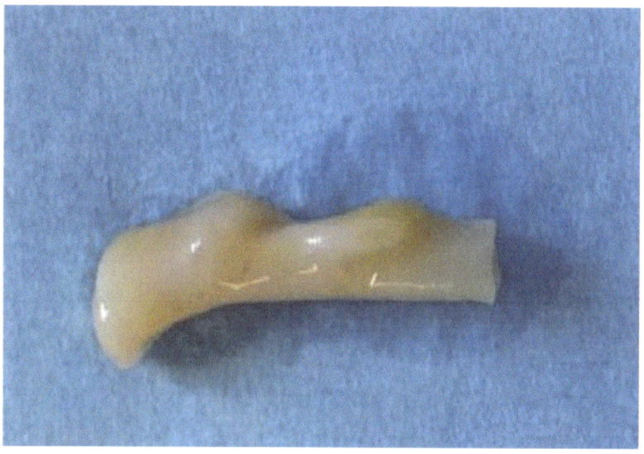

Figure 4. Example of pre-in vivo experimental implant of a decellularized artery. Protocol study of the authors.

There is not a unique decellularization protocol but every tissue need specific reactions and solutions. Protocols usually require a combination of physical, chemical and enzyme processes: the first phases are dedicated to rupture the cell membranes in order to release the intracellular components, which are then separated, dispersed, and degraded by enzymatic detergents and solutions. The last phases are directed to the elimination of cellular debris, that remained within the matrix, and of the reagents, that could interfere with the subsequent recellularization of scaffold or cause adverse reactions into recipient organism [54].

Some examples of physical treatment are the fast cycle of freeze/thaw that induce effective rupture of cell membranes, or the application of mechanical forces or the agitation of the material in combination with chemical detergents to promote the solubilization of homogenous membranous components [55].

As regards chemical treatments, the commonly solutions are usually composed by acid or alkaline solutions which are used for decellularization of thin layers of tissue samples as the bladder submucosa [56]. This type of treatment is effective in removing cell debris and, at the same time, has an antibacterial and antifungal activity [57]. Despite this, it has been demonstrated that the use of chemical solvents for prolonged times may alter the structure of the matrix, causing the detachment of glycosaminoglycans (GAGs) from collagen fibers [58]. Ionic detergents are another category of chemical agents usually employed during decellularization and they are involvede, with effectiveness, in removing the nuclear and cytoplasmic fragments. Their disadvantage could be the possible denaturation of ECM proteins.

In the decellularization protocols are also used chelating agents such as EDTA, whose function is to bind and ions, which are physiologicaly essential for covalent bonds between cells and matrix components, such as collagen fibers and fibronectin. In this way their use facilitate the loss of interconnections of the cells, resulting in disintegration and removal of cellular material [59].

To improve the effectiveness of the decellularization process, it is necessary the use of enzyme solutions, mainly consisting of trypsin, which specifically cleaves the protein bonds and exerts its maximum activity at the temperature of 37°C and pH 8. In addition to trypsin, is also frequently the use of the endonucleases, which include the deoxyribonuclease and ribonuclease. These enzymes catalyze the hydrolysis of covalent bonds of, respectively, DNA and RNA. However, it is recommended to avoid the exceeding exposure of samples with the action of the enzyme solutions because they could damage in the structure of the extracellular matrix. All decellularization solutions are usually implemented with antibiotic or antifungal agents to prevent any bacterial or fungal contamination.

To obtain an effectively decellularized scaffolds, it is important to remove the antigenic components, such as cell surface receptors, cytoplasmic proteins and nucleic acids that, due to their immunogenic properties, could trigger a defensive reaction of the immune system of the recipient. In contrast, components of the extracellular matrix, such as collagen and elastic fibers, are widely conserved among individuals of the same or different species and, therefore, they usually do not evoke the immunogenic reaction of the host. A low intensity inflammatory response of the receiving was observed, at least, at histological level and there have been reports of significant rejection reactions, as described by Kasimir and colleagues, after implant of porcine valve prosthesis in pediatric patients [60].

4.5. Clinical relevance and rationale for the use of ECM as a biologic scaffold

The use of ECM derived from decellularized tissue is increasingly frequent in regenerative medicine and tissue engineering strategies, with recent applications including the use of three-dimensional ECM scaffolds prepared by whole organ decellularization [61, 62]. Clinical products such as surgical mesh materials composed of ECM are harvested from a variety of allogenic or xenogeneic tissue sources, including dermis, urinary bladder, small intestine, mesothelium, pericardium, and heart valves, and from several different species. The potential advantage of tissue specificity for maintaining selected cell functions and phenotype has been suggested by studies of cells and ECM isolated from tissues and organs such as the liver [63], respiratory tract, nerve [64], adipose [65], and mammary gland [66]. The ECM has been shown to influence cell mitogenesis and chemotaxis, direct cell differentiation [67-71], and induce constructive host tissue remodelling responses [72-74]. It is likely that the three-dimensional ultrastructure, surface topology, and composition of the ECM all contribute to these effects. There is also evidence that residual cellular material attenuates or fully negates the constructive tissue remodelling advantages of biologic scaffold materials *in vivo* [75]. Therefore, tissue processing methods, including decellularization, are critical determinants of clinical success [76]. It should be understood that every cell removal agent and method will alter ECM composition and cause some degree of ultrastructure disruption. Minimization of these undesirable effects rather than complete avoidance is the objective of decellularization

4.6. Scaffold Free Method

An alternative method based on a scaffold free technique was also described and provide the advancement in the cell technology field. The application of such procedure depends from the necessity to improve the bioengineered vessel in terms to overcome process like the chronic inflammation, thrombosis, rejection, and poor mechanical properties of allogeneic or xenogeneic and synthetic vessels that as previously reported have impaired their clinical applications [77]. In addition it has emerged due to the failure cell to cell interaction and the assembly and alignment of ECM components, and the complex host response to scaffolds, [78]. In scaffold-free tissue engineering approach, the fabrication of the tissue construct is anchored in the crucial capability of the cells to manufacture their own extracellular matrix [39]. In 1998, the first scaffold-free tissue-engineered human blood vessel was established by L' Heureux and has been replicated for further preclinical evaluation using rat and mice models in 2006 [77]. Years later, groups of scientists reported a fully biological self-assembly approaches by implementing rapid prototyping bioprinting method and stimulation via bioreactors for scaffold-free small diameter vascular reconstruction [80].

Similar to scaffold-based technique, in tissue engineering for scaffold-free blood vessels, the bioreactors were also used to provide specific biochemical and physical signals to regulate cell differentiation, ECM production, and tissue assembly by using chemical, mechanical, or electromagnetic stimulation techniques to produce *de novo* tissue with properties comparable to the damaged or desired tissues [81, 82]. There are many types of launched bioreactors however, in engineering the vascular tissues, designs of various bioreactors have been based on the expansion and recoil properties of blood vessels, and so the combinations of stress, strain, and perfusion stimulation in biomimetic bioreactors have successfully developed vascular tissues [83, 84]. In case of cell senescence problem, lifespan extension via telomerase expression in vascular cells (smooth muscle cells and endothelial cells) from elderly patients has been found as an effective strategy for engineering autologous blood vessels and eventually provides bypass conduit for atherosclerotic diseases. Human telomerase, composed of an RNA component and a reverse transcriptase (hTERT), maintains the telomere length at the ends of the chromosomes [85, 86].Absence of hTERT expression in mature somatic cells induces lack of telomerase activity thus its ectopic expression has been shown to restore telomerase activity, arrest telomere shortening and senescence in some cells [87]. While high cell population is essential in cell-based vessel biofabrication and the expansion process is lengthy, cell-based therapies are more promising in terms of efficacy despite the fact that they are more complex and costly than scaffold-based techniques. Therefore, many researchers have focused on this approach and the representative studies are presented in. Among the well-studied scaffold-free techniques are the coculture system, sheet-based engineering, decellularization, direct cell injection, bioprinting, and bio fabrication in a bioreactor system.

5. Consideration on cost/effectiveness

Nevertheless systematic studies about tissue engineering costs and clinical benefit are not still available, due to many differences in procedure, techniques and materials, is well known that is a medicine promising but expensive tool. In fact, since now most of the result are due to high cost basic research with low effect on daily clinical practice and, in most cases, not widely available. The scientific surgeon community agree on necessity to use the best achievement in cell tecnologies to obtain an ideal and easy way to replace damaged tissue but also the cost to obtain it must be controlled. If the industry employ large sum of money to obtain a product its cost could be not widely available and so a novel promising technology could have a low impact worldwide.

More than 70 tissue engineering related start-up companies spent more than $600 million/year, with only two FDA-approved tissue-engineered products [88]. Given the modest performance in clinically approved organs, tissue engineering still remain a promising field. Often is a lacking in experimental model that avoid a perfect matching with human clinical situation. The community of bioengineering technology is advocating the application of clinically driven methodologies in large animal models enabling clinical translation.

The employment of sophisticated tecnologies in cell treatment as in decellularization process, cell isolation and in vitro expansion, costs of growth factor and bioreactor and so on are the most principles obstacle to a low cost wide available bioengineered tissue.

The huge clinical needing and the necessity of the low risk application of engineered substitute still represent a limitation in prevision of clinical application about most of the promising experimental result. At the same time such encouraging result provide the base for further development and spending limitation.

Since now all in vitro advancement in vessel replacement technology must consider its success not only in term of ideal vessel production but also in terms of feasibility, economic and time saving procedure.

6. Conclusion

Critical reading of researches in the field of microvascular tissue engineering gave the general impression of progress in the search for an ideal replacement for small diameter vessels but the goal its aimed to is still lacking. In fact, even if several studies seems to be promising they must be completely proved in vivo in human clinical situation and in long term period as also they must obtain a therapeutic result within an acceptable cost for the community.

Tissue Engineering in the context of Regenerative Medicine has been hailed for many years as one of the most important topics in medicine in the twenty-first century. While the first clinically relevant efforts were mainly concerned with the generation of bioengineered skin substitutes, subsequently tissue engineering applications have been continuously extended to a wide variety of tissues and organs.

The advent of either embryonic or mesenchymal adult stem-cell technology has fostered many of the efforts to combine this promising tool with tissue engineering approaches and has merged the field into the term Regenerative Medicine. As a typical example in translational medicine, the discovery of a new type of cells called *telocytes* that have been described in many organs and have been detected by electron microscopy opens another gate to regenerative medicine.

Besides cell-therapy strategies, the application of gene therapy combined with tissue engineering has been investigated to generate tissues and organs. The vascularization of constructs plays a crucial role besides the matrix and cell substitutes.

http://dx.doi.org/10.5772/59148

Therefore, novel in vivo models of vascularization have evolved allowing axial vascularization with subsequent transplantation of constructs.

This article is intended to give an overview over some of the most recent developments and possible applications in regenerative medicine through the perspective of tissue engineering achievements and cellular research. The synthesis of bioengineering with innovative methods of molecular biology and stem-cell technology appears to be very promising.

Most studies indicate the use of absorbable biomaterials, in view of their good integration, with the hope of developing autogenous vessels to replace prostheses. However, not one of these products has yet been approved for clinical experimentation. Degradability is one of the characteristics which tend to dissuade surgeons at the crucial moment of implant. In addition, synthetic not degradable material could not offer adequate surface to maintain and adequate patency in long period. There are many gaps in the examined articles. The first problem, already examined by many authors, is variability in animal models, which hinders direct comparison of results. Homogeneous studies on mechanical studies are also lacking, since so many of them focus on tensile strength, and neglect compliance, which is an essential feature of vessels.

An effective model of an artificial vessel is very far from being achieved, particularly considering the field of microvascular graft. So its development must take into account the context in which it could be applied. Experimental models have already been super-ceded, if we think that the application of a bio-absorbable prosthesis means that cells must be able to reconstruct a new artery and that, in clinical microsurgery practice, this must be achieved in already damaged arteries

However, the procedures are time-consuming and very expensive, requiring dedicated laboratories able to guarantee sterility and suitability for in vivo re-implantation of cell cultures.

As regards urgent procedures, such as revascularisation of all types, the cell culture step should be avoided. The ideal choice would be ready-to-use materials, that actually are needing their improvement.

Author details

A. Pontini[1*], M.M. Sfriso[2], M.I. Buompensiere[2], V. Vindigni[1] and F. Bassetto[1]

*Address all correspondence to: alex.pontini@sanita.padova.it

1 Department of Neurosensorial Specialties, Institute of Plastic Reconstructive Surgery and Burn Unit, Padova University Hospital, Padova, Italy

2 Department of Molecular Medicine, Human Anatomy Section, Padova University Hospital, Padova, Italy

References

[1] Tu JV, Pashos CL, Naylor CD, et al. Use of cardiac procedures and outcomes in elderly patients with myocardial infarction in the United States and Canada. N Engl J Med 1997;336:1500–1505.

[2] McKee JA, Banik SS, Boyer MJ, et al. Human arteries engineered in vitro. EMBO Rep 2003;4:633–638.

[3] Wang X, Lin P, Yao Q, Chen C. Development of small-diameter vascular graft. World J Surg. 2007 Apr;31(4):682-9

[4] J. L. Platt, "Preface: future approaches to replacement of organs, " *American Journal of Transplantation*, 2004 vol. 4, no. 6, pp.5–6.

[5] B. Ogle, M. Cascalho, and J. L. Platt, "Fusion of approaches to the treatment of organ failure, " *American Journal of Transplantation* 2004vol. 4, supplement 6, pp. 74–77.

[6] J. Yang, M. Yamato, C. Kohno et al., "Cell sheet engineering: recreating tissues without biodegradable scaffolds, " *Biomaterials*, 2005 vol. 26, no. 33, pp. 6415–6422.

[7] N. L'Heureux, N. Dusserre, A. Marini, S. Garrido, L. de la Fuente, and T. McAllister, "Technology insight: the evolution of tissue-engineered vascular grafts—from research to clinical practice, " *Nature Clinical Practice Cardiovascular Medicine*, 2007 vol. 4, no. 7, pp. 389–395.

[8] J. R. Porter, T. T. Ruckh, and K. C. Popat, "Bone tissue engineering: a review in bone biomimetics and drug delivery strategies, 2009 *Biotechnology Progress*, vol. 25, no. 6, pp. 1539– 1560.

[9] W. Ji, Y. Sun, F. Yang et al., "Bioactive electrospun scaffoldsdelivering growth factors and genes for tissue engineering applications, " *Pharmaceutical Research*, 2011 vol. 28, no. 6, pp. 1259–1272.

[10] J. P. Vacanti and R. Langer, "Tissue engineering: the designand fabrication of living replacement devices for surgical reconstruction and transplantation, " 1999 *The Lancet*, vol. 354, supplement 1, pp. S32–S34,

[11] B. S. Kim and D. J. Mooney, "Development of biocompatible synthetic extracellular matrices for tissue engineering, " *Trends in Biotechnology*, 1998 vol. 16, no. 5, pp. 224–230

[12] Judee Grace Nemeno-Guanzon, Soojung Lee, Johan Robert Berg, 1, 2 Yong Hwa Jo, Jee Eun Yeo,, 3 BoMi Nam, Yong-Gon Koh, and Jeong Ik Lee Trends in Tissue Engineering for Blood Vessels J Biomed Biotechnol. 2012;2012:956345.

[13] McKee JA, Banik SS, Boyer MJ, et al. Human arteries engineered in vitro. EMBO Rep 2003;4:633–638.

[14] Guidoin R, Chakfé N, Maurel S, How T, Batt M, Marois M, Gosselin C. (1993). Expanded polytetrafluoroethylene arterial prostheses in humans: histopathological study of 298 surgically excised grafts. 1993 Biomaterials;14(9):678-93

[15] Bordenave L, Fernandez P, Rémy-Zolghadri M, Villars S et al. In vitro endothelialized ePTFE prostheses: clinical update 20 years after the first realization. Clin Hemorheol Microcirc; 2005 33(3):227-34

[16] Cooper GJ, Underwood MJ, Deverall PB Arterial and venous conduits for coronary artery bypass. A current review. Eur J Cardiothorac Surg. 1996;10(2):129-40

[17] Klopsch C, Steinhoff G. Tissue-engineered devices in cardiovascular surgery. Eur Surg Res. 2012;49(1):44-52

[18] Tillman BW, Yazdani SK, Neff LP, Corriere MA, Christ GJ, Soker S, Atala A, Geary RL, Yoo JJ. Bioengineered vascular access maintains structural integrity in response to arteriovenous flow and repeated needle puncture. J Vasc Surg. 2012 Sep;56(3): 783-93.

[19] Schmedlen RH, Elbjeirami WM, Gobin AS, West JL. Tissue engineered small-diameter vascular grafts. Clin Plast Surg. 2003 Oct;30(4):507-17.

[20] Mooney DJ Mazzoni CL, Breuer C, McNamara K, Hern D, Vacanti JP, Langer R. Stabilized polyglycolic acid fibre-based tubes for tissue engineering. Biomaterials; 1996 17(2):115-24

[21] Kim BS, Mooney DJ Engineering smooth muscle tissue with a predefined structure. J Biomed Mater Res; 1998, 41(2): 322-32.

[22] Yao Y, Wang J, Cui Y, Xu R, Wang Z, Zhang J, Wang KLi Y, Zhao Q, Kong D. Effect of sustained heparin release from PCL/chitosan hybrid small-diameter vascular grafts on anti-thrombogenic property and endothelialization. Acta Biomater. 2014 Jun;10(6):2739

[23] Laube HR, Duwe J, Rutsch W, Konertz W Clinical experience with autologous endothelial cell-seeded polytetrafluoroethylene coronary artery bypass grafts. J Thorac Cardiovasc Surg; 2000, 120(1):134-41.

[24] Kumar TR, Krishnan LK A stable matrix for generation of tissue-engineered nonthrombogenic vascular grafts. Tissue Eng. 2002, Oct;8(5):763-70

[25] Zhang Z, Wang Z, Liu S, Kodama M. (2004). Pore size, tissue ingrowth, and endothelialization of small-diameter microporous polyurethane vascular prostheses. Biomaterials; 2004. 25(1):177-87

[26] Matsuda T, Nakayama Y. (1996). Surface microarchitectural design in biomedical applications: in vitro transmural endothelialization on microporous segmented polyurethane films fabricated using an excimer laser. J Biomed Mater Res; 1996 31(2):235-42.

[27] Berger K, Sauvage LR, Rao AM, Wood SJ (1972). Healing of arterial prostheses in man: its incompleteness. Ann Surg; 1972 175(1):118-27

[28] Zilla P, Bezuidenhout D, Human P (2007). Human, Prosthetic vascular grafts: wrong models, wrong questions and no healing. Biomaterials; 2007 28(34):5009-2.

[29] Dixit P, Hern-Anderson D, Ranieri J, Schmidt CE. (2001). Vascular graft endothelialization: comparative analysis of canine and human endothelial cell migration on natural biomaterials. J Biomed Mater Res;2001, 56(4):545-55.

[30] Kidane AG, Salacinski H, Tiwari A, Bruckdorfer KR, Seifalian AM. Anticoagulant and antiplatelet agents: their clinical and device application(s) together with usages to engineer surfaces. Biomacromolecules; 2004 5(3):798-813.

[31] Tiwari A, Cheng KS, Salacinski H, Hamilton G, Seifalian AM.. Improving the patency of vascular bypass grafts: the role of suture materials and surgical techniques on reducing anastomotic compliance mismatch. Eur J Vasc Endovasc Surg; 200325(4): 287-95.

[32] Avci-Adali M, Ziemer G, Wendel HP. Induction of EPC homing on biofunctionalized vascular graftsfor rapid in vivo self-endothelialization--a review of current strategies. Biotechnol Adv; 201028(1):119-29

[33] Rathore A, Cleary M, Naito Y, Rocco K, Breuer C. Development of tissue engineered vascular grafts and application of nanomedicine Wiley Interdiscip Rev Nanomed Nanobiotechnol. 2012 May-Jun;4(3):257-72

[34] Atala AL et al. Principles of Regenerative Medicine. Elsevier: Burlington, Massachusetts. First ed ed. 2008

[35] L. Bacakova, E. Filova, F. Rypacek, V. Svorcik, and V. Stary, "Cell adhesion on artificial materials for tissue engineering, " *Physiological Research*, 2004 vol. 53, supplement 1, pp. S35–S45.

[36] L. Bacakova, E. Filova, D. Kubies et al., "Adhesion and growth of vascular smooth muscle cells in cultures on bioactive RGD peptide-carrying polylactides, " *Journal of Materials Science* 2007 vol. 18, no. 7, pp. 1317–1323.

[37] M. Parizek, K. Novotna, and L. Bacakova, "The role of smoothmuscle cells in vessel wall pathophysiology and reconstruction using bioactive synthetic polymers, " *Physiological Research*, 2011 vol. 60, no. 3, pp. 419–437

[38] M. Herring, A. Gardner, and J. Glover, "A single staged technique for seeding vascular grafts with autogenous endothelium, " *Surgery*, 1978, vol. 84, no. 4, pp. 498–504.

[39] H.M. Nugent and E. R. Edelman, "Tissue engineering therapy for cardiovascular disease, " *Circulation Research*, 2003, vol. 92, no.10, pp. 1068–1078.

[40] L. Buttafoco, P. Engbers-Buijtenhuijs, A. A. Poot, P. J. Dijkstra, I. Vermes, and J. Feijen, "Physical characterization of vascular grafts cultured in a bioreactor, " *Biomaterials*, 2006 vol. 27, no. 11, pp. 2380–2389.

[41] Martin, D.Wendt, and M. Heberer, "The role of bioreactorsin tissue engineering, " *Trends in Biotechnology*, 2004vol. 22, no. 2, pp. 80–86.

[42] M. C. Peters, P. J. Polverini, and D. J. Mooney, "Engineeringvascular networks in porous polymer matrices, " *Journal of Biomedical Materials Research*, 2002 vol. 60, no. 4, pp. 668–678.

[43] Kidane AG, Salacinski H, Tiwari A, Bruckdorfer KR, Seifalian AM. Anticoagulant and antiplatelet agents: their clinical and device application(s) together with usages to engineer surfaces. Biomacromolecules;2004, 5(3):798-813.

[44] Wang W, Jin B, Ouyang C, Li Yet al. Acute phase reaction of different macromolecule vascular grafts healing in rat muscle. Sheng Wu Gong Cheng Xue Bao ;2010, 26(1): 79-84.

[45] Khorasani MT, Shorgashti S. (2006). Fabrication of microporous polyurethane by spray phase inversion method as small diameter vascular grafts material. J Biomed Mater Res A;2006, 77(2):253-60.

[46] Hong Y, Ye SH, Nieponice A, Soletti L et al. A small diameter, fibrous vascular conduit generated from a poly(ester urethane)urea and phospholipid polymer blend. Biomaterials;2009, 30(13):2457-67.

[47] Sodian R, Hoerstrup SP, Sperling JS, Martin DP et al. Evaluation of biodegradable, three-dimensional matrices for tissue engineering of heart valves. ASAIO J;2000, 46(1):107-10.

[48] Watanabe M, Shin'oka T, Tohyama S, Hibino N et al. Tissue-engineered vascular autograft: inferior vena cava replacement in a dog model. Tissue Eng;2001, 20017(4): 429-39.

[49] Pandis L, Zavan B, Abatangelo G, Lepidi S, Cortivo R, Vindigni V.. Hyaluronan-based scaffold for in vivo regeneration of the rat vena cava: Preliminary results in an animal model. J Biomed Mater Res A2010, ;93(4):1289-96.

[50] Schmidt CE, Baier JM. Acellular vascular tissues: natural biomaterials for tissue repair and tissue engineering. Biomaterials;2000, 21(22):2215-31.

[51] Chlupác J, Filová E, Bacáková L. Blood vessel replacement: 50 years of development and tissue engineering paradigms in vascular surgery. Physiol Res; 2009, 58 Suppl 2:S119-39.

[52] Hinds MT, Rowe RC, Ren Z, Teach J, Wu PC et al. Development of a reinforced porcine elastin composite vascular scaffold. J Biomed Mater Res A;2006, 77(3):458-69.

[53] Pavcnik D, Obermiller J, Uchida BT, Van Alstine W et al. Angiographic evaluation of carotid artery grafting with prefabricated small-diameter, small-intestinal submucosa grafts in sheep. Cardiovasc Intervent Radiol. 2009, 32(1):106-13.

[54] Gilbert TW, Sellaro TL, Badylak SF. Decellularization of tissues and organs Biomaterials. 2006 Jul;27(19):3675-83

[55] Schenke-Layland K, Vasilevski O, Opitz F, König K, Riemann I, Halbhuber KJ, Wahlers T, Stock UA. Impact of decellularization of xenogeneic tissue on extracellular matrix integrity for tissue engineering of heart valves. J Struct Biol. 2003 Sep;143(3): 201-8

[56] Wilshaw SP, Kearney JN, Fisher J, Ingham E. Production of an acellular amniotic membrane matrix for use in tissue engineering Tissue Eng. 2006 Aug;12(8):2117-29.

[57] Petersen TH, Calle EA, Colehour MB, Niklason LE. Matrix composition and mechanics of decellularized lung scaffolds Cells Tissues Organs. 2012;195(3):222-31

[58] Teebken OE[1], Bader A, Steinhoff G, Haverich A.Tissue engineering of vascular grafts: human cell seeding of decellularised porcine matrix Eur J Vasc Endovasc Surg. 2000 Apr;19(4):381-6

[59] Grauss RW, Hazekamp MG, Oppenhuizen F, van Munsteren CJ, Gittenberger-de Groot AC, DeRuiter MC. Histological evaluation of decellularised porcine aortic valves: matrix changes due to different decellularisation methods. Eur J Cardiothorac Surg. 2005 Apr;27(4):566-71

[60] Kasimir MT, Rieder E, Seebacher G, Silberhumer G, Wolner E, Weigel G, Simon PComparison of different decellularization procedures of porcine heart valves Int J Artif Organs. 2003 May;26(5):421-7

[61] K.Weinzierl, A. Hemprich, and B. Frerich, "Bone engineeringwith adipose tisssue derived stromal cells, " *Journal of Cranio-Maxillofacial Surgery* 2006, vol. 34, no. 8, pp. 466–471.

[62] Y. Zhu, T. Liu, K. Song, X. Fan, X. Ma, and Z. Cui, "Adiposederived stem cell: a better stem cell than BMSC, " *Cell Biochemistry and Function* 2008, vol. 26, no. 6, pp. 664–675.

[63] S. H. Bhang, S. W. Cho, J. M. Lim et al., "Locally delivered growth factor enhances the angiogenic efficacy of adiposederived stromal cells transplanted to ischemic limbs, " *Stem Cells*, 2009 vol. 27, no. 8, pp. 1976–1986.

[64] K. Rubina, N. Kalinina, A. Efimenko et al., "Adipose stromal cells stimulate angiogenesis via promoting progenitor cell differentiation, secretion of angiogenic factors, and enhancing vessel maturation, " *Tissue Engineering A*, 2009 vol. 15, no. 8, pp. 2039–2050.

[65] T. J. Lee, S. H. Bhang, H. S. Yang et al., "Enhancement of longterm angiogenic efficacy of adipose stem cells by delivery of FGF2, " *Microvascular Research*, 2012vol. 84, no. 1, pp. 1–8.

[66] Sterodimas, J. de Faria, B.Nicaretta, and I. Pitanguy, "Tissue engineering with adipose-derived stem cells (ADSCs): current and future applications, " *Journal of Plastic, Reconstructive and Aesthetic Surgery*, 2010 vol. 63, no. 11, pp. 1886–1892.

[67] S. Levenberg, J. S. Golub, M. Amit, J. Itskovitz-Eldor, and R. Langer, "Endothelial cells derived from human embryonic stem cells, " *Proceedings of the National Academy of Sciences of the United States of America*, 2002 vol. 99, no. 7, pp. 4391–4396.

[68] M. Hristov, W. Erl, and P. C. Weber, "Endothelial progenitor cells: mobilization, differentiation, and homing, " *Arteriosclerosis, Thrombosis, and Vascular Biology*, 2003 vol. 23, no. 7, pp. 1185– 1189.

[69] M. T. Hinds, M. Ma, N. Tran et al., "Potential of baboon endothelial progenitor cells for tissue engineered vascular grafts, " *Journal of Biomedical Materials Research A*, 2008 vol. 86, no.3, pp. 804–812.

[70] X. Wu, E. Rabkin-Aikawa, K. J. Guleserian et al., "Tissueengineered microvessels on three-dimensional biodegradable scaffolds using human endothelial progenitor cells, " *American Journal of Physiology*, 2004 vol. 287, no. 2, pp. H480–H487.

[71] J. M. Hill, G. Zalos, J. P. J. Halcox et al., "Circulating endothelial progenitor cells, vascular function, and cardiovascular risk, " *The New England Journal of Medicine*, 2003 vol. 348, no. 7, pp. 593–600.

[72] Kawamoto, T. Asahara, and D. W. Losordo, "Transplantation of endothelial progenitor cells for therapeutic neovascularization, " *Cardiovascular RadiationMedicine*, 2002 vol. 3, no. 3-4, pp. 221–225.

[73] T. Shirota, H. He, H. Yasui, and T. Matsuda, "Human endothelial progenitor cell-seeded hybrid graft: proliferative and antithrombogenic potentials *in vitro* and fabrication processing, " *Tissue Engineering*, 2003, vol. 9, no. 1, pp. 127–136.

[74] S. Kaushal, G. E. Amiel, K. J. Guleserian et al., "Functionalsmall-diameter neovessels created using endothelial progenitor cells expanded *ex vivo*, " *Nature Medicine*, 2001, vol. 7, no. 9, pp. 1035–1040.

[75] A. Kocher, M. D. Schuster, M. J. Szabolcs et al., "Neovascularization of ischemic myocardium by human bone-marrowderived angioblasts prevents cardiomyocyte apoptosis, reduces remodeling and improves cardiac function, " *Nature Medicine*, 2001 vol. 7, no. 4, pp. 430–436.

[76] Assmus, V. Sch°achinger, C. Teupe et al., "Transplantation of progenitor cells and regeneration enhancement in acute myocardial infarction (TOPCARE-AMI), " *Circulation*, 2002, vol. 106, no. 24, pp. 3009–3017.

[77] N. L'Heureux, S. P^aquet, R. Labb′e, L. Germain, and F. A. Auger, "A completely biological tissue-engineered human blood vessel, " *The FASEB Journal*, 1998 vol. 12, no. 1, pp. 47–56.

[78] L. Germain, M. Remy-Zolghadri, and F. Auger, "Tissue engineering of the vascular system: from capillaries to larger blood vessels, " *Medical and Biological Engineering and Computing*, 2000, vol. 38, no. 2, pp. 232–240.

[79] H. Ozaki and H. Karaki, "Organ culture as a usefulmethod for studying the biology of blood vessels and other smoothmuscle tissues, " *Japanese Journal of Pharmacology*, 2002, vol. 89, no. 2, pp. 93–100.

[80] N. L'Heureux, N. Dusserre, G. Konig et al., "Human tissue engineered blood vessels for adult arterial revascularization, " *Nature Medicine*, 2006, vol. 12, no. 3, pp. 361–365.

[81] C. Norotte, F. S. Marga, L. E. Niklason, and G. Forgacs, "Scaffold-free vascular tissue engineering using bioprinting, " *Biomaterials*, 2009, vol. 30, no. 30, pp. 5910–5917

[82] S. Chaterji, K. Park, and A. Panitch, "Scaffold-free *in vitro* arterial mimetics: the importance of smooth muscle-endothelium contact, " *Tissue Engineering A*, 2010 vol. 16, no. 6, pp. 1901– 1912.

[83] Z. H. Syedain, L. A. Meier, J. W. Bjork, A. Lee, and R. T. Tranquillo, "Implantable arterial grafts from human fibroblasts and fibrin using a multi-graft pulsed flow-stretch bioreactor with noninvasive strength monitoring, " *Biomaterials*, 2011, vol. 32, no. 3, pp. 714–722.

[84] J. Zhao, L. Liu, J.Wei et al., "A novel strategy to engineer smalldiameter vascular grafts from marrow-derived mesenchymal stem cells, " *Artificial Organs*, 2012, vol. 36, no. 1, pp. 93–101.

[85] L. Bacakova, E. Filova, F. Rypacek, V. Svorcik, and V. Stary, "Cell adhesion on artificial materials for tissue engineering, " *Physiological Research*, 2004 vol. 53, supplement 1, pp. S35–S45.

[86] E. Oragui, M. Nannaparaju, and W. S. Khan, "The role of bioreactors in tissue engineering for musculoskeletal applications, " *The Open Orthopaedics Journal*, 2011, vol. 5, supplement 2, pp. 267–270.

[87] N. Plunkett and F. J. O'Brien, "IV.3. bioreactors in tissue engineering, " *Studies in Health Technology and Informatics*, 2010 vol. 152, pp. 214–230.

[88] T. M. Nakamura, G. B. Morin, K. B. Chapman et al., "Telomerase catalytic subunit homologs from fission yeast and human, " *Science*, vol. 277, 1997 no. 5328, pp. 955–959.

[89] Othman SF, Xu H, Mao JJ. Future role of MR elastography in tissue engineering and regenerative medicine. J Tissue Eng Regen Med. 2013

Chapter 11

Production of Tissue-Engineered Human 3D Bronchi *In Vitro*

Sara Bouhout, Jadson Moreira Pereira, Franck Simon, Stéphane Chabaud, Stéphane Bolduc, Massimo Conti, Alex Therien, Eric Rousseau, Véronique J. Moulin, Jean-Pierre Lavoie, Michel Rouleau and Francine Goulet

Additional information is available at the end of the chapter

http://dx.doi.org/10.5772/59192

1. Introduction

1.1. Respiratory disorders and goblet cells

Asthma is experienced during the life span in approximately 10% of the population [1]. Chronic obstructive pulmonary disease (COPD) is a complex inflammatory airway disease that results in airflow limitation that is not fully reversible. Many animal model systems have been developed that recapitulate various features of COPD but all present significant limitations [2-3]. The lack of understanding of the underlying mechanisms and mediators that drive the onset and progression of chronic inflammation, emphysema and changes in lung function have limited the development of useful models and subsequently effective treatments. Most of the normal human airway is lined by a pseudostratified epithelium of ciliated cells, secretory cells and 6–30% basal cells, the proportion of which varies along the proximal-distal axis [4]. Among several other bronchial cell types, goblet cells (GC) are important players in the development and maintenance of pulmonary disorders. Disturbance of their distribution and phenotypic features, hyperplasia, metaplasia, dysplasia and mucus hypersecretion are common pathological features of chronic respiratory diseases, including asthma, COPD, lung cancer, and cystic fibrosis. Despite numerous studies, the molecular basis for GC phenotypic changes remain elusive [5-6]. In asthmatic bronchi, the number of GCs is increased, due to the transdifferentiation of ciliated and club cells (Clara) to GCs, resulting in more abundant mucus production [7]. GCs secrete the mucins MUC5B and MUC5AC, which are responsible for the viscoelasticity and hydration of the mucus covering the ciliated escalator [7]. The sputum in asthmatic bronchi is often so dry that it can lead to airway obstruction. An intratracheal injection of IL-13 in mice models leads also to hyperproliferation of GCs, since this cytokine induces the pathophysiological features of asthma in a manner that is independent of immunoglobulin E and eosinophils [8]. MUC5AC is the predominant mucin in the epithelium, whereas MUC5B is predominant in the submucosal glands [9]. In cystic fibrosis, the primary defect is dehydration of the mucus layer resulting in a failure of mucus clearance leading to plaque accumulation on the airway surface causing persistent airway infections, [10]. Modulation of GCs function could contribute to regulate the symptoms and or the evolution of respiratory diseases. Bronchial models are needed to evaluate new molecules to inhibit or promote selective mucus secretion pathways.

1.2. Animal models

Several animal models have been proposed or used to assess physiopathologic mechanisms leading to pulmonary diseases. The rat [11-12], the mouse [13-14], the guinea pig [15], the dog [16] and the horse [17-20] have been exploited to establish potential mechanisms of control of respiratory diseases. Despite valuable data gathered from animal experimentation, the failure to translate promising drug candidates from animal models to humans has led to demand for more predictive models and tools based on the latest technologies [2]. The current paradigm where animal models of allergen sensitization and challenge are considered the gold standard falls some way short of human asthma [3]. The use of human tissue-engineered bronchial models (BMs) may facilitate the design of studies that shed light on pathogenesis, which in turn can lead to the development of worthwhile therapeutic interventions [21].

1.3. Tissue-engineered airway models *in vitro*

In vitro models are proposed in the literature. One of the *in vitro* approach consists in a model of trachea and bronchi, ventilated through an endotracheal tube during controlled mechanical ventilation, assist control, pressure support, and continuous positive airway pressure, with a dry and humidified ventilator circuit [22]. This model is original and clinically useful, but such experimentation requires specific equipment and is mostly limited to assess conditions of administration of nebulizers and metered-dose inhalers. In terms of cell culture models, several cell populations are commercially available. For example, Clonetics™ human primary airway cells can be useful to produce airway cell monolayers in the study of respiration, mucin production, cilia formation and airway model cell culture [23]. However, two-dimensional cultures exclude the possibility to reproduce intercellular interactions with other cell populations. Three-dimensional models of different degrees of complexity present such advantage, allowing two or more cell types, e.g. epithelial and mesenchymal cells, to interact in culture [24-28]. Some airway models include immune cells [29-30], but ideally, an autologous culture system is required.

1.4. Human tissue-engineered three-dimensional BM

We were the first to establish the methods for HBEC and HBEF isolation from bronchial biopsies taken during standard bronchoscopy [31]. Using these two cell types and a native collagen matrix, we produced the first three-dimensional human normal and asthmatic BMs in culture, using the tissue-engineering approach [27-28]. We produced tubular and disk-shaped BMs. Cell culture is often criticized as it doesn't provide the support for the maintenance of cells phenotype. Freshly isolated HBEC lose their ciliated phenotype and similarly, GCs are not detectable in primary cultures. However, three-dimensional tissue culture systems significantly differ from two-dimensional (monolayer) cell culture, providing an environment that enhances cell differentiation and organization, matrix remodeling, cell-cell and cell-matrix interactions *in vitro* [24,27-28,32-36]. The main challenge remains to define the best and specific conditions that apply to each cell type *in vitro*. In the case of bronchi, some essential factors, e.g. retinoic acid, combined to specific culture conditions, such as the air-liquid interface, stimulate and maintain functional and differentiated properties of HBEC in tissue-engineered BMs in vitro [27-28]. This chapter presents the potential of tissue-engineered BMs, as an alternative to animal use, to study respiratory diseases *in vitro*.

2. Experimental approach

2.1. Collagen matrix and bronchial cell source

Our models are produced using bovine type I collagen as a matrix. It is a natural substrate, so HBEC attach, migrate and proliferate on it. HBFC seeded in this matrix grow, synthesize de novo human collagen and contract microfibers by gathering them into fibers. In the disk-shaped BMs, cells and collagen fibers become aligned horizontally, in the same plan with the tension induced by the peripheral anchorage that prevents circumferential matrix contraction. Human asthmatic and healthy bronchial cells were isolated from biopsies of volunteers taken during standard bronchoscopy [31]. The asthmatic donors (aged from 20 to 50 years-old), had never taken anti-inflammatory drugs, were non-smokers and their asthma required only an inhaled β2 agonist agent on demand. The study was approved by the local Ethics Committee and subjects had given informed written consent.

2.2. How do we produce human tissue-engineered BMs?

Step 1.Preparation of the mesenchymal layer of the BMs

A mixture of bovine Type I collagen (2,0 mg/ml) is prepared by dissolving the powder overnight at 4°C, in sterile 0.017 M acetic acid. A solution of 0.84 ml of DMEM 2.7X pH 7,4, containing 200 IU/ml penicillin G and 50 μg/ml of gentamicin, pH 8,0, is mixed with a second solution containing 1.43 ml of the stock collagen solution and 30 μl of NaOH 0.7 N to stabilize the final pH at 7,4. It is important to mention that collagen polymerization is pH-dependent and that it occurs at a pH above 7.0. However, cells should not be mixed with an acidic collagen solution. Thus, 0,56 ml of fetal calf serum (FCS) is mixed with 0.15 ml of a HBFC suspension (1×10^6 cells/ml; Figure 1B). Non-asthmatic and asthmatic HBFC are seeded individually in corresponding mesenchymal layers to produce non-asthmatic and asthmatic BMs, respectively. The cell suspension is slowly mixed with the neutralized collagen solution. This mixture (pH 7,4) is quickly distributed in a bacteriological Petri dish (35-mm diameter) already containing the peripheral anchorage (sterile ring of Whatman paper), to produce the mesenchymal layers for each BM. The anchorage method prevents collagen lattice to contract diametrically [27]. The dish must not be shaken until collagen polymerization has occured (20-30 min), at room temperature. The mesenchymal layers of the BMs are covered with 2 ml of DMEM supplemented with 10% FCS, 100 IU/ml penicillin G and 25 ug/ml gentamicin following collagen polymerization and cultured in this medium until their epithelialization. All mesenchymes are kept in an 8% CO_2 atmosphere at 37°C thereafter.

Step 2.Epithelialization of the BMs under submerged culture conditions.

Four days later, the epithelialization is performed by seeding HBEC (8 x 10^5 cells/ BM; Figure 1A), on the mesenchymal layers. Again, non-asthmatic and asthmatic HBEC are seeded on the corresponding mesenchymal layers. The constructs are maintained under submerged culture conditions in medium supplemented with 10% FCS, 1,8 x 10^{-6} M (or 24,3 ug/ml) adenine, 1,6 x 10^{-9} M (or 10 ng/ml) human EGF, 8,6 x 10^{-7} M (or 5 ug/ml) bovine crystallized insulin, 6,2 x 10^{-4} M (or 5 ug/ml) human transferrin, 2 x 10^{-9} M (or ug/ml) 3,3',5', triiodo-L-thyronin, 1,1 x 10^{-6} M (or 0.4 ug/ml) hydrocortisone, until a confluent epithelial cell layer is obtained. The culture media are changed daily, (Figure 1C).

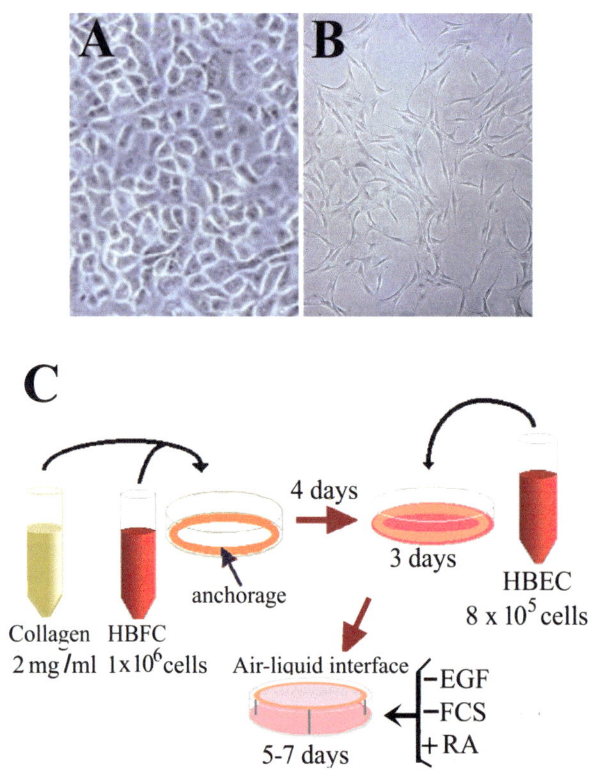

Figure 1. Photomicrographs taken under phase contrast microscope of non-asthmatic HBEC (A) and HBFC (B), in primary culture. (C): Steps of production of human tissue-engineered BMs *in vitro*.

Step 3.Culture at the air-liquid interface.

On the fourth day after epithelialization, the BMs are raised at the air-liquid interface in serum-free medium supplemented with 1,8 x 10^{-6}M adenine, 5 ug/ml bovine crystallized insulin, 6,2 x 10^{-4} M human transferrin, 2 x 10^{-9} M 3,3',5', triiodo-L-thyronin, 1,1 x 10^{-6} M hydrocortisone, and 5 x 10^{-8} M retinoic acid (RA). EGF and FCS are not added in the medium since these growth factors enhance the production of gelatinases by the cells, which degrade the collagen matrix of the mesenchymal layer in culture at the air-liquid interface. The BMs are placed on Petri dishes containing an internal elevated support (Falcon No 3037), or on any other type of air-liquid support. They are cultured at the air liquid interface for 5 to 7 days or more, according to each experiment. The culture media must be changed daily. Each experiment is always done 3 times on at least 3 samples per group of BMs tested.

2.3. Histological analyses

Human bronchial models were fixed with formol 10% and then embedded in paraffin. The 4 μm thick sections were stained using the hematoxylin-eosin staining.

2.4. Scanning electron microscope analyses

The samples were fixed with a solution of 2% glutaraldehyde in 0.1 M sodium cacodylate buffer, pH 7.5 for 24 hrs, post-fixed with 1% osmium tetroxyde, dehydrated with ethanol and coated with gold (Sputtercoater, Nanotech, England). Photographs were performed using a JEOL JSM-35CF scanning electron microscope (Université Laval).

2.5. Technical tips to isolate pure populations of HBEC and HBFC to produce various BMs

The methods established to produce non-and asthmatic BMs could be applied to develop other BMs, using bronchial cells isolated from 5-10 small biopsies (≤ 1 mm^2) taken from tumoral, cystic fibrosis, MPOC or other tissue types. HBEC and HBFC can be isolated from the same bronchial biopsies [31]. They are initially plated in the same dishes in primary culture, and colonies of both cell types can be observed under phase contrast microscope after 7-10 days. To purifie HBEC, some culture dishes containing at least one colony of HBEC were selected. The addition of lethally irradiated but living 3T3 cells in the selected dishes, allowed the progressive elimination of the HBFC. Then, the colonies of HBEC slowly reached confluence, while the 3T3 cells progressively detached from the dishes. The lethally irradiated 3T3 play three roles; 1) they still secrete some growth factors as feeder layer, 2) they increase cell density to support small colonies of HBEC in primary culture and 3) they contribute to eliminate HBFC that would overgrow HBEC. This methodological approach leads to reproducible and excellent results in obtaining pure HBEC and HBFC populations. This approach has great potential to isolate HBEC and HBFC from various bronchial tissues. The best medium to maintain bronchial cells viability after long-term storage in liquid nitrogen is composed of 90% FCS and 10% DMSO.

2.6. Ciliogenesis in bronchial epithelia grown on three-dimensional mesenchymes

To screen and target mechanisms that initiate and promote structural and functional disorders in bronchi, tissue-engineered three-dimensional BMs are excellent tools. Ciliogenesis, mucus secretion, pseudo-stratified epithelium and collagen remodeling have been observed in all human non-asthmatic and asthmatic BMs, produced under defined culture conditions [27-28,31]. Non-asthmatic BMs show well-differentiated pseudo-stratified epithelia (Figure 2A). In asthmatic BMs, partial desquamation occurs after 7-10 days in culture at the air-liquid interface, in a fashion that mimics clinical histopathological observations [28]. In addition, the asthmatic cells show an irregular distribution of cilia, with variable lengths, and some asthmatic cells don't become ciliated (Figure 2B).

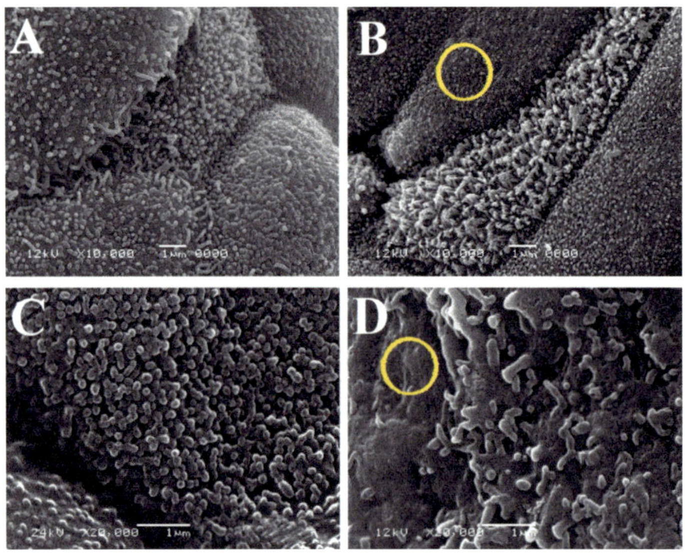

Figure 2. Photomicrographs taken under scanning electron microscopy, of BM's epithelia produced with non-asthmatic (A,C) and asthmatic (B,D) HBEC after 10 days of culture at the air-liquid interface. Ciliogenesis occurred in all BMs, but some asthmatic HBEC were devoided of cilia (yellow circles) and showed sparsely distributed shorter cilia, compared to normal HBEC. (Magnification: A-B: X 10,000; C-D: X 20,000).

Mucus secretion was previously shown in GCs observed on histological sections of BMs stained with periodic-acid-Schiff (PAS) [28]. The PAS staining reveals mucus-secreting cells on histological sections (Figure 3C-D, G-H). To eliminate cross-reaction with endogenous glycogen in situ, some sections were digested before PAS staining using a solution of 0.5% maltase (Fisher) in phosphate-buffered saline (PBS) for 30 min at 37°C and washed 10 min with distilled water. At least 10 tissue sections were analyzed for each BM tested. We observed the entire sections to take representative pictures. GCs can be seen in both, non-and asthmatic BMs cultured at the air-liquid interface (Figure 3B-D and E-H respectively). GCs were observed under scanning electron microscope in non-and asthmatic epithelia of BMs, cultured at the air-liquid interface for at least a week (Figure 4).

http://dx.doi.org/10.5772/59192

2.7. Mucus secretion GCs in human bronchial models *in vitro* in response to IL-13

The effects of cytokines, such as 10 ng/ml of IL-13 (PeproTec Inc., Rocky Hill, NJ, USA) [37], was assessed on the induction of GC metaplasia in non-asthmatic and asthmatic BMs (Figure 3B,D,F,H). PAS staining was performed to determine the percentage of mucine-positive cells on histological sections (Figure 3D,H).

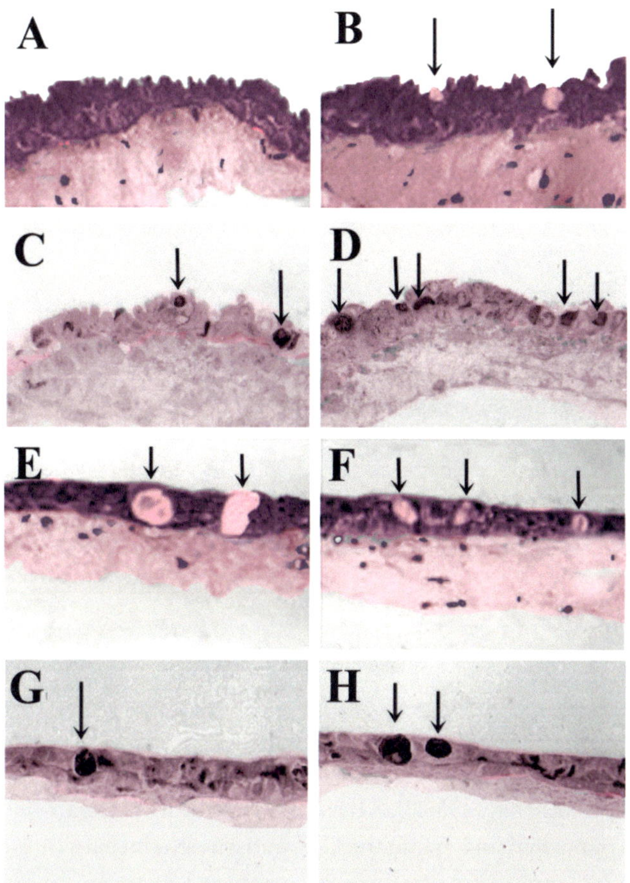

Figure 3. Photomicrographs of histological sections of non-asthmatic (A-D) and asthmatic (E-H) cultured for 10 days at the air-liquid interface without (A,C,E,G) and with (B,D,F,H) IL-13. Sections A-B, E-F were stained with hematoxylin-eosin and sections C-D,G-H were stained using the PAS method. On all BM sections, GCs are pointed with arrows. Magnification: 20 X

The isolation and growth of several populations of asthmatic HBEC (n > 30) have confirmed that they retain their morphological properties in culture. Asthmatic epithelia (Figure 3 E-H) are not as well organized as non-asthmatic controls (Figure 3 A-D) and partial desquamation occurs after 12 days of culture at the air-liquid interface (Figure 2B, D and Figure 4). Similarly, mesenchymal layers populated with asthmatic HBFC (Figure 3 E-H) are thinner than non-asthmatic mesenchymes (Figure 3 A-D), notably due to increased levels of collagenases secretion [28].

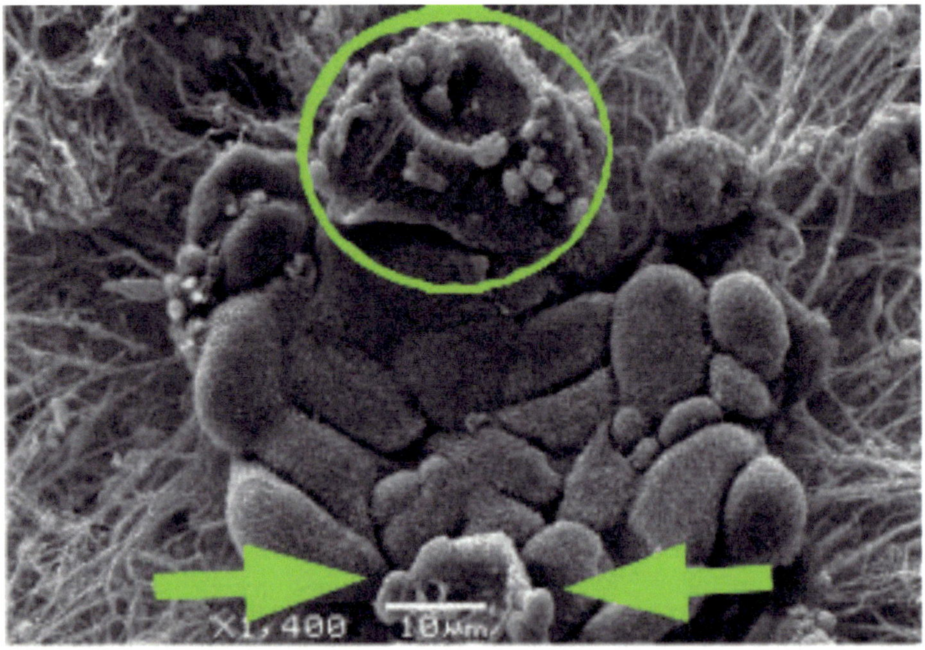

Figure 4. Photomicrograph taken under scanning electron microscopy, of BM's epithelia produced with asthmatic HBEC, after 10 days of culture at the air-liquid interface, showing an islet of HBEC including two GCs (green circle and arrowheads). (Magnification: X 1,400).

Our experiments confirmed that HBEC grown at the air-liquid interface and exposed to IL-13 generate GCs metaplasia *in vitro* [36-38]. After 7 days of culture at the air-liquid interface, IL-13 stimulates growth of normal and asthmatic GCs in tissue-engineered human BMs *in vitro* (10 ± 0,9% and 23 ± 1,3 % more GCs respectively, compared to corresponding controls grown without IL-13). The experiment was stopped before epithelial cell desquamation occurred in the asthmatic BMs, while non-and asthmatic HBEC were confluent on all BMs, to ensure that mucus quantification was not overestimated in normal BMs, due to higher cell number.

ELISA assays were also performed to quantify MUC5AC in the culture (enzyme-linked immunosorbent assay (ELISA) kit cat no: E90756Hu, Biomatik Cambridge, Ontario, Canada). MUC5AC synthesis increased significantly in non-and asthmatic BMs exposed to IL-13, as soon as after 3 days of culture at the air-liquid interface (Figure 5). The mucin synthesis increased up to 30% more in asthmatic than in non-asthmatic epithelia. Thus, the intrinsic properties of the cells can be preserved and expressed in the three-dimensional BMs.

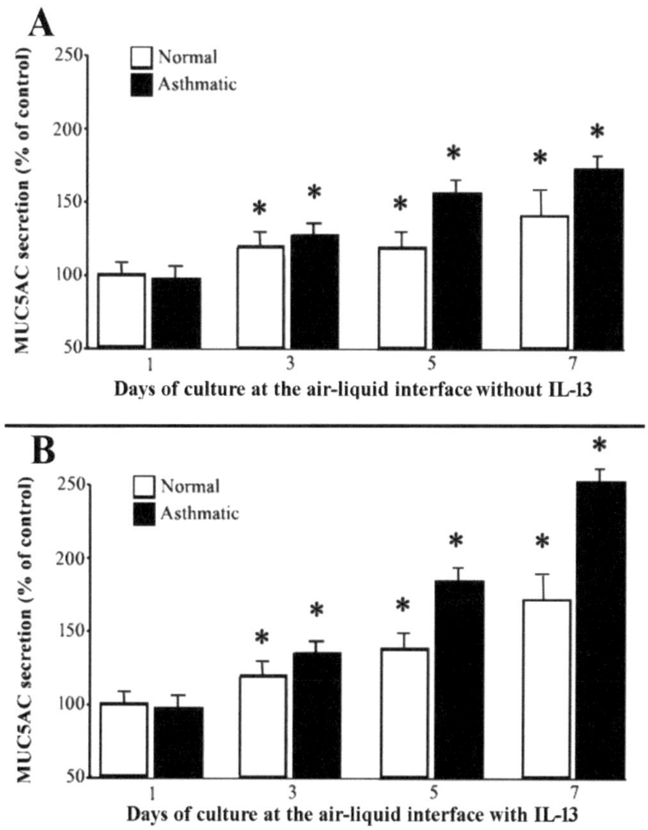

Figure 5. Synthesis of MUC5AC was assessed after 1, 3, 5 and 7 days of culture without (A) and with (B) IL-13, at the air-liquid interface. Values are relative to control ($p < 0.05$, control or non-exposed versus exposed to IL-13). We performed Student's t-test or one-way ANOVA and Tukey post hoc test considering significant $p < 0.05$. Data are presented as mean±s.d.

2.8. What advantages can be exploited using human tissue-engineered BMs?

The culture of human tissue-engineered BMs allows several advantages. First, histological analyses quality control and the quantification of any cell type can be performed on several groups of healthy, asthmatic or other pathological tissue-engineered BMs, compared to animal experimentation that is complex and expensive. The tissue-engineering approach offers an alternative to animal use, to study physiological pathways, outside the complex microenvironment that prevails *in vivo*. Cell layers organization and the quality of the tissues (ciliogenesis, cell morphology and functional parameters) can be monitored on BMs produced with cells of several donors. The tissue-engineered BMs can be produced in batches of many samples and it allows testing of several conditions, while lowering inter-individual differences. In addition, the models can be modified in various ways. For instance, the interactions epithelium-mesenchyme could be studied in hybrid constructs, using normal HBEC combined to pathological HBFC, and inversely. Isolating collagen can be a laborious task, but native type I collagen-based scaffolds remain the best matrix to develop viable tissue-engineered tissues and grafts. Many arguments support this affirmation; 1) it is the major component of all connective tissues, 2) it can be recognized and consequently colonized by mesenchymal cells, 3) cells synthesize and remodel collagen scaffolds (Figure 6), including collagens of other species (e.g. bovine collagen), since only a few amino acids differ from one source to another, 4) collagen network of fibers can trap growth factors and bioactive molecules that are then released in the culture medium, for the benefit of the surrounding cells, 5) interestingly, normal epithelial cells attach, migrate and adopt proper organization on collagen gels, but remain on its surface, 6) this type of matrix allows direct visualization under microscope of the cells that populate it, or proliferate on its surface in culture and 7) collagen gels can be produced in various shapes (disk, tubular, spherical) and several collagen layers can be superimposed, knitted or lyophilized and rehydrated, to reproduce the form of almost any type of organ. All these advantages can be fully exploited to establish new biomaterials for fundamental and clinical applications. Finally, airway smooth muscle is a major target in asthma management. A better understanding of the mechanisms underlying airway smooth muscle contraction and airway hyperresponsiveness is a priority, a prerequisite for advances in therapeutic control of asthma. Therefore, we are currently developing a construct containing smooth muscle cell layer (SMCs) as a third cell layer, to evaluate the effects of various bronchodilators on the contractile properties of reconstructed bronchi *in vitro* (Dr. J.-P. Lavoie and collaborators). SMCs frequently lose their typical features, including their contractile properties, in monolayer grown on plastic, but their phenotype is maintained in three-dimensional matrix. Then, exploiting animal models, such as the pseudostratified epithelium of the genetically tractable mouse trachea, could confirm observations made on BMs, and enable crucial discoveries regarding the pathogenesis of airway diseases [1]. Thus, the outcomes of experiments performed *in vitro* can be confirmed in animal models thereafter.

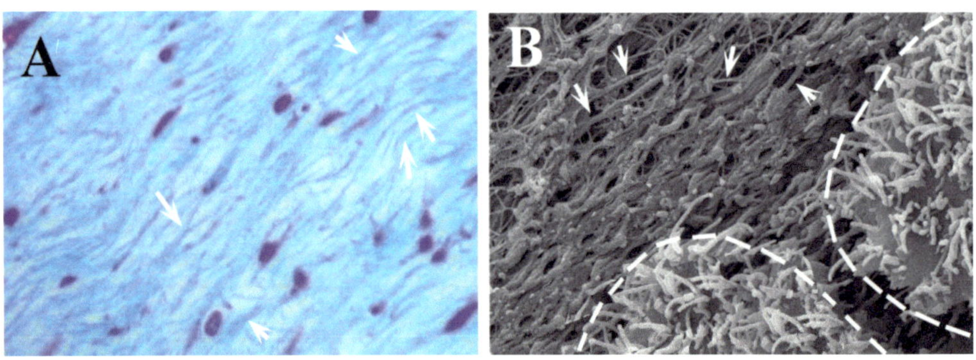

Figure 6. Photomicrographs of the collagen fibers (arrows) that compose the mesenchymal layer of the BMs, cultured for 12 days at the air-liquid interface. (A): Histological section of the mesenchymal layer of a BM stained with hematoxylin-eosin (HBFC in purple surrounded by blue fibers) and (B): View under scanning electron microscopy of the surface of the mesenchyme showing the collagen fibers network that supports epithelial cells (two asthmatic ciliated HBEC are circled with dashed lines). Note the high density and the alignment of the matrix fibers in response to the tension induced by the peripheral anchorage in the tissues, and the partial desquamation of the asthmatic epithelium that allowed mesenchyme visualization. (Magnification: A: 20 X, B: X 6,000)

http://dx.doi.org/10.5772/59192

2.9. What are the limits of this approach?

The quality of the tissues that can be developed using our approach cannot be denied. However, the stability of BMs in culture, especially asthmatic BMs, limits the duration of experiments. In addition, restrained availability of human bronchial tissue donors, combined to limited size of bronchial biopsies, can delay experimentation. Animal experimentation is very expensive, time consuming, complex, and sometimes ethically questionable, depending on the type of study. Tissue-engineered BMs offer a valuable alternative. Animal cells are more easily accessible and can be used to establish useful three-dimensional models and compare them with human materials.

Another challenge associated with tissue engineered BMs is the development of more complex tissues, including more than two types of cells. For instance, the addition of a layer of bronchial SMCs would certainly add functional attributes to our normal and asthmatic bilayered BMs. Many teams have made attempts to isolate and purify bronchial SMCs. It is very difficult to obtain SMCs populations that are not mixed with other mesenchymal cell types such as fibroblasts. But is it necessary to obtain 100% pure SMCs? This question and many others need to be assessed to understand the signaling pathways that support epithelium-mesenchymal interactions. Tissue-engineered three-dimensional BMs might provide answers to some of them, since they favor growth factors exchanges and intercellular crosstalk.

Pleural mesothelial cells are pluripotent cells [41], which could influence the morphology and the state of differentiation of bronchial epithelia and the functional properties of mesenchymes. They have been shown to clonally generate fibroblasts and smooth muscle cells in murine models. This supports the possibility that they may also modulate lung injury-repair by re-activation of developmental programs in the adult, reflecting an altered recapitulation of development, with implications for regenerative biology of the lung [41]. These cells could be included in some normal and pathologic BMs, to study their capacity to affect the phenotypic features and the functional state of HBEC and other cell populations. However, such experiments would involve the development of more complex multilayered BMs, which may not be easily maintained in culture.

3. Conclusion

We propose a tissue-engineering approach to reconstruct bronchi *in vitro*. Using type I bovine collagen, we have characterized human BMs that share several important features with native non-and asthmatic human bronchi. Other models are under development to target key molecules involved and new discoveries could be validated *in vivo* thereafter, as it can be achieved for other bioengineered tissues [38-40]. The creation of new human BMs can be performed with cells isolated from various donors, or pathologies.

Acknowledgements

The authors gratefully acknowledge the financial support of Mr. and Mrs. Andreas and Elektra Paterakis, Crete, Greece, Dr. Stéphane Bolduc for covering the publication costs, Dr. Richard Janvier for electron microscopy analyses at Université Laval, Dr. Amandine Vargas for scientific discussion, Mrs Lina Robayo for research assistance, Mr. Mario Lavigne and Mrs Nancy Dufour for technical assistance.

Author details

Sara Bouhout[1*], Jadson Moreira Pereira[1], Franck Simon[1], Stéphane Chabaud[1], Stéphane Bolduc[1], Massimo Conti[2], Alex Therien[3], Eric Rousseau[4], Véronique J. Moulin[1], Jean-Pierre Lavoie[5], Michel Rouleau[1] and Francine Goulet[1]

*Address all correspondence to: sara.ucergy@hotmail.fr

1 Centre de Recherche en Organogenèse Expérimentale de l'Université Laval / LOEX-Centre de Recherche du Centre Hospitalier Universitaire de Québec/FRQS, Québec, (QC), Canada

2 Institut Universitaire de Cardiologie et de Pneumologie de Québec, Centre de Pneumologie, Hôpital Laval, Chemin Sainte-Foy, Québec, Canada

3 Merck & Co., Inc., Kenilworth, NJ, USA

4 Département de Physiologie et Biophysique, Faculté de Médecine et des Sciences de la Santé, Université de Sherbrooke, Sherbrooke, Québec, Canada

5 Département des Sciences Cliniques, Faculté de Médecine Vétérinaire, St-Hyacinthe, (QC) , Canada

The authors declare no conflict of interest.

References

[1] Fuhlbrigge AL, Adams RJ, Guilbert TW, Grant E, Lozano P, Janson SL, Martinez F, Weiss KB, Weiss ST. The burden of asthma in the United States: level and distribution are dependent on interpretation of the national asthma education and prevention program guidelines. American Journal of Respiratory and Critical Care Medicine 2002;166(8):1044-9. DOI: 10.1164/rccm.2107057 http://www.atsjournals.org/doi/full/10.1164/rccm.2107057#.U_I5qdF0xjo

[2] Abbott-Banner KH, Holmes A, Adcock I, Rao NL, Barrett E, Knowles R. Models of respiratory disease symposium. Journal of Inflammation (London) 2013;10(Suppl 1): I1. DOI:10.1186/1476-9255-10-S1-I1. http://www.journal-inflammation.com/content/10/S1/I1 (accessed 14 August 2013)

[3] Holmes AM, Solari R, Holgate ST. Animal models of asthma: value, limitations and opportunities for alternative approaches. Drug Discovery Today 2011;16(15-16): 659-670. DOI: 10.1016/j.drudis.2011.05.014.

[4] Rock JR, Randell SH, Hogan BLM. Airway basal stem cells: a perspective on their roles in epithelial homeostasis and remodeling. Disease Models & Mechanisms 2010; 3(9-10): 545–56. DOI: 10.1242/dmm.006031. http://dmm.biologists.org/content/3/9-10/545.full (accessed 10 August 2010).

[5] Park JA. Tschumperlin DJ. Chronic Intermittent Mechanical Stress Increases MUC5AC Protein Expression. American Journal of Respiratory Cell and Molecular Biology 2009;41: 459-66. DOI: 10.1165/rcmb.2008-0195OC http://www.ncbi.nlm.nih.gov/pmc/articles/PMC2746990/ (accessed 29 January 2009)

[6] Zhang X, Zhang Y, Tao B, Wang D, Cheng H, Wang K, Zhou R, Xie Q, Ke Y. Docking protein Gab2 regulates mucin expression and goblet cell hyperplasia through TYK2/STAT6 pathway. The FASEB Journal 2012;26(11): 4603-13 DOI: 10.1096/fj.12-211755 http://www.fasebj.org/content/26/11/4603.long (accessed 2 August 2012)

[7] Lambrecht BN, Hammad H. The airway epithelium in asthma. Nature Medicine 2012;18: 684-92. DOI:10.1038/nm.2737 http://www.nature.com/nm/journal/v18/n5/full/nm.2737.html (accessed 4 May 2012)

[8] Wills-Karp M, Luyimbazi J, Xu X, Schofield B, Neben TY, Karp CL, Donaldson DD. Interleukin-13: Central mediator of allergic asthma. Science 1998;282(5397): 2258–61. DOI: 10.1126/science.282.5397.2258 http://www.sciencemag.org/content/282/5397/2258.long

[9] Hays SR, Fahy JV. Characterizing mucous cell remodeling in cystic fibrosis: relationship to neutrophils. American Journal of Respiratory and Critical Care Medicine 2006;174(9): 1018-24. DOI: 10.1164/rccm.200603-310OC http://www.ncbi.nlm.nih.gov/pmc/articles/PMC2648101/ (accessed 17 August 2006)

[10] Knowles MR, Boucher RC. Mucus clearance as a primary innate defense mechanism for mammalian airways. Journal of Clinical Investigation 2002;109(5): 571–7. DOI: 10.1172/JCI15217 http://www.ncbi.nlm.nih.gov/pmc/articles/PMC150901/

[11] Tamaoka M, Hassan M, McGovern T, Ramos-Barbon D, Jo T, Yoshizawa Y, Tolloczko B, Hamid Q, and Martin JG. The epidermal growth factor receptor mediates allergic airway remodelling in the rat. The European respiratory journal 2008;32: 1213-23. DOI: 10.1183/09031936.00166907 http://erj.ersjournals.com/content/32/5/1213.long (accessed 24 July 2008)

[12] Siddiqui S, Novali M, Tsuchiya K, Hirota N, Geller BJ, McGovern TK, Risse PA, Jo T, Zeroual MA, and Martin JG. The modulation of large airway smooth muscle phenotype and effects of epidermal growth factor receptor inhibition in the repeatedly allergen-challenged rat. American journal of physiology Lung cellular and molecular physiology 2013;304: L853-862. DOI: 10.1152/ajplung.00047.2012 http://ajplung.physiology.org/content/304/12/L853 (accessed 15 June 2013)

[13] Stumm CL, Halcsik E, Landgraf RG, Camara NO, Sogayar MC, Jancar S. Lung remodeling in a mouse model of asthma involves a balance between TGF-β1 and BMP-7. PLoS One. 2014;9(4): e95959. DOI: 10.1371/journal.pone.0095959 http://www.plosone.org/article/info%3Adoi%2F10.1371%2Fjournal.pone.0095959 (accessed 29 April 2014)

[14] Cleary RA, Wang R, Wang T, Tang DD. Role of Abl in airway hyperresponsiveness and airway remodeling. Respiratory Research 2013;14: 105. DOI: 10.1186/1465-9921-14-105 http://respiratory-research.com/content/14/1/105 (accessed 11 October 2013)

[15] Birrell MA, Bonvini SJ, Dubuis E, Maher SA, Wortley MA, Grace MS, Raemdonck K, Adcock JJ, Belvisi MG. Tiotropium modulates transient receptor potential V1 (TRPV1) in airway sensory nerves: A beneficial off-target effect? Journal of Allergy and Clinical Immunology 2014;133(3): 679-87.e9. DOI: 10.1016/j.jaci.2013.12.003 http://www.jacionline.org/article/S0091-6749(13)01856-3/fulltext (accessed 7 February 2014)

[16] Cormier Y, Whittom R, Boulet LP, Sériès F. Effect of inflammation on peripheral airway reactivity in dogs. Clinical Science (London) 1993; 84(1): 73-8.

[17] Leclere M, Lavoie-Lamoureux A, Gélinas-Lymburner E, David F, Martin JG, Lavoie JP. Effect of antigenic exposure on airway smooth muscle remodeling in an equine model of chronic asthma. American Journal of Respiratory Cell and Molecular Biology 2011; 45(1): 181-187. DOI: 10.1165/rcmb.2010-0300OC

[18] Leclere M, Lavoie-Lamoureux A, Lavoie JP. Heaves, an asthma-like disease of horses. Respirology 2011;16(7): 1027-46. DOI: 10.1111/j.1440-1843.2011.02033.x http://onlinelibrary.wiley.com/doi/10.1111/j.1440-1843.2011.02033.x/full (accessed 26 September 2011)

[19] Lavoie JP, Lefebvre-Lavoie J, Leclere M, Lavoie-Lamoureux A, Chamberland A, Laprise C, Lussier J. Profiling of differentially expressed genes using suppression subtractive hybridization in an equine model of chronic asthma. PLoS One. 2012;7(1): e29440. DOI: 10.1371/journal.pone.0029440 http://www.plosone.org/article/info%3Adoi%2F10.1371%2Fjournal.pone.0029440 (accessed 3 January 2012)

[20] Leclere M, Lavoie-Lamoureux A, Joubert P, Relave F, Setlakwe EL, Beauchamp G, Couture C, Martin JG, Lavoie JP. Corticosteroids and antigen avoidance decrease airway smooth muscle mass in an equine asthma model. American Journal of Respiratory Cell and Molecular Biology 2012;47(5): 589-96. DOI: 10.1165/rcmb.2011-0363OC

http://www.atsjournals.org/doi/full/10.1165/rcmb.2011-0363OC#.U_JELNF0xjo (accessed 21 June 2012)

[21] Kumar RK, Foster PS. Are mouse models of asthma appropriate for investigating the pathogenesis of airway hyper-responsiveness? Frontiers in Physiology 2012;3: 312. DOI: 10.3389/fphys.2012.00312 http://www.ncbi.nlm.nih.gov/pmc/articles/PMC3459343/

[22] Goode ML, Fink JB, Dhand R, Tobin MJ. Improvement in Aerosol Delivery with Helium–Oxygen Mixtures during Mechanical Ventilation. American Journal of Respiratory and Critical Care Medicine 2001;163(1): 109-14. DOI: 10.1164/ajrccm.163.1.2003025 http://www.atsjournals.org/doi/full/10.1164/ajrccm.163.1.2003025#.U_JKitF0xjo (accessed 1 January 2001)

[23] Pennino D, Bhavsar PK, Effner R, Avitabile S, Venn P, Quaranta M, Marzaioli V, Cifuentes L, Durham SR, Cavani A, Eyerich K, Chung KF, Schmidt-Weber CB, Eyerich S. IL-22 suppresses IFN-γ-mediated lung inflammation in asthmatic patients. Journal of Allergy and Clinical Immunology 2013;131(2): 562-70. DOI: 10.1016/j.jaci.2012.09.036 http://www.atsjournals.org/doi/full/10.1164/ajrccm.163.1.2003025#.U_JKitF0xjo (accessed 19 November 2012)

[24] Zhang S, Smartt H, Holgate ST, Roche WR. Growth factors secreted by bronchial epithelial cells control myofibroblast proliferation: an *in vitro* co-culture model of airway remodeling in asthma.Laboratory Investigation 1999; 79(4): 395-405.

[25] Sugihara H, Toda S, Miyabara S, Fujiyama C, Yonemitsu N. Reconstruction of alveolus-like structure from alveolar type II epithelial cells in three-dimensional collagen gel matrix culture. The American Journal of Pathology 1993;142(3): 783-792. http://www.ncbi.nlm.nih.gov/pmc/articles/PMC1886800/pdf/amjpathol00075-0123.pdf

[26] Du Y, Lo E, Ali S, Khademhosseini A. Directed assembly of cell-laden microgels for fabrication of 3D tissue constructs. Proceedings of the National Academy of Sciences of the Unites States of America 2008;105(28): 9522-9527. DOI: 10.1073/pnas.0801866105 http://www.pnas.org/content/105/28/9522.full

[27] Paquette JS, Tremblay P, Bernier V, Auger FA, Laviolette M, Germain L, Boutet M, Boulet LP, Goulet F. Production of tissue-engineered three-dimensional human bronchial models. In Vitro Cellular & Developmental Biology – Animal 2003;39(5): 213-220. DOI: 10.1290/1543-706X(2003)039<0213:POTTHB>2.0.CO;2

[28] Paquette J-S, Moulin V, Tremblay P, Bernier V, Laviolette M, Auger FA, Boulet L-P, Goulet F. Tissue-engineered human asthmatic bronchial equivalents. European Cells & Materials 2004;7: 1-11. http://www.ecmjournal.org/journal/papers/vol007/pdf/v007a01.pdf

[29] Choe MM, Sporn PH, Swartz MA. An *in vitro* airway wall model of remodeling. American Journal of Physiology-Lung Cellular Molecular Physiology 2003;285(2):

L427-33. DOI: 10.1152/ajplung.00005.2003 http://ajplung.physiology.org/content/285/2/L427

[30] Chakir J, Pagé N, Hamid Q, Laviolette M, Boulet LP, Rouabhia M. Bronchial mucosa produced by tissue engineering: a new tool to study cellular interactions in asthma. Journal of Allergy and Clinical Immunology 2001;107(1): 36-40.

[31] Goulet F, Boulet LP, Chakir J, Tremblay N, Dubé J, Laviolette M, Boutet M, Xu W, Germain L, Auger FA. Morphologic and functional properties of bronchial cells isolated from normal and asthmatic subjects. American Journal of Respiratory Cell and Molecular Biology 1996;15(3): 312-318. DOI: 10.1165/ajrcmb.15.3.8810634

[32] Black A, Berthod F, L'Heureux H, Germain L, Auger FA. *In vitro* reconstruction of a human capillary-like network in a tissue engineered skin equivalent. The FASEB Journal 1998;12: 1331-1340. http://www.fasebj.org/content/12/13/1331.long

[33] Proulx S, Uwamaliya J, Carrier P, Alexandre Deschambeault A, Audet C, Giasson CJ, Guérin SL, Auger FA, Germain L. Reconstruction of a human cornea by the self-assembly approach of tissue engineering using the three native cell types. Molecular Vision 2010;16: 2192-201. http://www.molvis.org/molvis/v16/a235/ (Accessed 29 October 2010)

[34] L'Heureux N, Stoclet JC, Auger FA, Lagaud GJ, Germain L, Andriantsitohaina RA. Human tissue-engineered vascular media: a new model for pharmacological studies of contractile responses. The FASEB Journal 2001;15: 515-24. DOI: 10.1096/fj.00-0283com http://www.fasebj.org/content/15/2/515.full.pdf

[35] Auger FA, Rouabhia M, Goulet F, Berthod F, Moulin V, Germain L. Tissue-engineered human skin substitutes developed from collagen-populated hydrated gels: clinical and fundamental applications. Medical & Biological Engineering and computing 1998;36(6): 801-812. DOI: 10.1007/BF02518887

[36] Tremblay P, Cloutier R, Lamontagne J, Belzile AM, Larkin AM, Chouinard L, Chabaud S, Laverty S, Lussier B, Goulet F. Potential of skin fibroblasts for application to anterior cruciate ligament tissue engineering. Cell Transplantation 2011;20: 535-542. DOI: 10.3727/096368910X536482

[37] Yasuo M, Fujimoto K, Tanabe T, Yaegashi H, Tsushima K, Takasuna K, Koike T, Yamaya M, Nikaido T. Relationship between Calcium-Activated Chloride Channel 1 and MUC5AC in Goblet Cell Hyperplasia Induced by Interleukin-13 in Human Bronchial Epithelial Cells. Respiration 2006;73: 347–59. DOI: 10.1159/000091391 http://www.karger.com/Article/Pdf/91391 (accessed 6 February 2006)

[38] Kuperman DA, Huang X, Koth LL, Chang GH, Dolganov GM, Zhu Z, Elias JA, Sheppard D, Erle, DJ. Direct effects of interleukin-13 on epithelial cells cause airway hyperreactivity and mucus overproduction in asthma. Nature Medicine 2002, 8, 885–889. DOI: 10.1038/nm734

[39] Dabbagh K, Takeyama K, Lee HM, Ueki IF, Lausier JA, Nadel JA. IL-4 induces mucin gene expression and goblet cell metaplasia *in vitro* and *in vivo*. Journal of Immunology 1999;162: 6233-7. http://www.jimmunol.org/content/162/10/6233.full.pdf±html (accessed 18 August 2014)

[40] Ezeamuzie CI, Sukumaran J, Philips E. Effect of wortmannin on human eosinophil responses *in vitro* and on bronchial inflammation and airway hyperresponsiveness in Guinea pigs *in vivo*. American Journal of Respiratory and Critical Care Medicine 2001;164(9): 1633-9. DOI: 10.1164/ajrccm.164.9.2101104 http://www.atsjournals.org/doi/full/10.1164/ajrccm.164.9.2101104#.U_KRKdF0xjo

[41] Batra H, Antony VB. The pleural mesothelium in development and disease. Frontiers in physiology DOI: 10.3389/fphys.2014.00284 http://journal.frontiersin.org/Journal/10.3389/fphys.2014.00284/full (accessed 01 August 2014)

Chapter 12

Experimental Design for the Innervation of Tooth Forming from Implanted Cell Re-associations

T. Kökten, H. Lesot and S. Kuchler-Bopp

Additional information is available at the end of the chapter

http://dx.doi.org/10.5772/59340

1. Introduction

Tooth engineering may refer to either dental tissue regeneration or to the reformation of a complete organ attached to surrounding bone [1–4]. In case of whole tooth organ engineering the use of embryonic dental cells allowed to search for the simplest experimental protocol to be developed in order to maintain and express odontogenic cell potentialities. This was achieved after implantation under the skin or in the kidney capsule or in the jaw of adult mice of cultured re-associations between pellets of dissociated single epithelial and mesenchymal cells [5–7]. The main points that actually need further research concern the identification of non dental cells able to engage in whole tooth formation [8–12], anchoring of the newly formed tooth root to surrounding alveolar bone [3,9,13], and innervation of the engineered organ [14].

The innervation of dental and peridental tissues has been extensively investigated during tooth development in the mouse [15–19]. The timing and pattern of tooth innervation are controlled by diffusible signaling molecules and ultimately by epithelial-mesenchymal interactions [15,20]. Tooth innervation is particularly important in signaling damage in the crown, but also may interfere with several functions including the regulation of blood flow, tissue homeostasis, immune cell function, inflammation, and healing [21,22]. The innervation of engineered tooth germs does not occur spontaneously after implantation, but requires immunodepressed conditions: cyclosporin A (CsA) treatment of the host mice or implantation under the skin of immunocompromised Nude mice [23]. New questions arising from this previous work are discussed in this chapter. They concern a) the nature of relationships between axons and other cells in the dental pulp (odontoblasts, glial cells and blood vessels: endothelial cells and pericytes) in the case of cell re-associations implanted in Nude mice, b) the differential innervation of the dental and peridental mesenchymes, and c) experimental attempts to avoid the use of CsA or related molecules to achieve dental pulp innervation [23]. For this purpose, specific cell re-associations were designed and analyzed, after *in vitro* culture and further implantation in adult mice, by means of histology, transmission electron microscopy and immunofluorescence. After two weeks of implantation, the cell re-associations reached a stage corresponding to a first lower molar at post-natal day 4 (PN4), when taking into account the crown development and stage of matrix deposition and mineralization [24,25]. However, axons reached the odontoblast layer, which in physiological conditions, occurs at PN7 only [23]. For this reason, the status of innervation in cell re-associations implanted for two weeks was compared to the situation in mouse first lower molar at PN7.

2. Cellular network mediating sensory function in the dental mesenchyme

Interactions between axons and other cells have been investigated in the human dental pulp [26]. The situation in tooth forming from implanted cell re-associations is not known. To analyze it, re-associations between dissociated dental embryonic epithelial and mesenchymal cells were cultured for seven days, then trigeminal ganglia were put in contact with the re-associations and cultured overnight before implantation in Nude mice (for detailed method and scheme of the protocol, see [23]). This protocol was selected because it allows the development of a whole tooth organ (Figure 1A-E; [24]), in conditions where the cellular heterogeneity in the dental mesenchyme can be maintained as it is in physiological conditions [25]. The localization of axons and glial cells, their relationships, and also with neighbor cells will be discussed in three complementary contexts: 1) with odontoblasts (Figures 2 and 4), 2) with microvessels present in the odontoblast layer (Figure 5) and 3) with blood vessels in the dental and peridental mesenchymes (Figures 6 and 8). These observations will be compared to what exists in the mesenchyme of a molar tooth at a similar developmental stage (PN7).

2.1. Innervation in the odontoblast layer

After implantation for two weeks in Nude mice, cell differentiation was achieved in the crown part of the tooth (Figure 1A-C) as it occurs after implantation in ICR mice (Charles River CD-1, [24]) or during molar development (Figure 1F-H, J). In implanted cell re-associations as in molar at PN7, odontoblasts were functional and secreted predentin and dentin constituents in the crown (compare Figure 1C, E with Figure 1G, H) and root portions (compare Figure 1A, E with Figure 1J). The cells were elongated with a nucleus in basal position (Figure 1B, G, H). Odontoblast cell processes were present in the dentinal tubules (Figure 1B, C, G, H). These cell processes are involved in the secretion and permanent mecanosensing in the matrix (for review, see [27–29]). The innervation was analyzed using antibodies to peripherin and neurofilament-200 (NF200) [18,30]. After two weeks of implantation of cell re-associations in Nude mice, axons positive for peripherin (Figure 2B) reached the layer of odontoblasts, which are positive for nestin (Figure 2A-E) or vimentin (Figure 2F). The same was observed in the first molar at PN7 (Figure 2G-L). Similarly, antibodies to NF200 allowed the detection of axons in the dental mesenchyme and reaching the odontoblasts of teeth forming from implanted cell re-associations (Figure 2A, C) as in molars at PN7 (Figure 2I). NF200 is a marker for myelinated

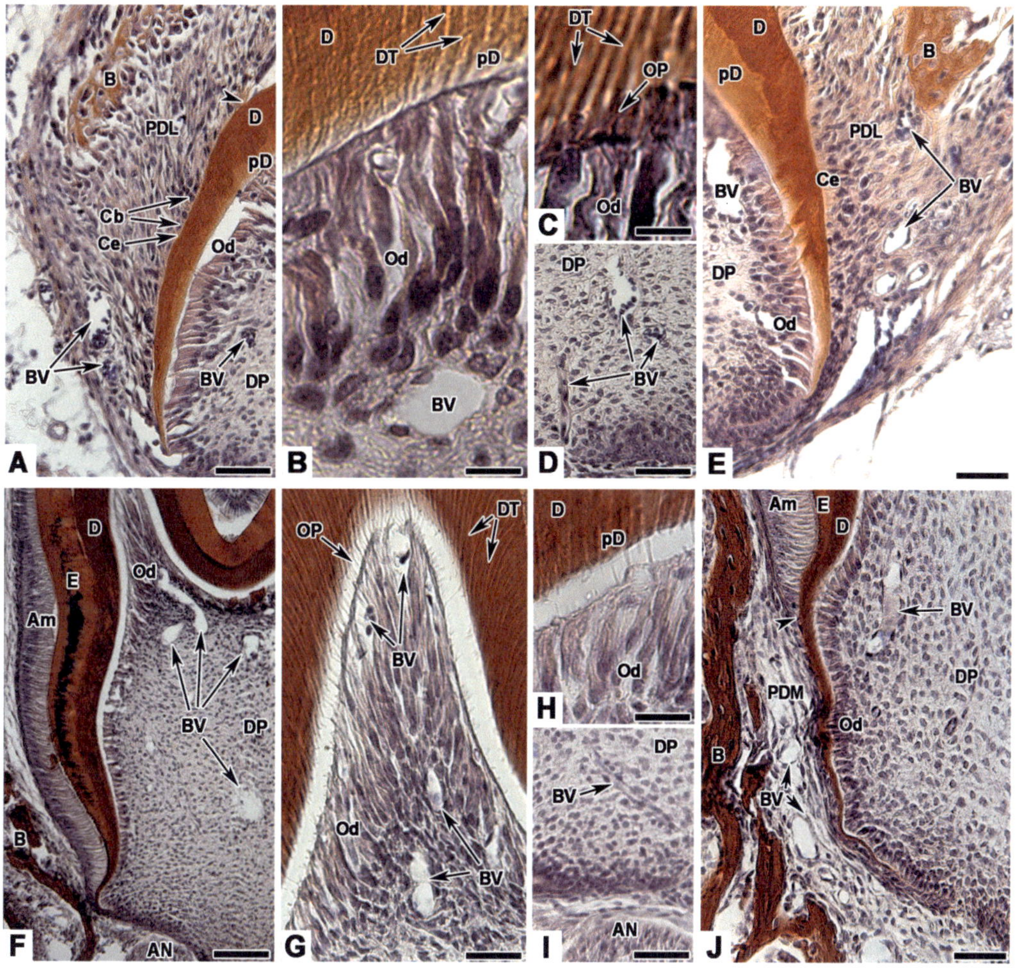

Figure 1. Histology of cultured cell re-associations implanted for two weeks under the skin of Nude mice (A-E) and of the first lower molar of ICR mice at post-natal day 7 (PN7) (F-J). The protocols for cell dissociation, re-association, culture, implantation and innervation have been detailed elsewhere [23]. All procedures were designed in compliance with the recommendations of the European Union (2010/63/EU) for the care and use of laboratory animals [23]. For histology, samples were fixed in Bouin-Hollande and demineralized in 15% EDTA before their inclusion in paraffin. Serial sections (7 μm) were stained with Mallory's stain. After implantation of cell re-associations for two weeks under skin, crown was developed and root initiated (A, E), as in the developing molar at PN7 (F, J). Arrowheads showed the limit between crown and root in implanted cell re-associations (A), as in molar at PN7 (J). In implanted cell re-associations as in molars at PN7, blood vessels (BV) were present in the dental pulp (DP) (compare D, E with F, I, J) and reached the odontoblast (Od) layer (compare B with G). In both cases, odontoblasts were elongated and polarized, as seen from the position of their nucleus, opposite to the secretory pole (compare B with G, H). These cells were also functional, secreted predentin (pD) and dentin (A, B, E and H). Dentinal tubules (DT) and odontoblast cell processes (OP) were observed in the dentin of implanted cell re-associations as in molars at PN7 (compare B, C with G, H). In both types of samples, ameloblasts (Am) were elongated, polarized and secreted enamel (J, see [23]). Newly formed bone was present at the periphery of implanted cell re-associations (A, E) corresponding to alveolar bone in molar at PN7 (F, J). In contact with the external surface of root dentin, cementoblasts (Cb) were observed in implanted cell re-associations (A). These cells were functional and deposited cementum (Ce) (A, E). Periodontal ligament (PDL) fibers were attached to the root by cementum and extended until reaching newly formed bone (A, E). AN: Alveolar nerve; B: Bone; D: Dentin; E: Enamel. Scale bars=80 μm (F); 40 μm (A, D, E, F, G, I, J) and 20 μm (B, C, H).

A-fibers and previous observations by transmission electron microscopy showed that such fibers could not be detected in implanted re-associations [23], probably due to the too short period of implantation. For this reason and despite the positive reaction to anti-NF200 antibodies, all further detections of axons were performed using antibodies to peripherin.

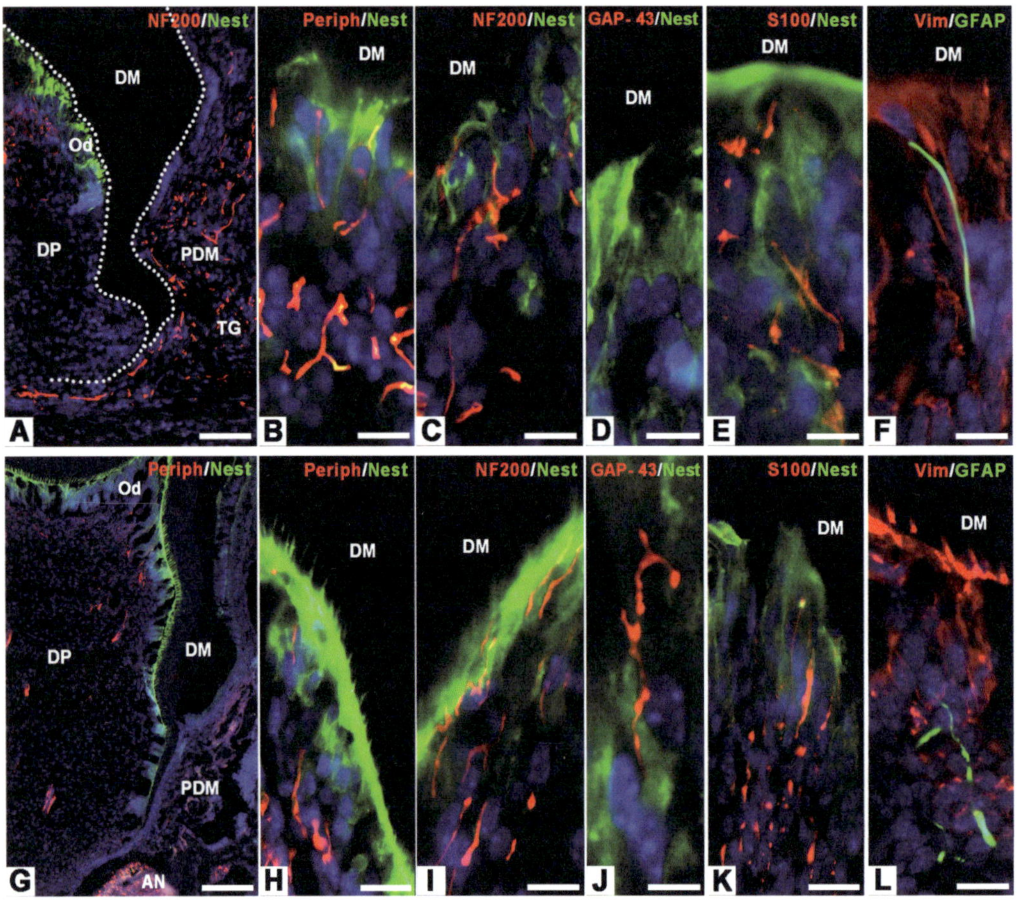

Figure 2. Innervation in the odontoblast layer of cell re-associations implanted for two weeks in Nude mice (A-F), compared to first lower molar at PN7 (G-L). Odontoblasts (Od) were labeled with anti-nestin (A-E and G-K) or anti-vimentin antibodies (F, L). Axons were labeled for peripherin (B, G, H) or NF200 (A, C, I) and glial cells using antibodies to either S100-beta (E, K) or GFAP (F, L). Growth cones were labeled for GAP-43 (D, J). In case of implanted re-associations, axons coming from the trigeminal ganglion (TG) entered the dental mesenchyme and reached the odontoblast layer, as seen after staining for NF200 (A). Similarly, the dental mesenchyme of first lower molar at PN7 was innervated by alveolar nerve (AN) as shown after staining for peripherin (G). Higher magnifications showed that axons positive for either peripherin or NF200 were present in the odontoblast layer, but did not enter the dentinal tubules, neither in implanted cell re-associations (B, C), nor in molars at PN7 (H, I). Growth cones were present in the odontoblast layer of molars at PN7 (J), but not in the case of implanted cell re-associations (D). S100-beta positive glial cells were detected in the odontoblast layer of implanted cell re-associations (E), as in intact molars at PN7 (K). The same was observed for GFAP positive glial cells (compare F with L). DM: dental matrix; DP: dental pulp; PDM: peridental mesenchyme. Scale bars=80 μm (A, G) and 20 μm (B-F and H-L).

http://dx.doi.org/10.5772/59340

Axons had not yet reached the odontoblast processes present in predentin/dentin, neither in cell re-associations implanted for two weeks (Figure 2A-C), nor in the molar at PN7 (Figure 2H, I). In the mouse first lower molar, the presence of axons in dentinal tubules was only detected at PN10, and still very few were observed there [23]. Nagahama et al., [31] reported that a subodontoblastic nerve plexus develops at about PN11 in the first molar, and dentinal innervation still continue to increase in the crown after tooth eruption (PN20). In the crown of mature human primary teeth, nerve endings terminate in or near the odontoblast layer, with a small number penetrating into dentin [32]. Odontoblasts and neighbor axons have been suggested to form a mechanosensory complex (for review see [29]). This raises the question of who is doing what when axons are also present in dentinal tubules [33]. Although Carda and Peydró [33] reported the existence of membrane densifications, similar to synapses in between odontoblast processes and nerve endings, there is a general agreement in the literature that true synapses do not exist in between the two cell types. Instead, appositions with a narrow extracellular gap have been reported [34,35]. Such apposition was also observed by transmission electron microscopy in implanted re-associations (Figure 3A, arrows in A') as well as in molars at PN7 (Figure 3E, arrows in E'). The question on how signals sensed by odontoblasts are transmitted to axons remains unsolved [29].

GAP-43 is an intracellular growth-associated protein. It was suggested to participate in neuronal pathfinding and branching during development and regeneration, by integrating plasma membrane and cytoskeletal responses to extracellular signals [36]. Staining for GAP-43 allows the visualization of growth cone, at the tip of axons [37,38]. More surprisingly, GAP-43 is also expressed by Schwann cells in rat molars [39]. *In vitro* and *in vivo* observations showed that, when related to Schwann cells, GAP-43 was expressed by their precursors and by non-myelinating mature cells [40]. Double stainings for GAP-43 and nestin did not show the presence of GAP-43 in the odontoblast layer in implanted re-associations (Figure 2D), while it was present there in the first molar at PN7 (Figure 2J). GAP-43, in the re-associations, was only detected in the peridental mesenchyme as discussed below. The reasons for this absence of GAP-43 in most parts of the dental mesenchyme from implanted cell re-associations will have to be further investigated. It might have important consequences on later stages of tooth innervation [41]. It will thus be necessary to go further by looking at the situation when cultured cell re-associations are implanted in CsA-treated mice. Indeed, CsA can interfere with GAP-43 expression [42].

Different glial cell types have been observed in the dental and peridental mesenchymes in rodents as in human [43,44]. Glial cells have multiple functions during the development of the peripheral nervous system (PNS) and in repair process [45]. During early PNS development, axonal signals are critical for Schwann cell migration, survival and proliferation [46,47]. In the present study, glial cells were searched for, using antibodies to either S100-beta or glial fibrillary acidic protein (GFAP) (for review, see [46]). In the implanted re-associations, S100-beta and GFAP were detected in the odontoblast layer (Figure 4A, B), as in the intact molar at PN7 (Figure 4F, G). The two antigens showed associations with axons, as seen after double-stainings using anti-peripherin antibodies, in the re-associations (Figure 4A, B) as in the molar at PN7 (Figure 4F, G). Furthermore, S100-beta and GFAP positive cells showed only partial

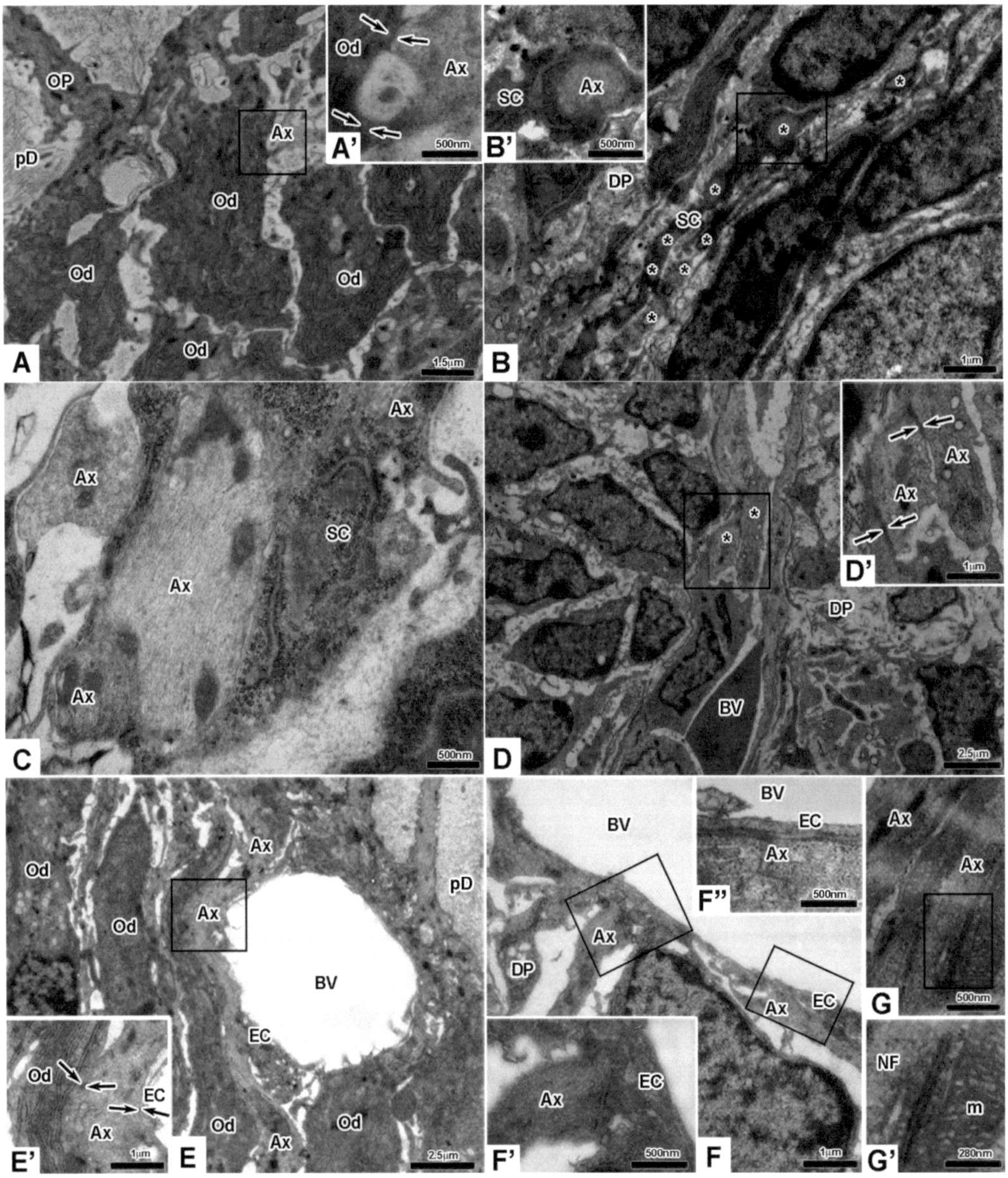

Figure 3. Transmission electron microscopy of cell re-associations implanted for two weeks in Nude mice (A-D') and first lower molar at PN7 (E-G'). Unmyelinated axons (Ax) were present in the odontoblast layer of implanted cell re-associations (A, A'). Arrows in A' showed contacts between an axon and an odontoblast (Od). In the dental pulp (DP) of teeth forming from re-associations, Schwann cells (SC) in contact with or wrapping around axons were seen (B, B', C). In the apical part of the pulp (D, D') unmyelinated axons were present in proximity of blood vessel (BV) and arrows in D' showed contacts between axons. In the odontoblast layer of molar at PN7, an axon showed contact with an endothelial cell (EC) and with an odontoblast (E, arrows in E'). In the dental pulp, axons were in contact with endothelial cells (F, F', F''). Also, a contact between two axons was visualized in molar at PN7 (G, G'). Ax and *: unmyelinated axon; m: mitochondria; NF: neurofilament; OP: odontoblast process; pD: predentin.

http://dx.doi.org/10.5772/59340

co-distribution in the re-associations (arrowheads in Figure 4C-E), as in the molar at PN7 (arrowheads in Figure 4H-J). More GFAP-positive than S100-beta-positive cells were associated with axons. More cells were positive for S100-beta than for GFAP in implanted cell re-associations (Figure 4E) as in intact molars (Figure 4J). These complementary observations indicate that several populations of glial cells are present in implanted cell re-associations, as in a developing molar, in physiological conditions.

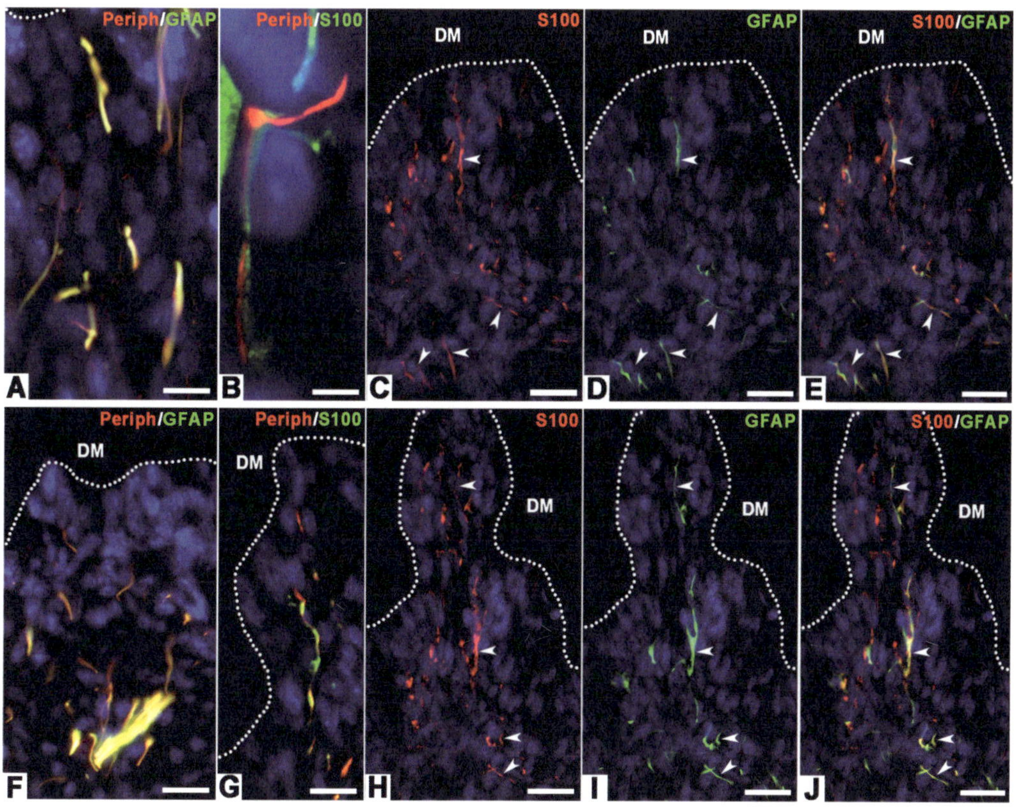

Figure 4. Glial cells in the odontoblast layer of cell re-associations implanted for two weeks in Nude mice (A-E), compared to first lower molars at PN7 (F-J). Axons were visualized with anti-peripherin antibody (A-B and F-G) and glial cells using antibodies to either S100-beta (B, C, E and G, H, J) or GFAP (A, D, E and F, I, J). In implanted cell re-associations as well as in molar at PN7, axons and glial cells were detected in the odontoblast layer (compare A, B with F, G). In the odontoblast layer of implanted cell re-associations, GFAP positive glial cells were associated with axons (A), as in the first lower molar at PN7 (F). The same results were observed for S100-beta positive glial cells (compare B with G). Double stainings for S100-beta and GFAP showed the presence of distinct glial cell types (C-E) as in intact molars at PN7 (H-J). More cells were positive for S100-beta than for GFAP and the two antigens showed only partial co-distribution in implanted cell re-associations (arrowheads in C-E), as in molar at PN7 (arrowheads in H-J). DM: dental matrix. Scale bars=20 μm (A, C-J) and 8 μm (B).

2.2. Blood vessels-axons relationships in the odontoblast layer

Microvessels are present in the odontoblast layer in re-associations implanted for two weeks in either Nude mice (Figure 5; [23]), or in ICR mice [48]. These microvessels in the odontoblast layer were detected after staining for CD34, collagen type IV or CD146 in implanted cell re-associations (Figure 5A-H) as in molars at PN7 (Figure 5I-P). All three antigens were found to co-distribute, as shown after double stainings for CD146 and collagen type IV (Figure 5Q-S) or for CD34 and collagen type IV (Figure 5T-V). The same was observed in molars at PN7 (data not shown). CD146, a marker of endothelial cells and pericytes, was detected in microvessels from the odontoblast layer in implanted re-associations (Figure 5A, D, F, H), as in molars at PN7 (Figure 5I, L, N, P). However, staining for α-SMA, a marker of smooth muscle cells, was detected in pericytes associated with large vessels in the apical part of the dental mesenchyme [25], but not in the odontoblast layer, neither in implanted cell re-associations nor in molars at PN7 (Figure 6G, K).

The timing of capillaries development and their entering the odontoblast layer (Figure 3E, E') has been correlated with the mineral requirement for dentinogenesis [49]. The relative roles of odontoblasts and capillaries present in this cell layer in the transport of calcium during dentin mineralization has been discussed previously [50]. In the odontoblast layer, double stainings for peripherin and either CD34, or collagen IV, or CD146 showed that very frequently axons were detected in close proximity with capillaries (Figure 5B-D). Such neurovascular relationships were also documented in molars at PN7 (Figure 5J-L). Transmission electron microscopy confirmed these observations and further showed that there exist direct interactions (Figure 3E, E'). Tabata et al., [51] suggested that the innervation of capillaries might be involved in the regulation of blood flow as in larger vessels. Although most pulp axons are sensory ones, both sensory and sympathetic axons can make contacts with dental pulp vessels [34]. Sympathetic axons have been detected in the odontoblast layer, although not extending in dentinal tubules [52].

Double stainings were performed to compare the distribution of glial cells as visualized after staining for S100-beta or GFAP, with that of microvessels visualized after staining for CD34 or CD146 (Figure 5). In implanted re-associations, S100-beta positive cells showed more proximity with microvessels (Figure 5E, F) than GFAP positive glial cells (Figure 5G, H). The same was observed in molars at PN7 (compare Figure 5M, N with Figure 5O, P). This confirms the differential patterning of S100-beta versus GFAP-positive glial cells as discussed above. Furthermore, the stainings for S100-beta in the odontoblast layer of implanted re-associations or molars at PN7 (Figure 5F, M) are in good agreement with observations made in the dental pulp of adult human tooth where the S100-beta positive cells were ensheating blood vessels [26]. Due to the angle of sectioning, several layers of nuclei corresponded to functional odontoblasts at the tip of the cusps (Figure 1G). Typical Schwann cells were not observed by transmission electron microscopy in the odontoblast layer of implanted re-associations or molars at PN7. This fits with previous report showing that in the mouse molar, Schwann cells were present near the base of the odontoblast layer from PN9 and were not detected in the odontoblast layer before PN50-60 [53].

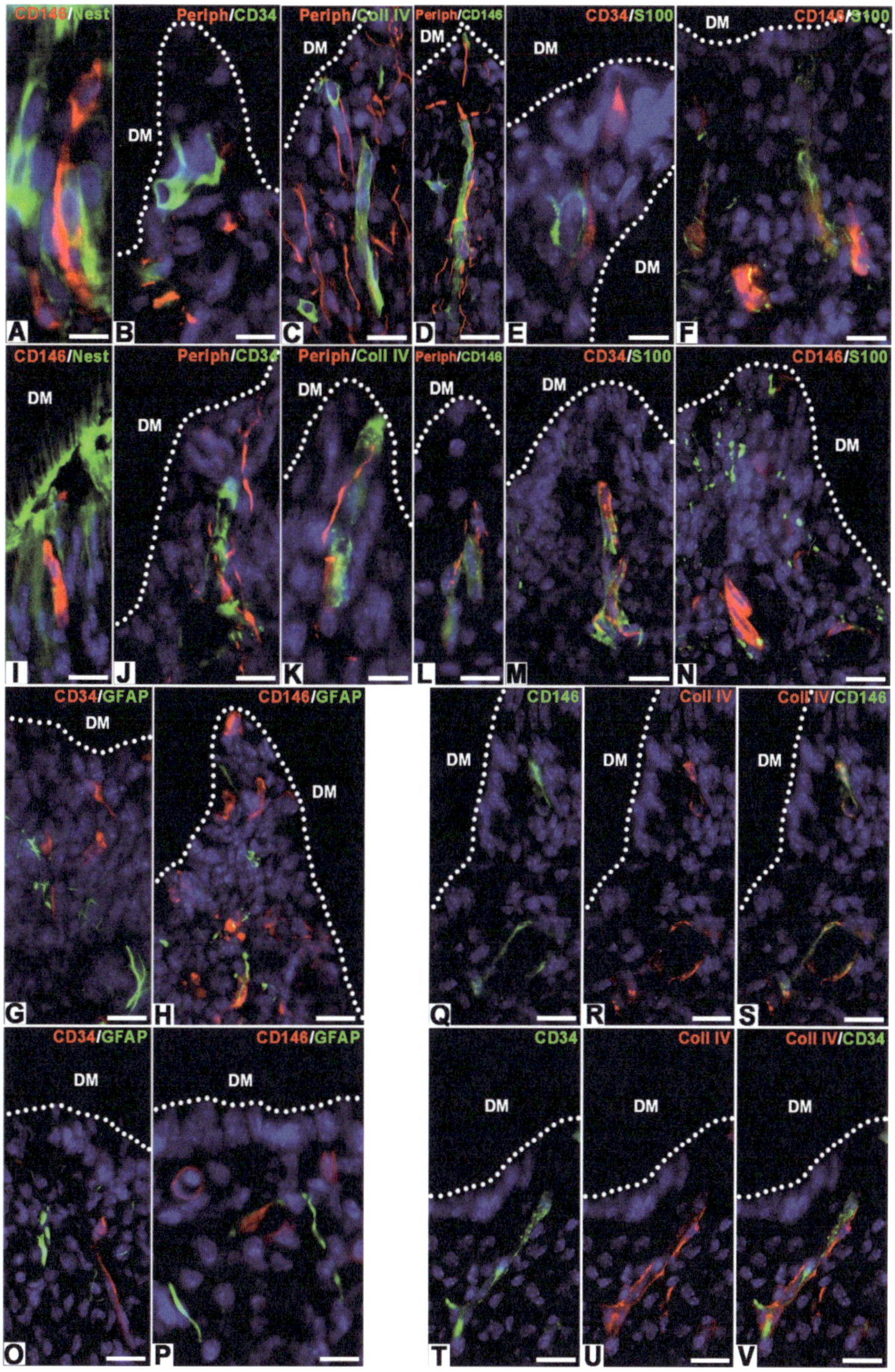

Figure 5. Neurovascular relationships in the odontoblast layer of cell re-associations implanted for two weeks in Nude mice (A-H and Q-V), compared to first lower molar at PN7 (I-P). Odontoblasts were labeled with an anti-nestin antibody (A, I). Blood vessels were either labeled for CD146 (A, D, F, H, I, L, N, P, Q, S), CD34 (B, E, G, J, M, O, T, V),

collagen type IV (C, K, R, S, U, V). Axons were labeled with anti-peripherin antibody (B-D, J-L) and glial cells using either anti-S100-beta (E, F, M, N) or anti-GFAP antibodies (G, H, O, P). Blood vessels were present in the odontoblast layer in implanted cell re-associations (A), as in the first lower molar at PN7 (I). In both cases also, axons were detected in the odontoblast layer and showed close proximity with blood vessels (compare B-D with J-L). CD146 (Q) and collagen type IV (R) were detected in the same blood vessels (S) and the same was observed when comparing the localization of CD34 and collagen type IV (T-V). In the odontoblast layer of implanted cell re-associations, as in molar at PN7, glial cells were detected after staining for S100-beta (compare E, F with M, N) or for GFAP (compare G, H with O, P). In both cases, a certain proportion of glial cells were detected next to blood vessels positive with CD34 (compare E, M with G, O) or CD146 (compare F, N with H, P). DM: dental matrix. Scale bars=20 μm.

2.3. Blood vessels-axons relationships in the central and apical parts of the dental mesenchyme

The progressive vascularization of the dental mesenchyme has been studied during mouse molar development [48,54] as well as in implanted cell re-associations [25,48]. The vascularization in implanted intact molars or cell re-associations showed similar density but was slightly retarded in re-associations. The implantation of cell re-associations in GFP mice showed that all blood vessels in the implant originated from the host [48].

In the developing molar, pericytes positive for α-SMA were detected only from PN4 in the dental mesenchyme. They were observed in association with large blood vessels at the apical part of the pulp both in the intact molar (Figure 1I; Figure 6K-N) as in cell re-associations implanted for two weeks (Figure 1D; Figure 6G-J; [25]). Pericytes are involved in the development, maturation, stabilization, and remodeling of capillaries and small vessels [55–57]. These cells are also involved in the regulation of capillary blood flow [58]. Still pericytes were suggested to possibly act as mesenchymal stem cells (MSCs) [59–61]. A study performed with mouse incisor as a model showed that pericytes, together with other MSCs-like cells, could participate in tooth growth and repair, giving rise to odontoblasts and odontoblast-like cells [62].

Transmission electron microscopy and double immunostainings for peripherin and CD146 showed close proximity between axons and blood vessels in the central part of the dental pulp of implanted cell re-associations (Figure 3D; Figure 6A), as in intact molars (Figure 3F, F′, F′′; Figure 6D). Such relationships have also been reported in the case of rat and human teeth by scanning electron microscopy, which showed a dense network of interactions between axons and vessels [63,64]. Transmission electron microscopy also showed contacts between axons in implanted cell re-associations (Figure 3D′) as in molar at PN7 (Figure 3G, G′). Blood vessels in the dental pulp are innervated by sensory and sympathetic nerve fibers [17,21]. Both sensory and sympathetic axons would be positive for peripherin. However during development, sensory nerves enter the dental mesenchyme at PN3-4, while sympathetic nerves penetrate this tissue much later, at PN9 [17]. Sympathetic axons are mainly detected in the deeper pulp along the blood vessels [65]. They are involved in vasoregulation [22,66]. Due to their late entering in the dental mesenchyme in physiological conditions, sympathetic nerves should be absent from the re-associations implanted for two weeks.

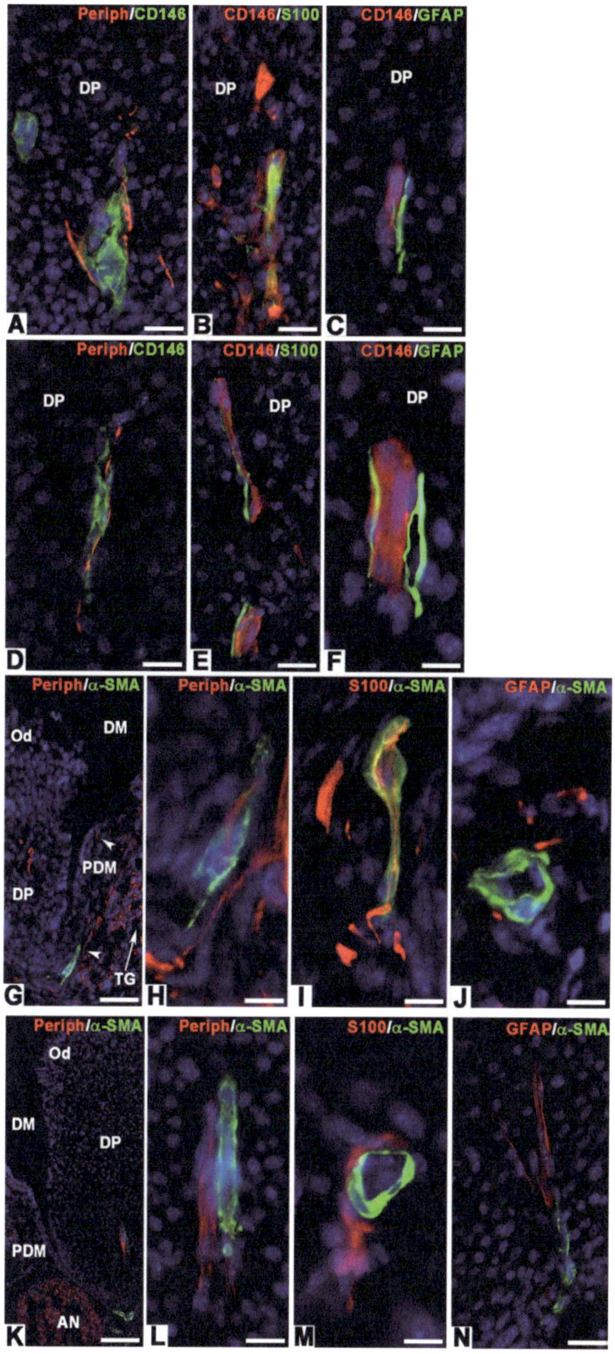

Figure 6. Neurovascular relationships in the central (A-F) and apical parts (G-N) of the dental mesenchyme of cell re-associations implanted for two weeks in Nude mice (A-C, G-J), compared to first lower molar at PN7 (D-F, K-N). Blood vessels were labeled with an anti-CD146 antibody (A-F) in the central part of the dental mesenchyme and with an anti-α-SMA antibody in its apical part (G-N). Axons were detected with an anti-peripherin antibody (A, D, G, H, K, L) and

glial cells using antibodies to either S100-beta (B, E, I, M) or GFAP (C, F, J, N). In implanted re-associations, axons originating from the trigeminal ganglion (TG) extended in the peridental mesenchyme (arrowheads) and dental pulp (G). In the central part of the dental mesenchyme, close relationships were observed between axons and blood vessels in implanted cell re-associations (A) as in the first lower molar at PN7 (D). In both types of samples, more vicinity was observed between blood vessels and S100-beta positive cells than with GFAP positive glial cells (compare B, E with C, F). In implanted cell re-associations pericytes positive for α-SMA surrounded large blood vessels and were detected only in the apical part of the dental mesenchyme (K). In both cases, axons were also detected in close proximity with α-SMA positive blood vessels (H, L). In implanted cell re-associations as in molars at PN7, some S100-beta positive (compare I with M), and GFAP positive glial cells (compare J with N) were detected next to pericytes. However, S100-beta positive cells showed more proximity with pericytes than GFAP positive ones (compare I, M with J, N). AN: alveolar nerve; DM: dental matrix; DP: dental pulp; Od: odontoblast; PDM: peridental mesenchyme. Scale bars=80 μm (K); 40 μm (G) and 20 μm (A-F, H-J, L-N).

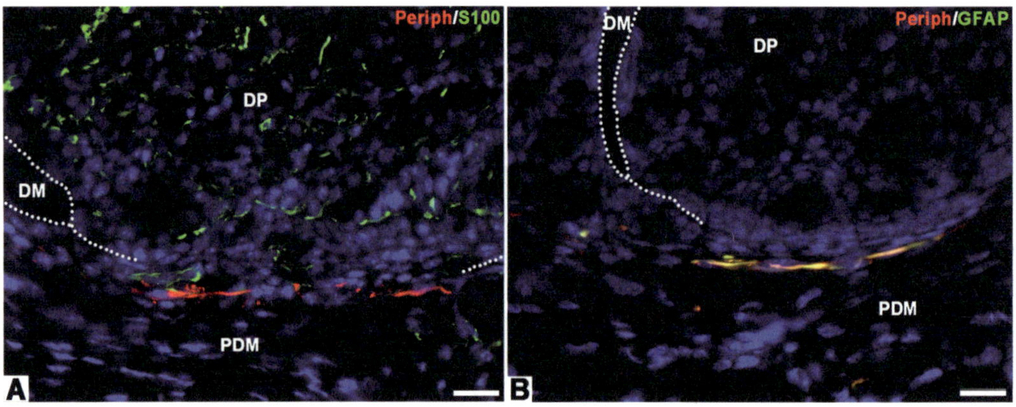

Figure 7. Innervation of cell re-associations implanted for two weeks in ICR mice without cyclosporine A treatment (A, B). Axons were labeled with anti-peripherin antibody (A, B) and glial cells using antibodies to either S100-beta (A) or GFAP (B). In such experimental conditions, axons remained at the border between the peridental mesenchyme (PDM) and dental pulp (DP) but did not enter the dental pulp (A, B). Similarly, GFAP positive glial cells co-localized with axons, at the limit between the peridental mesenchyme and dental pulp of the forming tooth (B). S100-beta positive glial cells showed a broader pattern, being present in close proximity with axons in the peridental mesenchyme, as well in the dental pulp (A). DM: dental matrix. Scale bars=20 μm.

Neurovascular bundles present in the dental mesenchyme have been suggested to represent a niche for stem cells [59,67]. Immunostainings performed in the human dental pulp showed that STRO-1 positive cells were present in the walls of blood vessels and perineurium surrounding the nerve bundles, while absent in the layer of mature odontoblasts [59]. These authors suggested that the potentialities of this minor stem cell population should be tested to search for a possible ability to differentiate into functional neuronal-like cells. Recently, this was supported by cultures of murine dental pulp stem cells (DPSC) under neuroinductive conditions [68]. Although STRO-1 cannot be searched for in mouse tissues, the immunostaining for peripherin remained negative when cell re-associations (Figure 7A, B) or even intact molars at PN4 were implanted for two weeks in ICR mice in the absence of a trigeminal ganglion [14]. Thus, if dental pulp stem cells can differentiate into neuronal-like cells *in vitro*, this potential was neither expressed in implanted intact molars at PN4 [14] nor in teeth forming after implantation of cultured cell re-associations [23]. Higuchi et al. [69] have reported the re-innervation of Embryonic Day (ED)18 rat molar germs after they were implanted under the skin of adult Wistar rats for eight weeks. Unfortunately, the early stages when the dental pulp first started to be re-innervated were not analyzed [69]. Nevertheless, these results suggest that, from a certain stage, immunodepressed conditions may no longer be necessary to allow dental pulp innervation.

When cell re-associations were implanted in Nude mice, both S100-beta and GFAP positive cells were detected in the dental pulp, which sometimes could be in close proximity with blood vessels (Figure 6B, C). The same was observed in molars at PN7 (Figure 6E, F). Both in implanted re-associations (Figure 6I, J) and in intact molars (Figure 6M, N), S100-beta positive cells showed more proximity with pericytes than GFAP positive cells. Transmission electron microscopy showed the presence of Schwann cells in the dental pulp of re-associations implanted in CsA-treated mice [23]. When cell re-associations were implanted in ICR mice without CsA treatment, the innervation of the dental mesenchyme was not possible: the dental mesenchyme remained negative for peripherin (Figure 7A, B; [23]). Nevertheless, glial cells expressing S100-beta (Figure 7A), but not GFAP (Figure 7B), were present in the dental mesenchyme. These results agree with previous observations showing that, in human pulp tissues, S100-beta and GFAP are expressed by different cell types, showing distinct patterns [26]. The origin of glial cells present in the dental pulp is not known. Implantation experiments will have to be performed in GFP mice to determine whether S100-beta positive cells have an endogenous origin, come from the host or have a mixed origin, as already observed for other dental pulp cells [25]. Glial cells in dental tissues from rodents appeared quite heterogeneous [43] and in the human dental pulp as well [26]. Schwann cells ensheating axons were observed in the dental pulp of cell reassociations (Figure 3B, B′, C). Schwann cells are derived from neural crest cells, and were considered to consist in two types whether myelinating or not, both being involved in the maintenance of axons [70,71]. Differences were also found when comparing Schwann cells in the dental and peridental mesenchymes [43]. These cells show a remarkable plasticity during reparative processes, as being able to dedifferentiate and participate in re-innervation by directing axonal regrowth and re-myelinisation [45]. Recently, culture conditions were determined, which allowed human dental pulp stem cells to undergo Schwann-like cell differentiation and to support neural outgrowth *in vitro* [72].

3. Innervation of the periodontium

As a component of the periodontium, the periodontal ligament mediates the attachment of the tooth to surrounding alveolar bone (Figure 1J). The periodontal ligament fibers are the main constituent of the ligament. This tissue is also vascularized (Figure 1J), innervated (Figure 8G) and contains a dense network of mechanoreceptors [73,74]. The periodontal ligament develops in parallel with root formation [75]. During development, the vascularization of the dental sac (prospective peridental mesenchyme) largely precedes the vascularization of the dental mesenchyme [48]. Similarly, nerves come close to the condensing dental mesenchyme long before the periodontium differentiates and becomes innervated [15,76,77].

In the mesenchyme surrounding the forming tooth in implanted re-associations, blood vessels and axons were detected, but showed only a limited degree of associations (Figure 8A, B). The same was observed in molars at PN7 (Figure 8F, G). In implanted re-associations, S100-beta positive and also GFAP-positive cells were detected in association with axons (Figure 8D,E), as in molars at PN7 (Figure 8I, J). In implanted re-associations, there were co-localizations of peripherin and GAP-43 in the peridental mesenchyme, but not in the dental pulp (Figure 8C). However the situation was different in the molar at PN7, where such co-localizations were found in the peridental and dental mesenchymes (Figure 8H).

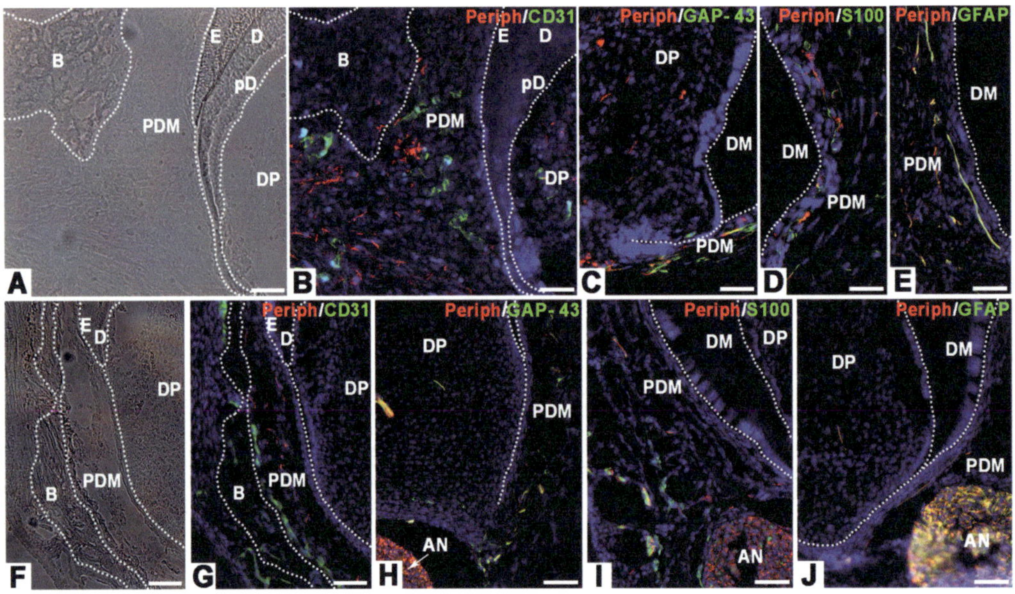

Figure 8. Innervation of the peridental mesenchyme of cell re-associations implanted for two weeks in Nude mice (A-E), compared to first lower molars at PN7 (F-J). Axons were visualized using an antibody to peripherin (B-E, G-J) and their growth cones with an anti-GAP-43 antibody (C, H). Glial cells were labeled using antibodies to either S100-beta (D, I) or GFAP (E, J). Blood vessels were labeled with an anti-CD31 antibody (B, G). Newly formed bone was detected at the periphery of implanted cell re-associations (A, B), analogous to the alveolar bone surrounding the molar at PN7 (F, G). In both cases, axons and blood vessels were observed in the peridental mesenchyme (PDM) between bone and dental tissues (compare B with G). Also, growth cones were detected, which co-distributed with axons in the peridental mesenchyme of implanted cell re-associations (C), as in intact molar at PN7 (H). GAP-43 and peripherin also co-distributed in the dental pulp of molar at PN7 (H). However, GAP-43 was not detected in the dental pulp (DP) of tooth forming from implanted cell re-associations (C). In the peridental mesenchyme of implanted cell re-associations as well as in molars at PN7, different glial cell types were detected after staining for S100-beta (compare D with I) or GFAP (compare E with J). In both cases, these different glial cell types were observed next to axons in the peridental mesenchyme (compare D, E with I, J). AN: alveolar nerve; B: bone; D: dentin; DM: dental matrix; E: enamel; pD: predentin. Scale bars=40 μm (A-C and E-J) and 20 μm (D).

Besides sensory axons, the periodontal ligament also contains sympathetic neurons, which are involved in bone remodeling [78,79]. However, this could not be investigated in the present experimental design. Implantation of cell re-associations would have to be performed in the jaw, which still raises major difficulties to correctly position the implant, with prospective roots having a correct orientation in the jaw [48]. When cell re-associations had been cultured for eight days, the crown morphology of the forming teeth is well visible (Figure 10B). However, the shape of the material to be implanted is more or less spherical (Figure 10B). It is thus very difficult to avoid a rotation of the samples during their implantation [2], although in some instances, it could be very well achieved [9].

The engineering of tooth root and periodontium (Figure 1E) is now considered as the main point in tooth engineering [80]. Indeed, complementary approaches to address this question are in progress [81–83]. The presence of axons in the peridental mesenchyme of engineered teeth has been observed after implantation in ICR mice i.e. in conditions where the dental mesenchyme cannot be innervated [23]. This indicates that not only the kinetics but also the conditions allowing innervation (immunodepressed conditions) are different in the dental and peridental mesenchymes.

4. Attempts to replace cyclosporin A treatment by using stem cells

The entering of axons in the dental mesenchyme of cell re-associations implanted under the skin of adult mice requires an artificially created immunodepression: treatment of host mice with CsA or implantation in Nude mice [23]. This requirement is not a consequence of a possible change in the mesenchymal cell heterogeneity/behavior during tissue and cell dissociation steps prior to re-association, culture, and implantation. Indeed, the same was observed when implanting intact PN4 molars [14]. The importance of immunomodulation, as observed for the re-innervation in the clinical context of face transplantation [84], has also been reported in a reparative process [85]. The next question was: how to avoid the use of CsA, which actually represents a heavy constraint in view of clinical application [23]?

Stem cells have been suggested to facilitate pulp innervation, possibly by chemoattraction [86]. Before stem cells were taken into account, dental pulp cells were known to interfere with neuron survival and differentiation *in vitro*, which was correlated with their expressing a wide range of neurotrophic factors [87,88]. These included Nerve growth factor (NGF), brain-derived neurotrophic factor (BDNF), glial-derived neurotrophic factor (GDNF), and chemokine ligand (CXCL)12 [88]. Other studies led to propose a potential role of mesenchymal stem cells (MSCs) in immunomodulation, by exerting an immunosuppressive effect on cells from both innate and acquired immunity systems [89–91]. Dental and peridental MSCs also showed such properties [92–95]. Some of these MSCs might originate from pericytes [96], although other origins have also been suspected [62]. The number of pericytes as visualized using antibodies to α-SMA was very limited in re-associations implanted for two weeks (Figure 6G). MSCs from different origins have been shown to stimulate axonal sprouting and tissue innervation [97]. Bone marrow also contains mesenchymal stem cells (BM-MSCs) exhibiting immunomodulatory and regulatory properties [98,99]. BM-MSCs have been shown to stimulate neurite outgrowth, probably due to their expressing NGF and BDNF under specific experimental conditions [100]. Since bone marrow cells can be easily prepared in large quantities, they were tested for their possible effect in stimulating dental pulp innervation, to possibly avoid the use of CsA.

For this purpose, bone marrow derived cells (BMDCs) [8] were used at passage two [101]. Re-associations between mixed BMDCs/dental mesenchymal cells and an intact dental epithelium from the cap stage at ED14 (Figure 9G) were cultured for eight days *in vitro* (Figure 9H, I) and analyzed for tissue organization and cell differentiation (Figure 10A, C). BMDCs were prepared from GFP mice (Figure 9D) in order to follow their fate in cultured re-associations (Figure 11). Histology of cultured re-associations demonstrated that teeth could develop *in vitro* when mixed BMDCs/mesenchymal cells at a ratio of 50% were re-associated with an ED14 competent dental epithelium and cultured for 8 days (Figure 10A-C), as in the absence of BMDCs (Figure 11F; [24]). Although initially mixed with dental mesenchymal cells, GFP-labeled BMDCs were only detected in the dental mesenchyme and at the periphery of the forming tooth after 6 days of culture (Figure 11A). After 8 days, all BMDCs were located at the periphery of the forming tooth (Figure 11B). This suggests that, in these experimental conditions, BMDCs do not directly participate in the late stages of tooth formation (i.e. odontoblast differentiation). This is further supported by histology showing rather small teeth forming, which can be related to the lower number of dental mesenchymal cells used for the re-associations, when mixed 50:50 with BMDCs (Figure 10D).

To investigate the possible re-innervation of such re-associations, a trigeminal ganglion (TG) was deposited on top of cells/tissue re-associations cultured for 7 days (Figure 9I), and maintained *in vitro* overnight prior to implantation under the skin of ICR mice (Figure 9J). Implants were maintained for two weeks in order to allow direct comparison with previous results obtained after implantation in CsA-treated mice or Nude mice [23]. Histology showed the presence of functional odontoblasts secreting predentin/dentin (Figure 10G) and thus inducing the differentiation of functional ameloblasts secteting enamel (Figure 10E, F). Double stainings for CD31 and peripherin showed the presence of blood vessels in the dental and peridental mesenchymes (Figure 11C, D, 7:7 samples (100%)). Few axons only were detected in the dental mesenchyme (Figure 11E, 3:7 samples (42%)), while the control re-associations without BMDCs remained negative for peripherin (Figure 11F). However, in these preliminary experiments, axons did not reach the odontoblast layer (Figure 11E).

From the literature, several mechanisms have been proposed to try to explain how exogenous stem cells might stimulate innervation. They include cell replacement, neurotrophic support or immunomodulation and may vary with different target tissues. In the experimental conditions reported here, the possibility of differentiation towards neural cells can be rejected since, already before implantation, bone marrow cells no longer remained present in the dental mesenchyme, which developed during the *in vitro* culture of the re-associations (Figure 11B). Furthermore, the axons present in the dental mesenchyme remained GFP negative (Figure 11E). Nevertheless, these preliminary experiments will have to be completed a) by increasing their number, b) by further testing BMDCs at different passages, and c) by testing sub-populations of these BMDCs.

http://dx.doi.org/10.5772/59340

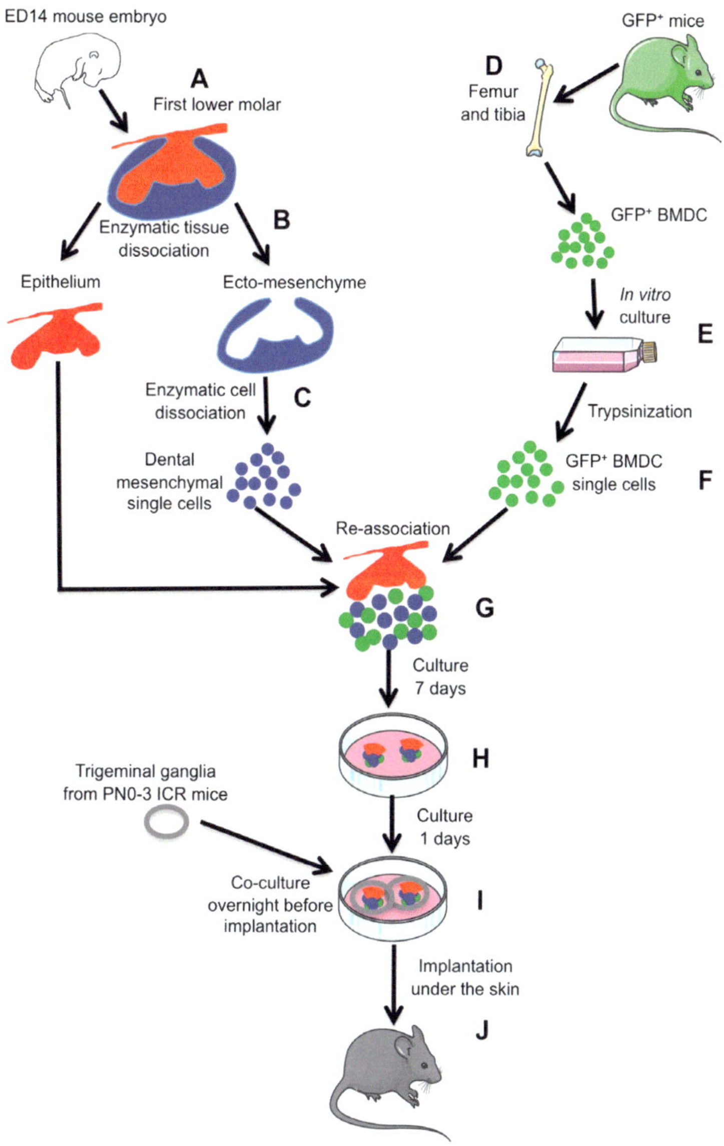

Figure 9. Schematic representation of the experimental procedures to test the effects of bone marrow derived cells on the innervation of implanted cells/tissue re-associations. The mandibular first molars were dissected from ICR embryos at ED14 (A). The dental epithelium and ecto-mesenchyme were dissociated by enzymatic treatment (B). Then dental ecto-mesenchyme was dissociated into single cells (C) and mixed with bone marrow derived cells (50:50) (G). For this purpose, bone marrow derived cells were prepared from femur and tibia bones (D), dissected from adult GFP positive mice (C57BL/6 from the IGBMC, Illkirch, France). The bone marrow derived cells (BDMCs) were cultured *in vitro* and used at passage 2 (E). After trypsinization, BDMCs single cells (F) were mixed with the dental mesenchymal single cells and put in contact with an ED14 intact dental epithelium (G). These re-associations were cultured for 7 days *in vitro* (H) and then a trigeminal ganglion was put on the top of each re-association for a further co-culture overnight (I). These re-associations were implanted between the skin and muscles behind the ears in adult ICR mice and maintained for two weeks (J).

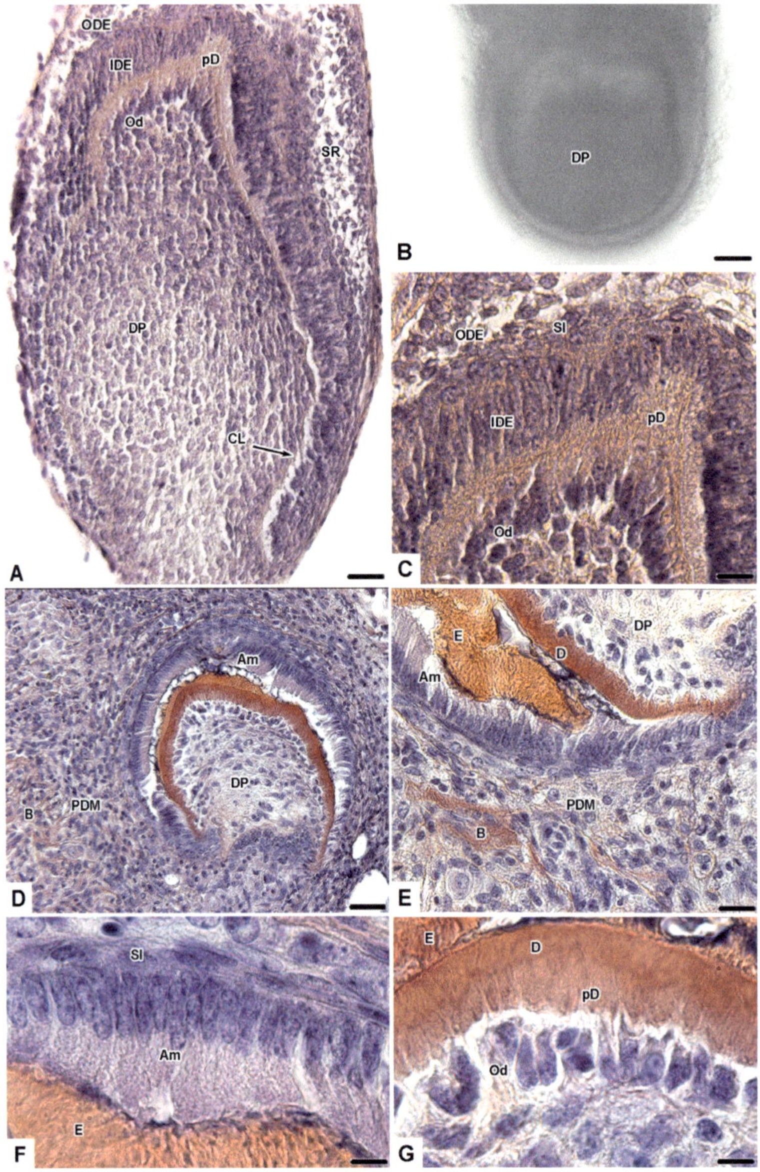

Figure 10. Histology of re-associations between an enamel organ and mixed dental mesenchymal cells and bone marrow derived cells prior to implantation (A-C) and implanted for two weeks under the skin of ICR mice (D-G). Re-associations between mixed BMDCs/dental mesenchymal cells (50:50) and an intact dental epithelium from the cap stage at embryonic days 14 (ED14) were cultured before implantation for two weeks under the skin of ICR mice. After eight days in culture, cells/tissue re-associations showed a characteristic dental epithelial histogenesis, with the presence of the inner (IDE) and outer dental epithelium (ODE) (A,C), the stellate reticulum (SR) (A) and the stratum intermedium (SI) (C). Odontoblasts (Od) were differentiated and induced the differentiation of ameloblasts, which elongated in the IDE (A). Odontoblasts secreted predentin (pD) (A, C). After implantation (D-G), crown (D) and root (E) were developed and newly formed bone (B) was present next to the tooth (E). Ameloblasts (Am) were elongated, polarized and secreted enamel (F). Ameloblasts were in contact with the stratum intermedium (F). Odontoblasts were elongated and polarized with their nucleus opposite to the secretory pole (G). They were functional, secreting predentin/dentin and

http://dx.doi.org/10.5772/59340

dentinal tubules were visible (G). CL: cervical loop; D: dentin; DP: dental pulp; E: enamel; PDM: peridental mesenchyme. Scale bars = 80 μm (B); 40 μm (A, D); 20 μm (C, E) and 8 μm (F, G).

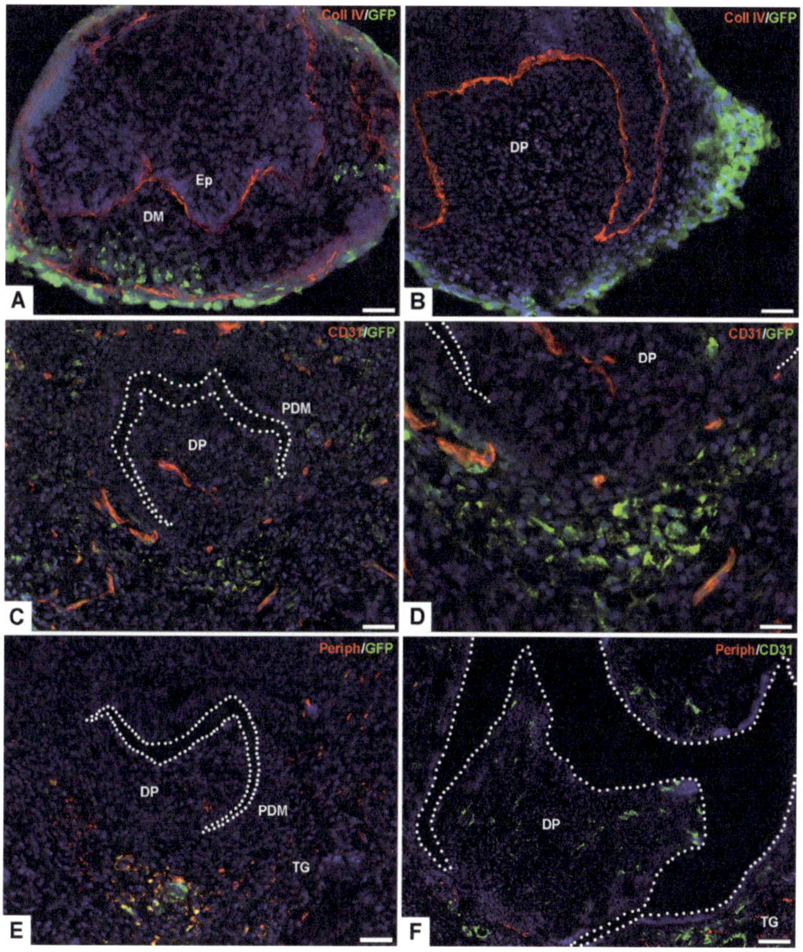

Figure 11. Vascularization and innervation of re-associations between mixed dental cells and bone marrow derived cells and an intact dental epithelium cultured for six days (A), or eight days (B), or cultured and implanted for two weeks with a trigeminal ganglion (C-E), and control re-associations without bone marrow-derived cells (F). Implantations were performed under the skin of ICR mice. The epithelial-mesenchymal junction was labeled with an antibody against collagen type IV (A, B). Blood vessels were labeled with an antibody to CD31 (C, D, F) and axons with an antibody to peripherin (E, F). GFP positive bone marrow derived cells (BMDCs) were labeled with a specific antibody to GFP (A-E). After six days in culture, BMDCs were present in the dental mesenchyme (DM) and at the periphery of the forming tooth (A). After eight days in culture, GFP-positive BMDCs were only detected at the periphery of the forming tooth (B). After two weeks of implantation, blood vessels were present in the peridental tissues and could enter the dental pulp (DP) in cell re-associations with BMDCs (C, D), as in cell re-associations without BMDCs (F). In implanted cell re-associations with BMDCs, axons from the trigeminal ganglion (TG) were detected in the dental pulp (E), while in control re-associations (without BMDCs) axons were observed only at the limit between the peridental mesenchyme (PDM) and dental pulp (F). After implantation for two weeks, GFP positive BMDCs were detected only in the periphery of the forming tooth and not in the dental pulp (C-E). Ep: epithelium; PDM. Scale bars=40 μm (A-C, E, F) and 20 μm (D).

5. Conclusions and prospects

All together, the results presented here show how far it is possible to reproduce the innervation of the dental and peridental mesenchymes in teeth forming from cultured and implanted cell re-associations, as it may occur during odontogenesis. As long as immunodepressed conditions can be maintained (implantation in Nude mice), it is possible within two weeks of implantation to reproduce the innervation of the dental mesenchyme, with axons reaching odontoblasts. Glial cells were present as well, although further work will be necessary to determine their origin. Still some differences exist when comparing the situation in implanted re-associations with developing molars. It will be necessary to determine how far these might result from differences in the kinetic of innervation in the two conditions and whether this would have further consequences after longer implantation period. Longer implantations would also allow determining whether myelinisation can be achieved. Using a method described by Honda et al., [102], implantation in the jaw, next to the alveolar nerve, will have to be tested to try to better approach the physiological situation and see whether tooth innervation can occur in these conditions. At the same time, this might allow the sympathetic innervation of the tooth, thus avoiding parallel experiments where cell re-associations would have to be grown in contact with a superior cervical ganglion.

The method used here will have to be adjusted when new non-dental mesenchymal cell sources will become available to engage in whole tooth engineering. Indeed, dental embryonic cells represent a very convenient tool to design and try to simplify experimental conditions for tooth organ engineering. However, such cells would not be available in a clinical context, so that several other cell sources are being tested (for review see [2,80,103]). Dental embryonic cells were also a convenient model to investigate the rapid loss of odontogenic potentialities with aging and after *in vitro* culture [104]. Despite the presence of a morphogenetic center, the primary enamel knot, the enamel organ at the cap stage has no more odontogenic potential [105]. Previous studies offered the possibility to use younger epithelial sources, when still retaining odontogenic inductive potential as tested for tooth engineering [8]. Obviously the odontogenic potentialities of dental pulp cells rapidly decrease during development and are quite poor in adult tissues. This can be seen from reparative processes, which are quite limited physiologically and lead to osteodentin instead of tubular dentin in most experimental conditions [62,106]. For this reason, the design of experimental conditions where non dental cells can be induced to give rise to odontoblasts secreting tubular dentin [8,59] or to ameloblasts secreting enamel may require important technical adjustments [107–109]. Investigating the innervation of the mesenchyme and more specifically of the odontoblast layer forming in these conditions will be a future challenge.

Abbreviations

α-SMA: alpha smooth muscle actin; BDNF: brain-derived neurotrophic factor; BM-MSC: bone marrow mesenchymal stem cell; BMDC: bone marrow derived cell; CD34: cluster of differen-

http://dx.doi.org/10.5772/59340

tiation 34; CsA: cyclosporin A; CXCL: chemokine ligand; DPSC: dental pulp stem cell; ED: embryonic day; GAP-43: growth-associated protein-43; GDNF: glial-derived neurotrophic factor; GFAP: glial fibrillary acidic protein; GFP: green fluorescent protein; MSC: mesenchymal stem cell; NF200: neurofilament 200; NGF: nerve growth factor; PN: post-natal; PNS: peripheral nervous system; TEM: transmission electron microscopy; TG: trigeminal ganglion.

Acknowledgements

The authors thank Prof. Jean-Christophe Maurin for help with trigeminal ganglia dissection, Jean-Luc Weickert and Dr. Nadia Messaddeq for their help with transmission electron microscopy and Hervé Gegout for the histology. Tunay Kökten was funded by the Dental Faculty, University of Strasbourg.

Author details

T. Kökten[1,2], H. Lesot[1,2*] and S. Kuchler-Bopp[1,2]

*Address all correspondence to: Lesot@unistra.fr

1 INSERM UMR S1109, "Osteoarticular and Dental Regenerative Nanomedicine" Laboratory, FMTS, Faculty of Medicine, Strasbourg Cedex, France

2 Dental Faculty, Université de Strasbourg, Strasbourg, France

References

[1] Huang GT. Dental pulp and dentin tissue engineering and regeneration: advancement and challenge. Front. Biosci. (Elite Ed). 2011;3 788-800.

[2] Kuchler-Bopp S, Keller L, Poliard A, Lesot H. Tooth Organ Engineering: Biological Constraints Specifying Experimental Approaches. In: Eberli D, éditeur. Tissue Engineering for Tissue and Organ Regeneration [Internet]. InTech; 2011 http://www.intechopen.com/books/tissue-engineering-for-tissue-and-organ-regeneration/tooth-organ-engineering-biological-constraints-specifying-experimental-approaches.

[3] Oshima M, Mizuno M, Imamura A, Ogawa M, Yasukawa M, Yamazaki H, Morita R, Ikeda E, Nakao K, Takano-Yamamoto T, Kasugai S, Saito M, Tsuji T. Functional tooth regeneration using a bioengineered tooth unit as a mature organ replacement regenerative therapy. PLoS ONE. 2011;6(7) e21531.

[4] Hynes K, Menicanin D, Gronthos S, Bartold PM. Clinical utility of stem cells for periodontal regeneration. Periodontol. 2000. 2012;59(1) 203-227.

[5] Hu B, Nadiri A, Kuchler-Bopp S, Perrin-Schmitt F, Peters H, Lesot H. Tissue engineering of tooth crown, root, and periodontium. Tissue Eng. 2006;12(8) 2069-2075.

[6] Nakao K, Morita R, Saji Y, Ishida K, Tomita Y, Ogawa M, Saitoh M, Tomooka Y, Tsuji T. The development of a bioengineered organ germ method. Nat. Methods. 2007;4(3) 227-230.

[7] Honda MJ, Fong H, Iwatsuki S, Sumita Y, Sarikaya M. Tooth-forming potential in embryonic and postnatal tooth bud cells. Med. Mol. Morphol. 2008;41(4) 183-192.

[8] Ohazama A, Modino SAC, Miletich I, Sharpe PT. Stem-cell-based tissue engineering of murine teeth. J. Dent. Res. 2004;83(7):518-522.

[9] Yen AHH, Sharpe PT. Stem cells and tooth tissue engineering. Cell Tissue Res. 2008;331(1) 359-372.

[10] Nakahara T. Potential feasibility of dental stem cells for regenerative therapies: stem cell transplantation and whole-tooth engineering. Odontology. 2011; 99(2) 105-111.

[11] Otsu K, Kumakami-Sakano M, Fujiwara N, Kikuchi K, Keller L, Lesot H, Harada H. Stem cell sources for tooth regeneration: current status and future prospects. Front Physiol. 2014;5 36.

[12] Zhang Y, Chen Y. Bioengineering of a human whole tooth: progress and challenge. Cell Regeneration. 2014;3(1) 8.

[13] Keller L, Kökten T, Kuchler-Bopp S, Lesot H. Tooth organ engineering. In "Stem Cell Biology and Tissue Engineering in Dental Sciences. " (A. Vishwakarma, P. Sharpe, S. Shi, X.-P. Wang, M. Ramalingam, Eds), in press.

[14] Keller L, Kuchler-Bopp S, Lesot H. Whole-Tooth Engineering and Cell Sources. In: Huang GT-J, Thesleff I, éditeurs. Stem Cells in Craniofacial Development and Regeneration. John Wiley & Sons, Inc.; 2013 p431-446. http://onlinelibrary.wiley.com/doi/10.1002/9781118498026.ch24/summary.

[15] Mohamed SS, Atkinson ME. A histological study of the innervation of developing mouse teeth. J. Anat. 1983;136(Pt 4) 735-749.

[16] Matsuo S, Ichikawa H, Henderson TA, Silos-Santiago I, Barbacid M, Arends JJ, Jacquin MF. trkA modulation of developing somatosensory neurons in oro-facial tissues: tooth pulp fibers are absent in trkA knockout mice. Neuroscience. 2001;105(3) 747-760.

[17] Moe K, Kettunen P, Kvinnsland IH, Luukko K. Development of the pioneer sympathetic innervation into the dental pulp of the mouse mandibular first molar. Arch. Oral Biol. 2008;53(9) 865-873.

[18] Moe K, Sijaona A, Shrestha A, Kettunen P, Taniguchi M, Luukko K. Semaphorin 3A controls timing and patterning of the dental pulp innervation. Differentiation. 2012;84(5) 371-379.

[19] Luukko K, Kettunen P. Coordination of tooth morphogenesis and neuronal development through tissue interactions: Lessons from mouse models. Exp. Cell Res. 2014;325(2) 72-77.

[20] Kettunen P, Løes S, Furmanek T, Fjeld K, Kvinnsland IH, Behar O, Yagi T, Fujisawa H, Vainio S, Taniguchi M, Luukko K. Coordination of trigeminal axon navigation and patterning with tooth organ formation: epithelial-mesenchymal interactions, and epithelial Wnt4 and Tgfbeta1 regulate semaphorin 3a expression in the dental mesenchyme. Development. 2005;132(2) 323-334.

[21] Olgart L. Neural control of pulpal blood flow. Crit. Rev. Oral Biol. Med. 1996;7(2) 159-171.

[22] Byers MR, Suzuki H, Maeda T. Dental neuroplasticity, neuro-pulpal interactions, and nerve regeneration. Microsc. Res. Tech. 2003;60(5) 503-515.

[23] Kökten T, Bécavin T, Keller L, Weickert J-L, Kuchler-Bopp S, Lesot H. Immunomodulation stimulates the innervation of engineered tooth organ. PLoS ONE. 2014;9(1) e86011.

[24] Lechguer AN, Couble ML, Labert N, Kuchler-Bopp S, Keller L, Magloire H, Bleicher F, Lesot H. Cell Differentiation and Matrix Organization in Engineered Teeth. J. Dent. Res. 2011;90(5) 583-589.

[25] Keller LV, Kuchler-Bopp S, Lesot H. Restoring physiological cell heterogeneity in the mesenchyme during tooth engineering. Int. J. Dev. Biol. 2012;56(9) 737-746.

[26] Farahani RM, Simonian M, Hunter N. Blueprint of an ancestral neurosensory organ revealed in glial networks in human dental pulp. J. Comp. Neurol. 2011;519(16) 3306-3326.

[27] Linde A, Goldberg M. Dentinogenesis. Crit. Rev. Oral Biol. Med. 1993;4(5) 679-728.

[28] Ruch JV, Lesot H, Bègue-Kirn C. Odontoblast differentiation. Int. J. Dev. Biol. 1995;39(1) 51-68.

[29] Magloire H, Maurin JC, Couble ML, Shibukawa Y, Tsumura M, Thivichon-Prince B, Bleicher F. Topical review. Dental pain and odontoblasts: facts and hypotheses. J. Orofac. Pain. 2010;24(4) 335-349.

[30] Vang H, Chung G, Kim HY, Park S-B, Jung SJ, Kim J-S, Oh SB. Neurochemical properties of dental primary afferent neurons. Exp. Neurobiol. 2012;21(2) 68-74.

[31] Nagahama SI, Cunningham ML, Lee MY, Byers MR. Normal development of dental innervation and nerve/tissue interactions in the colony-stimulating factor-1 deficient osteopetrotic mouse. Dev. Dyn. 1998;211(1) 52-59.

[32] Egan CA, Hector MP, Bishop MA. On the pulpal nerve supply in primary human teeth: evidence for the innervation of primary dentine. Int. J. Paediatr. Dent. 1999;9(1) 57-66.

[33] Carda C, Peydró A. Ultrastructural patterns of human dentinal tubules, odontoblasts processes and nerve fibres. Tissue Cell. 2006 ;38(2) 141-150.

[34] Byers MR. Dental sensory receptors. Int. Rev. Neurobiol. 1984;25 39-94.

[35] Ibuki T, Kido MA, Kiyoshima T, Terada Y, Tanaka T. An ultrastructural study of the relationship between sensory trigeminal nerves and odontoblasts in rat dentin/pulp as demonstrated by the anterograde transport of wheat germ agglutinin-horseradish peroxidase (WGA-HRP). J. Dent. Res. 1996;75(12) 1963-1970.

[36] Shen Y, Mani S, Donovan SL, Schwob JE, Meiri KF. Growth-associated protein-43 is required for commissural axon guidance in the developing vertebrate nervous system. J. Neurosci. 2002;22(1) 239-247.

[37] Fried K, Risling M. Growth-associated protein (GAP-43)-like immunoreactivity in primary and permanent tooth pulp nerve fibers of the cat. Brain Res. 1992;572(1-2) 19-26.

[38] Ishizuka H, Hiura A. A light and electron microscopic study on pulpal nerve fibers in the lower incisor of the mouse. Arch. Histol. Cytol. 1992;55(2) 167-178.

[39] Maeda T, Byers MR. Different localizations of growth-associated protein (GAP-43) in mechanoreceptors and free nerve endings of adult rat periodontal ligament, dental pulp and skin. Arch. Histol. Cytol. 1996;59(3) 291-304.

[40] Curtis R, Stewart HJ, Hall SM, Wilkin GP, Mirsky R, Jessen KR. GAP-43 is expressed by nonmyelin-forming Schwann cells of the peripheral nervous system. J. Cell Biol. 1992;116(6) 1455-1464.

[41] Mani S, Shen Y, Schaefer J, Meiri KF. Failure to express GAP-43 during neurogenesis affects cell cycle regulation and differentiation of neural precursors and stimulates apoptosis of neurons. Mol. Cell. Neurosci. 2001;17(1) 54-66.

[42] Ibarra A, Hernández E, Lomeli J, Pineda D, Buenrostro M, Martiñón S, Garcia E, Flores N, Guizar-Sahagun G, Correa D, Madrazo I. Cyclosporin-A enhances non-functional axonal growing after complete spinal cord transection. Brain Res. 2007;1149 200-209.

[43] Byers MR, Maeda T, Brown AM, Westenbroek RE. GFAP immunoreactivity and transcription in trigeminal and dental tissues of rats and transgenic GFP/GFAP mice. Microsc. Res. Tech. 2004;65(6) 295-307.

[44] Farahani RM, Sarrafpour B, Simonian M, Li Q, Hunter N. Directed glia-assisted angiogenesis in a mature neurosensory structure: pericytes mediate an adaptive re-

sponse in human dental pulp that maintains blood-barrier function. J. Comp. Neurol. 2012;520(17) 3803-3826.

[45] Kim HA, Mindos T, Parkinson DB. Plastic fantastic: Schwann cells and repair of the peripheral nervous system. Stem Cells Transl. Med. 2013;2(8) 553-557.

[46] Jessen KR, Mirsky R. The origin and development of glial cells in peripheral nerves. Nat. Rev. Neurosci. 2005;6(9) 671-682.

[47] Pereira JA, Lebrun-Julien F, Suter U. Molecular mechanisms regulating myelination in the peripheral nervous system. Trends Neurosci. 2012;35(2) 123-134.

[48] Nait Lechguer A, Kuchler-Bopp S, Hu B, Haïkel Y, Lesot H. Vascularization of engineered teeth. J. Dent. Res. 2008;87(12) 1138-1143.

[49] Yoshida S, Ohshima H. Distribution and organization of peripheral capillaries in dental pulp and their relationship to odontoblasts. Anat. Rec. 1996;245(2) 313-326.

[50] Bishop MA. An investigation of pulp capillaries and tight junctions between odontoblasts in cats. Anat. Embryol. 1987;177(2) 131-138.

[51] Tabata S, Ozaki HS, Nakashima M, Uemura M, Iwamoto H. Innervation of blood vessels in the rat incisor pulp: a scanning 1. electron microscopic and immunoelectron microscopic study. Anat. Rec. 1998;251(3) 384-391.

[52] Shimeno Y, Sugawara Y, Iikubo M, Shoji N, Sasano T. Sympathetic nerve fibers sprout into rat odontoblast layer, but not into dentinal tubules, in response to cavity preparation. Neurosci. Lett. 2008;435(1) 73-77.

[53] Corpron RE, Avery JK. The ultrastructure of intradental nerves in developing mouse molars. Anat. Rec. 1973;175(3) 585-606.

[54] Rothová M, Feng J, Sharpe PT, Peterková R, Tucker AS. Contribution of mesoderm to the developing dental papilla. Int. J. Dev. Biol. 2011;55(1) 59-64.

[55] Hirschi KK, D'Amore PA. Pericytes in the microvasculature. Cardiovasc. Res. 1996;32(4) 687-698.

[56] Mendes LF, Pirraco RP, Szymczyk W, Frias AM, Santos TC, Reis RL, Marques AP. Perivascular-like cells contribute to the stability of the vascular network of osteogenic tissue formed from cell sheet-based constructs. PLoS ONE. 2012;7(7) e41051.

[57] Tigges U, Welser-Alves JV, Boroujerdi A, Milner R. A novel and simple method for culturing pericytes from mouse brain. Microvasc. Res. 2012;84(1) 74-80.

[58] Hamilton NB, Attwell D, Hall CN. Pericyte-mediated regulation of capillary diameter: a component of neurovascular coupling in health and disease. Front. Neuroenergetics. 2010;2.

[59] Shi S, Gronthos S. Perivascular niche of postnatal mesenchymal stem cells in human bone marrow and dental pulp. J. Bone Miner. Res. 2003;18(4) 696-704.

[60] Caplan AI. All MSCs are pericytes? Cell Stem Cell. 2008;3(3) 229-230.

[61] Crisan M, Yap S, Casteilla L, Chen C-W, Corselli M, Park TS, Andriolo G, Sun B, Zheng B, Zhang L, Norotte C, Teng P-N, Traas J, Schugar R, Deasy BM, Badylak S, Buhring H-J, Giacobino J-P, Lazzari L, Huard J, Péault B. A perivascular origin for mesenchymal stem cells in multiple human organs. Cell Stem Cell. 2008;3(3) 301-313.

[62] Feng J, Mantesso A, De Bari C, Nishiyama A, Sharpe PT. Dual origin of mesenchymal stem cells contributing to organ growth and repair. Proc. Natl. Acad. Sci. U.S.A. 2011;108(16) 6503-6508.

[63] Zhang JQ, Nagata K, Iijima T. Scanning electron microscopy and immunohistochemical observations of the vascular nerve plexuses in the dental pulp of rat incisor. Anat. Rec. 1998;251(2) 214-220.

[64] Iijima T, Zhang J-Q. Three-dimensional wall structure and the innervation of dental pulp blood vessels. Microsc. Res. Tech. 2002;56(1) 32-41.

[65] Pohto P, Antila R. Innervation of blood vessels in the dental pulp. Int. Dent. J. 1972;22(2) 228-239.

[66] Hildebrand C, Fried K, Tuisku F, Johansson CS. Teeth and tooth nerves. Prog. Neurobiol. 1995;45(3) 165-222.

[67] Zhao H, Feng J, Seidel K, Shi S, Klein O, Sharpe P, Chai Y. Secretion of shh by a neurovascular bundle niche supports mesenchymal stem cell homeostasis in the adult mouse incisor. Cell Stem Cell. 2014;14(2) 160-173.

[68] Ellis KM, O Carroll DC, Lewis MD, Rychkov GY, Koblar SA. Neurogenic potential of dental pulp stem cells isolated from murine incisors. Stem Cell Res. Ther. 2014;5(1) 30.

[69] Higuchi K, Santiwong P, Tamaki H, Terashima T, Nakayama H, Notani T, Iseki H, Baba O, Takano Y. Development and terminal differentiation of pulp and periodontal nerve elements in subcutaneous transplants of molar tooth germs and incisors of the rat. Eur. J. Oral Sci. 2008;116(4) 324-333.

[70] Bhatheja K, Field J. Schwann cells: origins and role in axonal maintenance and regeneration. Int. J. Biochem. Cell Biol. 2006;38(12) 1995-1999.

[71] Woodhoo A, Sommer L. Development of the Schwann cell lineage: from the neural crest to the myelinated nerve. Glia. 2008 ;56(14) 1481-1490.

[72] Martens W, Sanen K, Georgiou M, Struys T, Bronckaers A, Ameloot M, Phillips J, Lambrichts I. Human dental pulp stem cells can differentiate into Schwann cells and promote and guide neurite outgrowth in an aligned tissue-engineered collagen construct in vitro. FASEB J. 2014;28(4) 1634-1643.

[73] Byers MR. Sensory innervation of periodontal ligament of rat molars consists of unencapsulated Ruffini-like mechanoreceptors and free nerve endings. J. Comp. Neurol. 1985;231(4) 500-518.

[74] Maeda T, Ochi K, Nakakura-Ohshima K, Youn SH, Wakisaka S. The Ruffini ending as the primary mechanoreceptor in the periodontal ligament: its morphology, cytochemical features, regeneration, and development. Crit. Rev. Oral Biol. Med. 1999;10(3) 307-327.

[75] Lungová V, Radlanski RJ, Tucker AS, Renz H, Míšek I, Matalová E. Tooth-bone morphogenesis during postnatal stages of mouse first molar development. J. Anat. 2011;218(6) 699-716.

[76] Pearson AA. The early innervation of the developing deciduous teeth. J. Anat. 1977;123(Pt 3) 563-577.

[77] Luukko K. Immunohistochemical localization of nerve fibres during development of embryonic rat molar using peripherin and protein gene product 9.5 antibodies. Arch. Oral Biol. 1997;42(3) 189-195.

[78] Elefteriou F, Campbell P, Ma Y. Control of bone remodeling by the peripheral sympathetic nervous system. Calcif. Tissue Int. 2014;94(1) 140-151.

[79] Kondo M, Kondo H, Miyazawa K, Goto S, Togari A. Experimental tooth movement-induced osteoclast activation is regulated by sympathetic signaling. Bone. 2013;52(1) 39-47.

[80] Volponi AA, Pang Y, Sharpe PT. Stem cell-based biological tooth repair and regeneration. Trends Cell Biol. 2010;20(12) 715-722.

[81] Dangaria SJ, Ito Y, Luan X, Diekwisch TGH. Successful periodontal ligament regeneration by periodontal progenitor preseeding on natural tooth root surfaces. Stem Cells Dev. 2011;20(10) 1659-1668.

[82] Maeda H, Fujii S, Tomokiyo A, Wada N, Akamine A. Periodontal tissue engineering: defining the triad. Int. J. Oral Maxillofac. Implants. 2013;28(6) 461-471.

[83] Han J, Menicanin D, Gronthos S, Bartold PM. Stem cells, tissue engineering and periodontal regeneration. Aust. Dent. J. 2014;59 (Suppl 1) 117-130.

[84] Siemionow M, Gharb BB, Rampazzo A. Pathways of sensory recovery after face transplantation. Plast. Reconstr. Surg. 2011;127(5) 1875-1889.

[85] Iohara K, Murakami M, Takeuchi N, Osako Y, Ito M, Ishizaka R, Utunomiya S, Nakamura H, Matsushita K, Nakashima M. A Novel Combinatorial Therapy With Pulp Stem Cells and Granulocyte Colony-Stimulating Factor for Total Pulp Regeneration. Stem Cells Transl. Med. 2013;2(10) 521-533.

[86] Arthur A, Shi S, Zannettino ACW, Fujii N, Gronthos S, Koblar SA. Implanted adult human dental pulp stem cells induce endogenous axon guidance. Stem Cells. 2009;27(9) 2229-2237.

[87] Nosrat IV, Widenfalk J, Olson L, Nosrat CA. Dental pulp cells produce neurotrophic factors, interact with trigeminal neurons in vitro, and rescue motoneurons after spinal cord injury. Dev. Biol. 2001;238(1) 120-132.

[88] Nosrat IV, Smith CA, Mullally P, Olson L, Nosrat CA. Dental pulp cells provide neurotrophic support for dopaminergic neurons and differentiate into neurons in vitro; implications for tissue engineering and repair in the nervous system. Eur. J. Neurosci. 2004;19(9) 2388-2398.

[89] Zhang W, Ge W, Li C, You S, Liao L, Han Q, Deng W, Zhao RCH. Effects of mesenchymal stem cells on differentiation, maturation, and function of human monocyte-derived dendritic cells. Stem Cells Dev. 2004;13(3) 263-271.

[90] Liu Y, Wang S, Shi S. The role of recipient T cells in mesenchymal stem cell-based tissue regeneration. Int. J. Biochem. Cell Biol. 2012;44(11) 2044-2050.

[91] Bernardo ME, Fibbe WE. Mesenchymal stromal cells: sensors and switchers of inflammation. Cell Stem Cell. 2013;13(4) 392-402.

[92] Wada N, Menicanin D, Shi S, Bartold PM, Gronthos S. Immunomodulatory properties of human periodontal ligament stem cells. J. Cell. Physiol. 2009;219(3) 667-676.

[93] Wada N, Gronthos S, Bartold PM. Immunomodulatory effects of stem cells. Periodontol. 2000. 2013;63(1) 198-216.

[94] Yamaza T, Kentaro A, Chen C, Liu Y, Shi Y, Gronthos S, Wang S, Shi S. Immunomodulatory properties of stem cells from human exfoliated deciduous teeth. Stem Cell Res. Ther. 2010;1(1) 5.

[95] Zhao Y, Wang L, Jin Y, Shi S. Fas ligand regulates the immunomodulatory properties of dental pulp stem cells. J. Dent. Res. 2012;91(10) 948-954.

[96] Feng J, Mantesso A, Sharpe PT. Perivascular cells as mesenchymal stem cells. Expert Opin. Biol. Ther. 2010;10(10) 1441-1451.

[97] Mahmood A, Wu H, Qu C, Xiong Y, Chopp M. Effects of treating traumatic brain injury with collagen scaffolds and human bone marrow stromal cells on sprouting of corticospinal tract axons into the denervated side of the spinal cord. J. Neurosurg. 2013;118(2) 381-389.

[98] Uccelli A, Pistoia V, Moretta L. Mesenchymal stem cells: a new strategy for immunosuppression? Trends Immunol. 2007;28(5) 219-226.

[99] Ramasamy R, Fazekasova H, Lam EW-F, Soeiro I, Lombardi G, Dazzi F. Mesenchymal stem cells inhibit dendritic cell differentiation and function by preventing entry into the cell cycle. Transplantation. 2007;83(1) 71-76.

[100] Miranda CO, Teixeira CA, Sousa VF, Santos TE, Liz MA, Marques AM, Pinto-do-Ó P, Sousa MM. Primary bone marrow mesenchymal stromal cells rescue the axonal phenotype of Twitcher mice. Cell Transplant. 2014;23(2) 239-252.

[101] Ball LM, Bernardo ME, Roelofs H, Lankester A, Cometa A, Egeler RM, Locatelli F, Fibbe WE. Cotransplantation of ex vivo expanded mesenchymal stem cells accelerates lymphocyte recovery and may reduce the risk of graft failure in haploidentical hematopoietic stem-cell transplantation. Blood. 2007;110(7) 2764-2767.

[102] Honda MJ, Ohara T, Sumita Y, Ogaeri T, Kagami H, Ueda M. Preliminary study of tissue-engineered odontogenesis in the canine jaw. J. Oral Maxillofac. Surg. 2006;64(2) 283-289.

[103] Huang GT, Gronthos S, Shi S. Mesenchymal stem cells derived from dental tissues vs. those from other sources: their biology and role in regenerative medicine. J. Dent. Res. 2009;88(9) 792-806.

[104] Keller L, Kuchler-Bopp S, Mendoza SA, Poliard A, Lesot H. Tooth engineering: searching for dental mesenchymal cells sources. Front Physiol. 2011;2 7.

[105] Mina M, Kollar EJ. The induction of odontogenesis in non-dental mesenchyme combined with early murine mandibular arch epithelium. Arch. Oral Biol. 1987;32(2) 123-127.

[106] Smith AJ, Lesot H. Induction and regulation of crown dentinogenesis: embryonic events as a template for dental tissue repair? Crit. Rev. Oral Biol. Med. 2001;12(5) 425-437.

[107] Nakagawa E, Itoh T, Yoshie H, Satokata I. Odontogenic potential of post-natal oral mucosal epithelium. J. Dent. Res. 2009;88(3) 219-223.

[108] Takahashi C, Yoshida H, Komine A, Nakao K, Tsuji T, Tomooka Y. Newly established cell lines from mouse oral epithelium regenerate teeth when combined with dental mesenchyme. In Vitro Cell. Dev. Biol. Anim. 2010;46(5) 457-468.

[109] Wang B, Li L, Du S, Liu C, Lin X, Chen Y, Zhang Y. Induction of human keratinocytes into enamel-secreting ameloblasts. Dev. Biol. 2010;344(2) 795-799.